T0206165

Vitamin C

OXIDATIVE STRESS AND DISEASE

Series Editors

Lester Packer, PhD
Enrique Cadenas, MD, PhD
University of Southern California School of Pharmacy
Los Angeles, California

Lipoic Acid: Energy Production, Antioxidant Activity and Health Effects, edited by Mulchand S. Patel and Lester Packer

Dietary Modulation of Cell Signaling Pathways, edited by Young-Joon Surh, Zigang Dong, Enrique Cadenas, and Lester Packer

Micronutrients and Brain Health, edited by Lester Packer, Helmut Sies, Manfred Eggersdorfer, and Enrique Cadenas

Adipose Tissue and Inflammation, edited by Atif B. Awad and Peter G. Bradford

Herbal Medicine: Biomolecular and Clinical Aspects, Second Edition, edited by Iris F.F. Benzie and Sissi Wachtel-Galor

Inflammation, Lifestyle, and Chronic Diseases: The Silent Link, edited by Bharat B. Aggarwal, Sunil Krishnan, and Sushovan Guha

Flavonoids and Related Compounds: Bioavailability and Function, edited by Jeremy P.E. Spencer and Alan Crozier

Mitochondrial Signaling in Health and Disease, edited by Sten Orrenius, Lester Packer, and Enrique Cadenas

Vitamin D: Oxidative Stress, Immunity, and Aging, edited by Adrian F. Gombart

Carotenoids and Vitamin A in Translational Medicine, edited by Olaf Sommerburg, Werner Siems, and Klaus Kraemer

Hormesis in Health and Disease, edited by Suresh I. S. Rattan and Éric Le Bourg

Liver Metabolism and Fatty Liver Disease, edited by Oren Tirosh

Nutrition and Epigenetics, edited by Emily Ho and Frederick Domann

Lipid Oxidation in Health and Disease, edited by Corinne M. Spickett and Henry Jay Forman

Diversity of Selenium Functions in Health and Disease, edited by Regina Brigelius-Flohé and Helmut Sies

Mitochondria in Liver Disease, edited by Derick Han and Neil Kaplowitz

Fetal and Early Postnatal Programming and its Influence on Adult Health, edited by Mulchand S. Patel and Jens H. Nielsen

Biomedical Application of Nanoparticles, edited by Bertrand Rihn

The Biology of the First 1,000 Days, edited by Crystal D. Karakochuk, Kyly C. Whitfield, Tim J. Green, and Klaus Kraemer

Hydrogen Peroxide Metabolism in Health and Disease, edited by Margreet C.M. Vissers, Mark Hampton, and Anthony J. Kettle

Glutathione, edited by Leopold Flohé

Vitamin C: New Biochemical and Functional Insights, edited by Qi Chen and Margreet C.M. Vissers

Cancer and Vitamin C, edited by Qi Chen and Margreet C.M. Vissers

For more information about this series, please visit:
https://www.crcpress.com/Oxidative-Stress-and-Disease/book-series/CRCOXISTRDIS

Vitamin C
New Biochemical and Functional Insights

Edited by

Qi Chen, PhD
Associate Professor
Department of Pharmacology, Toxicology and Therapeutics
University of Kansas Medical Center
Kansas City, Kansas, USA

Margreet C.M. Vissers, PhD
Research Professor
Department of Pathology and Biomedical Science
University of Otago, Christchurch, New Zealand

CRC Press
Taylor & Francis Group
Boca Raton London New York

CRC Press is an imprint of the
Taylor & Francis Group, an **informa** business

Cover: The image on the cover was created by Dr. Abel Ang and depicts the expanding knowledge of the functional roles of vitamin C and resulting health influences.

CRC Press
Taylor & Francis Group
6000 Broken Sound Parkway NW, Suite 300
Boca Raton, FL 33487-2742

First issued in paperback 2021

ISBN-13: 978-1-138-33799-2 (hbk)
ISBN-13: 978-1-03-217525-6 (pbk)
DOI: 10.1201/9780429442025

The Open Access version of chapter 1 was funded by Niddk Dir Ddb Mcns.

Visit the Taylor & Francis Web site at
http://www.taylorandfrancis.com

and the CRC Press Web site at
http://www.crcpress.com

This book is dedicated to the memory of

Lester Packer
August 28, 1929–July 27, 2018

With thanks for his tireless work and his endless enthusiasm for science,
and in particular, for his work as editor of the Oxidative Stress and
Disease series of books. This Vitamin C project (now two volumes) was initiated
on Professor Packer's invitation at the end of 2017.

CONTENTS

CONTENTS

PREFACE

Almost a century ago, ascorbic acid was identified as the compound present in fruits and vegetables that was responsible for the alleviation of scurvy, a much-feared disease commonly experienced during extended periods of poor nutrition. Prior to its identification, it had already been known as vitamin C—the "water-soluble C" factor that was antiscorbutic. The hexuronic acid identified by Albert Szent-Györgyi and colleagues in the early 1930s was named *ascorbic acid* due to its antiscurvy properties. Since that time, much has been learned about the chemistry, biochemistry, and biology of this compound, with more than 70,000 publications listed in the PubMed database. Research and discussion have frequently focused on determining and justifying the amount of dietary vitamin C needed on a daily basis. This topic is of relevance to the human condition, as we are one of the few species that cannot synthesize ascorbate and are therefore dependent on sufficient dietary intake. Most countries have determined the recommended daily intake based on the amount required to avoid deficiency. However, as outlined in many chapters, there is not a simple, unequivocal answer to the question of optimal daily intake of vitamin C. Rather, there are likely to be multiple answers that are a reflection of both intake and turnover under different clinical and health circumstances.

This discussion has been fueled by a better understanding of the pharmacokinetics of vitamin C around the body, its transport in the circulation and uptake into tissues, and the discovery of new enzymes that have a vitamin C requirement for optimal functioning. During the past 20 years, and since the publication of the last books on vitamin C (1999 and 2003), there has been a renaissance of interest in vitamin C. Much has been learned, with the identification of its role in the regulation of stress responses, gene expression, epigenetic programming, and metabolism. Vitamin C research is rapidly expanding—the number of papers published annually has increased steadily and has doubled in the past 20 years. The aim of this publication is to summarize our knowledge of vitamin C and to describe recent findings and new directions. Improved understanding of its new biological functions, pharmacokinetics, and biological chemistry has expanded the range of clinical settings in which it is appropriate to consider the body's status and daily vitamin C requirement. These insights will help us determine the appropriate management of the body's ascorbate status, on a day-to-day basis and in a clinical setting.

The depth and breadth of this topic are now substantial, and the scope is sufficiently broad such that two volumes have been generated to cover this extensive topic. This volume contains chapters summarizing the current knowledge of the chemistry and biochemistry of vitamin C, together with its biological functions in supporting many aspects of immunity and neurological function. A decision was made to create a separate volume to cover the topic of vitamin C in cancer. There has been a renewed interest in this research area with many new developments, and much of this information is contained in the second volume. The content of the two books is interdependent, and we hope that readers will value the information contained in both volumes.

The opening chapter of this book creates a context for both volumes. Levine and colleagues have provided a perspective and an overview of the extensive studies on the uptake and turnover of dietary vitamin C and have given an insightful summary of the ongoing discussion on the recommended daily intake. This thoughtful reflection makes an excellent introduction to this seemingly simple, but actually complex, topic.

The biology of vitamin C in the body is determined by its chemistry and biochemistry. A good understanding of the redox chemistry behind the turnover and pharmacokinetics of ascorbate in vivo is essential for the correct interpretation of experimental data. The chapters included in Part II provide an up-to-date overview of the chemistry of vitamin C (Chapter 2 by Vissers et al.), its uptake and transport throughout the body and into cells (Chapter 3 by Nydegger et al.), and its interaction with enzymes for which it is a cofactor (Chapter 5 by Vissers and Das). The distribution of the vitamin throughout the body has long been thought to contain clues to its function, and Chapter 4 by Lykkesfeldt and Chapter 6 by Bánhegyi et al. provide important and up-to-date information in this regard.

One of the most compelling areas of current clinical research is the potential contribution of vitamin C to the management of acute infectious diseases including sepsis. Recent studies have shown accelerated turnover of vitamin C in patients with acute sepsis, leading to severely depleted plasma levels. Significant clinical benefits including reduced mortality have been reported when healthy plasma levels are restored with gram quantities of vitamin C daily. This topic and the function of vitamin C in pneumonia and sepsis are covered in Chapter 7 by Carr. Immune cells are known to contain high concentrations of vitamin C, and in Chapter 8, Ang et al. provide an in-depth summary of the potential functional roles for ascorbate in white blood cells. Low vitamin C status has long been associated with impaired wound healing, but it is surprising how little robust information exists to support this claim, and Chapter 9 by Pullar and Vissers highlights the need for better, well-controlled clinical studies in this neglected clinical area.

The brain contains high levels of vitamin C and has specialized transport and retention such that levels are sustained even in times of body depletion. This is thought to reflect essential functions for vitamin C in the brain, but this area has been underexplored, most likely due to the difficulty of making measurements in the brain and the lack of good animal models of brain disorders. Therefore, we are very pleased to be able to include a section on the role of vitamin C in brain function. There is a close association between vitamin C deficiency and symptoms of malaise and depression. Chapter 10 by Dixit et al. and Chapter 11 by Hoffer detail the role for vitamin C in supporting many brain functions, and its possible impact on a number of brain pathologies, including neurodegenerative conditions, mood changes, and psychiatric diseases. In addition, the emerging role of vitamin C in supporting epigenetic reprogramming via the ten-eleven translocation (TET) enzymes and the potential impact of this on neurological function are introduced in Chapter 12 by Huff and Wang.

We thank the contributors for their generosity and enthusiasm in the preparation of chapters for this book. They have all taken time out of very busy schedules to provide an excellent overview of the current state of knowledge. The aim of this book is to reflect on and document the resurgence of interest in vitamin C. We trust that the content of this volume, and its companion book on cancer and vitamin C, will be of interest to research scientists, clinicians, students of medical science and nutrition, interested patient groups, and general readers.

QI CHEN
MARGREET C.M. VISSERS

CONTRIBUTORS

ABEL ANG
Mackenzie Cancer Research Group
Department of Pathology and Biomedical
    Science
University of Otago
Christchurch, New Zealand

GÁBOR BÁNHEGYI
Department of Medical Chemistry
Semmelweis University
and
MTA-SE Pathobiochemistry Research Group
Budapest, Hungary

ANITRA C. CARR
Department of Pathology and Biomedical
    Science
University of Otago
Christchurch, New Zealand

DAVID C. CONSOLI
Division of Diabetes, Endocrinology and
    Metabolism
Department of Medicine
Vanderbilt University Medical Center
Nashville, Tennessee

ANDREW B. DAS
Centre for Free Radical Research
Department of Pathology and Biomedical
    Science
University of Otago
Christchurch, New Zealand

SHILPY DIXIT
Division of Diabetes, Endocrinology and Metabolism
Department of Medicine
Vanderbilt University Medical Center
Nashville, Tennessee

IFECHUKWUDE C. EBENUWA
National Institute of Diabetes and Digestive
    and Kidney Diseases
National Institutes of Health
Bethesda, Maryland

GERGELY GYIMESI
Department of Nephrology and Hypertension
University Hospital Bern, Inselspital
and
Department of Biomedical Research
University of Bern
Bern, Switzerland

FIONA E. HARRISON
Division of Diabetes, Endocrinology and Metabolism
Department of Medicine
Vanderbilt University Medical Center
Nashville, Tennessee

MATTHIAS A. HEDIGER
Department of Nephrology and Hypertension
University Hospital Bern, Inselspital
and
Department of Biomedical Research
University of Bern
Bern, Switzerland

L. JOHN HOFFER
Lady Davis Institute for Medical Research
Division of General Internal Medicine
Jewish General Hospital
Montreal, Canada

TYLER C. HUFF
Department of Human Genetics
University of Miami Miller School
   of Medicine
Miami, Florida

MARK LEVINE
National Institute of Diabetes and Digestive
   and Kidney Diseases
National Institutes of Health
Bethesda, Maryland

JENS LYKKESFELDT
Faculty of Health and Medical Sciences
University of Copenhagen
Frederiksberg, Denmark

JÓZSEF MANDL
Department of Medical Chemistry
Semmelweis University
Budapest, Hungary

DAMIAN NYDEGGER
Department of Nephrology and Hypertension
University Hospital Bern, Inselspital
and
Department of Biomedical Research
University of Bern
Bern, Switzerland

KRISTA C. PAFFENROTH
Division of Diabetes, Endocrinology and
   Metabolism
Department of Medicine
Vanderbilt University Medical Center
Nashville, Tennessee

JULIET M. PULLAR
Centre for Free Radical Research
Department of Pathology and Biomedical Science
University of Otago
Christchurch, New Zealand

NICHOLAS SMIRNOFF
Biosciences
College of Life and Environmental Sciences
University of Exeter
Exeter, United Kingdom

ANDRÁS SZARKA
Laboratory of Biochemistry and Molecular Biology
Department of Applied Biotechnology and Food
   Science
Budapest University of Technology and Economics
Budapest, Hungary

HONGBIN TU
National Institute of Diabetes and Digestive and
   Kidney Diseases
National Institutes of Health
Bethesda, Maryland

PIERRE-CHRISTIAN VIOLET
National Institute of Diabetes and Digestive and
   Kidney Diseases
National Institutes of Health
Bethesda, Maryland

MARGREET C.M. VISSERS
Centre for Free Radical Research
Department of Pathology and Biomedical
   Science
University of Otago
Christchurch, New Zealand

GAOFENG WANG
Department of Human Genetics
University of Miami Miller School
   of Medicine
Miami, Florida

YAOHUI WANG
National Institute of Diabetes and Digestive and
   Kidney Diseases
National Institutes of Health
Bethesda, Maryland

JORDYN M. WILCOX
Division of Diabetes, Endocrinology and
   Metabolism
Department of Medicine
Vanderbilt University Medical Center
Nashville, Tennessee

Overview of Vitamin C

CHAPTER ONE

A "C Odyssey"

RECOMMENDED DIETARY ALLOWANCES AND OPTIMAL HEALTH: PARADIGM AND PROMISE OF VITAMIN C

Mark Levine, Pierre-Christian Violet, Ifechukwude C. Ebenuwa, Hongbin Tu, and Yaohui Wang

DOI: 10.1201/9780429442025-1

CONTENTS

INTRODUCTION: RECOMMENDED DIETARY ALLOWANCES AND THEIR LIMITATIONS

The nutritional biochemist Alfred E. Harper was a member of the Food and Nutrition Board (FNB) of the National Research Council/National Academy of Sciences for many years, and chair of the Food and Nutrition Board from 1978 to 1982. In a number of articles, he described the original intent of Recommended Dietary Allowances (RDAs) in the United States, briefly summarized as follows [1–3]. In 1940, to guide the U.S. government concerning national defense, a Committee on Food and Nutrition was established under the National Research Council, U.S. National Academy of Sciences, to advise the government on problems concerned with national defense. In 1941, the committee name was changed to the FNB. The allowances for specific nutrients from the FNB were intended to serve as a guide for planning adequate nutrition for U.S. civilians. Specifically, there was no intent to have RDAs as guides to perfect health, nor were they designed to attain ideal intakes. The RDA was stated to be not just "minimal sufficient to protect against actual deficiency disease" but sufficient "to ensure good nutrition and protection of all body tissues," and in the 1953 edition they are stated to be "nutrient allowances suitable for the maintenance of good nutrition in essentially the total population." The scientific bases for many RDAs were prevention of deficiency with a margin of safety, often determined from depletion-repletion studies or balance experiments. As

Harper wrote, "The RDA has been adopted and adapted by various organizations for many purposes, but they were devised for the planning and procurement of food supplies that would be nutritionally adequate for population groups. Therefore, any assessment of the adequacy, accuracy, and reliability of the RDA will be meaningful only if it is done in relation to their use for this primary purpose. To base such an assessment on their adequacy for other purposes would be like judging the adequacy of the design of the family car for use as a snowplow" [3].

As nutritional science has grown, and policy needs have changed, limitations of RDAs were recognized. Beginning in the 1990s, the FNB expanded nutritional intake concepts in the form of dietary reference intakes, commonly known as DRIs [4–6]. Unfortunately, because of data limitations, scientific bases for many dietary reference intakes remain as prevention of deficiency with a margin of safety, because these are the only data available.

To paraphrase Alfred E. Harper, when there is heavy snow, you need a snowplow. To achieve a nutrition goal, we have to be thoughtful in defining the nutrition problem that we are addressing, and if necessary, to think outside the box to solve the problem. Prevention of deficiency with a safety margin is not the only means to determine nutrient intake, nor, from a clinical vantage point, is it the preferred one. If there truly were a means to realize goals of ideal nutrient intake, there would be unprecedented possibilities to optimize health, prevent disease, and even treat disease. But specific methods and measures are essential to realize such lofty possibilities. Such specifics have taken decades to formulate, evolve, and solidify. What is so simple in concept has been so difficult to bring to fruition.

CONCENTRATION-FUNCTION HYPOTHESIS: PHYSIOLOGY/ PHARMACOKINETIC APPROACHES

Fundamental biochemical kinetics concepts can be the foundation for nutrient recommendations. Such concepts derive from work of Tatum and Beedle [7], David Perla and Jesse Marmorston [8], Roger Williams [9,10], and, perhaps surprisingly and independent of his later involvement with ascorbic acid in colds and cancer, Linus Pauling [11–13]. With these biochemical and experimental supports, a new hypothesis was proposed: bases of vitamin recommendations could be concentration-function relationships, or kinetics relationships, in cells, tissues, animals, and healthy humans [14–17]. Approaches were to conduct physiology and pharmacokinetic studies in healthy humans. Stated in another way, the overarching concept was that kinetics in situ would underlie vitamin recommendations in healthy people. Physiology and pharmacokinetic studies would provide essential data for the x-axis, those concentrations found in vivo, preferably in humans. With concentration data describing an x-axis, function, or the y-axis, could follow in relation to concentrations in vivo [18].

VITAMIN C AS A MODEL VITAMIN

For such work to proceed, a vitamin was selected arbitrarily: ascorbic acid. Initial efforts were focused on assay development and proof of concept. Accurate assay of ascorbic acid was an essential prerequisite, a foundation on which everything else was based. Since its discovery and isolation, ascorbic acid measurements had many uncertainties, due to limitations of sensitivity, specificity, stability, and confounding substance interferences [19]. The emergence of high-performance liquid chromatography (HPLC) techniques coupled to electrochemical detection provided a path forward [20]. A new assay was developed that addressed and solved these issues, using HPLC specifically coupled to coulometric (flow-through) electrochemical detection [21,22]. With this assay, the concept was tested and verified that kinetics relationships could be determined in situ for ascorbic acid. The experimental system was ascorbic acid–dependent norepinephrine biosynthesis in chromaffin granules, the secretory vesicles of adrenal medulla, isolated from bovine (cow) adrenal glands [23–25]. Norepinephrine synthesis is mediated by the enzyme dopamine β-monooxygenase [26,27]. Using this system, in situ kinetics were described for norepinephrine biosynthesis from dopamine, mediated by ascorbic acid as a cosubstrate [28]. To our knowledge, this was the first demonstration of kinetics in situ for any vitamin with the physiologic enzymatic substrate. Notably, the findings showed that the mechanisms of ascorbic acid action on norepinephrine biosynthesis in situ were more complex than its direct in vitro action as a cofactor/cosubstrate for isolated dopamine β-monooxygenase [22,27–29]. One strong implication of these findings was that an in vivo system, rather than an in vitro system, was the preferred basis for determining concentration-function relationships, on which vitamin recommendations could ultimately be based. An equally strong implication of these findings was that data from humans, rather than animals (i.e., cows), constituted the holy grail for determining concentration-function relationships.

VITAMIN C PHYSIOLOGY AND PHARMACOKINETICS IN HEALTHY HUMANS: TIGHT CONTROL AND UNDERLYING MECHANISMS

But, before proceeding with such clinical experiments, a key precondition was to know the x-axis range for ascorbic acid concentrations in humans. From a clinical perspective, this is not much different from having normal limits on a basic metabolic panel, a common test in clinical care. However, data were limited or unavailable that described whether and how a wide range of different doses of ascorbic acid modulated plasma and tissue concentrations: pharmacokinetic data [30–33]. Comprehensive pharmacokinetic data of this kind were unavailable not just for ascorbic acid but for all vitamins. Without pharmacokinetic information as a foundation, it would be impossible to consider biosynthetic consequences and clinical outcomes in relation to any ascorbic acid concentration. For clinicians, an analogy would be to try to manage diabetes without prior knowledge of normal and abnormal blood glucose concentrations.

For ascorbic acid, specific pharmacokinetic goals were to learn how seven different ascorbic acid doses over an 80-fold range impacted steady-state blood and tissue concentrations in men and women, while simultaneously characterizing concentration relationships and normal physiology as extensively as possible [34,35]. Approaches were to conduct pharmacokinetic experiments in healthy humans, using oral and intravenous ascorbic acid. Data were obtained using a depletion-repletion design, with correction for all possible unintended vitamin and mineral deficiencies [36]. Subjects were hospitalized as inpatients at the National Institutes of Health (NIH) Clinical Research Center for approximately 6 months, to facilitate dietary control, compliance, and pharmacokinetic

samplings over each of the seven different ascorbic acid doses. Subjects were first safely depleted of ascorbic acid using a vitamin C–restricted diet, with correction of any possible deficiencies in other vitamins [34–36]. When plasma concentrations were less than 10 μM, subjects were dose repleted in a stepwise manner. Subjects had to achieve steady state at each dose before advancing to the next highest dose. Daily doses were from 30 to 2500 mg. Half of the total daily dose was administered twice daily: before dinner, at least 4 hours after the past meal, or before breakfast in the morning, after overnight fasting. Ascorbic acid for oral administration was in a water solution, pH adjusted, with individualized doses that were routinely monitored for stability. Administration in this manner eliminated confounding effects of interferences from capsules or food components [37–39]. Ascorbic acid for intravenous injection was in individualized sterilized vials, pH adjusted, and routinely monitored for sterility and stability. The coulometric HPLC electrochemical measurement technique described earlier was utilized so that ascorbic acid could be measured in clinical samples. Measurements provided necessary sensitivity and specificity, without interferences from compounds in biological samples, and sample stability was accounted for during sample processing, assay, and long-term storage [21,40].

Inpatient studies were completed by 7 healthy men and 11 healthy women between ages 18 and 30 at the Clinical Research Center, NIH. Subjects were hospitalized for approximately 6 months each (Figure 1.1a). Both sex-specific and combined pharmacokinetic data showed that ascorbic acid concentrations were tightly controlled as a function of dose in all subjects [34,35] (Figure 1.1b). At a dose of 200 mg, there was nearly complete saturation of steady-state plasma and tissue concentrations, with higher oral doses having minimal additional effects. As doses increased above 100 mg, the fraction of the absorbed dose decreased, and the excess was excreted unchanged in urine. These pharmacokinetic data have subsequently been used by many countries to calculate intakes for ascorbic acid, including dietary reference intakes and recommended dietary allowances [6,41,42].

Some might conclude the following: mission accomplished, end of the story, time to investigate another vitamin. In contrast, our view is that these data unmasked possibilities that are the true

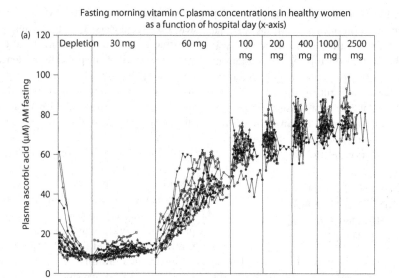

Fasting morning vitamin C plasma concentrations in healthy women
as a function of hospital day (x-axis)

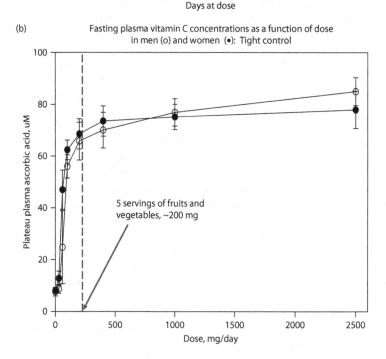

Fasting plasma vitamin C concentrations as a function of dose
in men (o) and women (•): Tight control

5 servings of fruits and
vegetables, ~200 mg

Figure 1.1. Fasting steady-state ascorbic acid plasma concentrations as a function of dose. (a) Concentrations in plasma in healthy women as a function of hospital day (x-axis) [35]. Number of days at each dose can be seen by comparing hospital day (x-axis) to each dose, listed at the top of the figure. Similar findings were observed for men [34,131]. (b) Fasting steady-state plasma concentrations as a function of dose for men and women: summary data. (From Levine, M. et al. 1996. *Proc. Natl. Acad. Sci. USA* 93, 3704–3709; Levine, M. et al. 2001. *Biofactors* 15, 71–74.)

beginning of understanding ascorbic acid, based on the clinical physiology and pharmacokinetic findings. The aggregate data showed that ascorbic acid concentrations were tightly controlled in humans as a function of oral doses. An analogy is again apparent based on glucose concentrations in humans. Glucose concentrations are tightly controlled as a function of glucose ingested, pancreatic hormones insulin and glucagon, gut hormones, gluconeogenesis in the liver, and insulin responsiveness of muscle and fat [43]. Plasma and tissue ascorbic acid concentrations appeared to have an analogous tight control. What were the mechanisms? In addition to dose ingested, the clinical data implicated that at least four potential physiologic processes were involved in tight control: intestinal absorption, tissue transport, renal reabsorption/excretion, and utilization. Each of these mechanisms can and has been probed, in varying degrees, for clinical relevance.

Bioavailability and Unexpected Consequences

In animals, intestinal absorption of ascorbic acid is mediated by sodium-dependent transporter SLC23a1, which is localized to the small intestine [44–46]. There may be additional mechanisms because knockout mice for SLC23a1 still absorb ascorbic acid, dehydroascorbic acid (oxidized ascorbic acid) is nearly as effective as ascorbic acid in preventing deficiency in guinea pigs, and dehydroascorbic acid appears to be absorbed in humans [47–49].

In humans, intestinal absorption of ascorbic acid was measured as part of the inpatient pharmacokinetic studies at the NIH. True bioavailability, or fractional absorption, was determined for each of seven different doses. Subjects were at steady state for each dose, which was administered one day by mouth and the following day by vein. Plasma samples were collected continuously over 36 hours. These data showed that fractional absorption declined as doses rose, especially above 100 mg daily [34,35,50] (Table 1.1).

TABLE 1.1
Bioavailability of oral ascorbate: Nonlinear tissue distribution model

Dose (mg)	Bioavailability (median%)
15[a]	89
30[a]	87
50[a]	85
100[a]	80
200	72
500	63
1250	46

[a] Amounts in foods.

Bioavailability experiments showed plasma and tissue concentrations were tightly controlled with oral administration of ascorbic acid. At doses at and above 100 mg, intravenous administration bypassed tight control of ascorbic acid concentrations in plasma that were seen with oral dosing, until the dose was excreted in urine. These data indicated that intravenous administration of ascorbic acid could produce pharmacologic concentrations that were not otherwise possible with oral dosing (Figure 1.2a–c). Using another clinical analogy, these findings are similar to plasma concentrations that are achieved with oral versus intravenous administration of many antibiotics.

Why might it matter that only intravenous administration produces pharmacologic ascorbic acid concentrations? Concentrations produced only with intravenous administration could have pharmacologic effects not found with oral dosing [51,52]. Many clinical trials open now have been designed to investigate precisely such effects in cancer treatment and in sepsis (see ClinicalTrials. gov for full listings).

Ascorbic acid in cancer treatment has a long and convoluted history, described elsewhere [17,51,53,54]. Briefly, ascorbic acid was proposed as a cancer treatment agent because of its effects on maintaining tissue collagen. The hypothesis was simple: cancers metastasized via collagen breakdown, and collagen could be strengthened, or its breakdown prevented, with ascorbic acid. In comparison to 200 mg of ascorbic acid, which produces near saturation of plasma and tissues, 10,000 mg of ascorbic acid was administered to patients with a variety of cancers by Ewan Cameron and his colleagues [13,55–57]. They reported treatment effects, improved well-being, and prolonged survival in some cases. Cameron's studies were criticized because of their retrospective design, lack of pathology confirmation, lack of controls, and potential confounding effects of endemic deficiency in the treatment population [51]. Ascorbic acid at the same dose had no effect on cancer treatment in two double-blind placebo-controlled trials, both performed at the Mayo Clinic, and ascorbic acid was dismissed as a cancer treatment agent [58–60]. The physiology and pharmacokinetic studies in healthy people provided a straightforward potential explanation. Ascorbic acid was administered only by mouth in the studies at Mayo Clinic, but both by mouth and

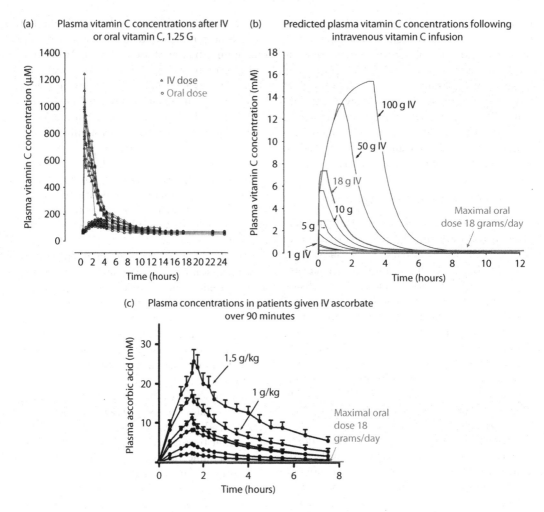

Figure 1.2. Plasma ascorbic acid concentrations as a function of oral or intravenous dosing. (a) Plasma concentration as a function of time in women who received 1,250 mg (1.25 G) of ascorbic acid either by mouth or by vein (intravenously). (b) Observed or modeled plasma ascorbic acid concentrations as a function of time at doses from 1 to 100 grams by mouth or by vein (intravenously) in healthy people. (c) Observed plasma ascorbic acid concentrations as a function of time in patients with cancer at doses approximately 0.7–100 g administered intravenously. ([b] Padayatty, S. J. et al. 2004. *Ann. Intern. Med.* 140, 533–537; [c] Hoffer, L. J. et al. 2008. *Ann. Oncol.* 19, 1969–1974.)

intravenously by Ewan Cameron and colleagues. Of note, no ascorbic acid concentrations in patients were measured in any of these cancer studies. The profound differences in ascorbic acid concentrations from oral versus intravenous dosing provided one rationale to reopen the investigation of ascorbic acid in cancer treatment [51,52]. A second rationale was provided from clinical case reports of three patients with aggressive, pathologically confirmed cancers who were cured with intravenous ascorbic acid [61]. Despite limitations of these cases, they provided firm incentive to investigate pharmacologic ascorbate. These studies revealed that pharmacologic ascorbate concentrations killed cancer cells *in vitro* and *in vivo*

without affecting normal cells and normal tissues [62]. Pharmacologic ascorbate acted by generating hydrogen peroxide in extracellular fluid [63] (Figure 1.3). Hydrogen peroxide diffuses into cells and is a prodrug for reactive oxygen species that are toxic to cancers but not normal tissues. The central role of hydrogen peroxide was shown because cancer killing effects are negated with catalase, which dismutates hydrogen peroxide to water and oxygen [62,64]. Reactive oxygen species are likely to form in the setting of hydrogen peroxide, pharmacologic ascorbate, and trace iron found intracellularly and/or on domains of proteins facing the extracellular fluid or in extracellular

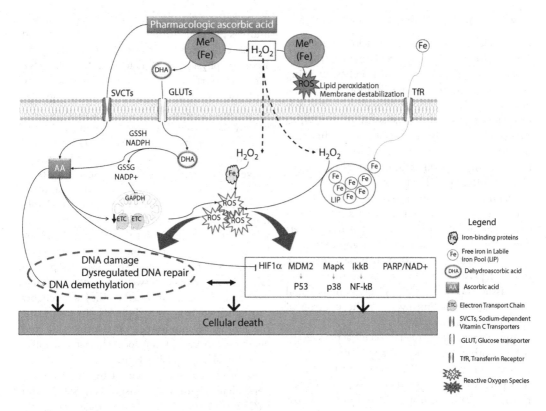

Figure 1.3. Multiple actions of pharmacologic ascorbic acid in cancer treatment. (Modified from Levine, M. and Violet, P. C. 2017. *Cancer Cell* 31, 467–469.)

fluid [63,64]. Subsequently, additional mechanisms have emerged that could explain ascorbic acid action in cancer treatment, including effects on hypoxia-inducible factor (HIF), regulation of DNA methylation, and effects of dehydroascorbic acid on ATP production in cancer cells [65–74,133] (Figure 1.3). Pharmacologic ascorbic acid has an exceptional safety profile and produces few adverse events in patients who are appropriately screened [54,75–77,132]. Small clinical trials indicate promise for pharmacologic ascorbate in metastatic pancreatic cancer, ovarian cancer, non-small cell lung cancer, glioblastoma, metastatic colon cancer, and metastatic gastric cancer [76–82]. Multiple clinical trials are open worldwide to test pharmacologic ascorbic acid in cancer treatment (see ClinicalTrials.gov for full list).

For more than 30 years, it has been described that ascorbic acid concentrations are low in critically ill patients [83–88]. Organ injury in sepsis is attenuated by ascorbic acid in animal models [88–90]. In one small single-center trial, requirement for vasopressors was decreased by pharmacologic ascorbic acid, and 28-day survival was improved

[91]. In another small single-center trial, which utilized corticosteroids and thiamine as well as pharmacologic ascorbate, in-hospital survival was increased [92], with a trend to increased survival in a third study [93]. Unfortunately, in sepsis studies, there can be inadvertent bias in single-center unblinded observational studies with relatively few participants [94–97]. These trials were single-center small-scale trials. Larger and prospective multiple-center clinical trials are warranted and are in progress to test whether ascorbic acid alone or ascorbic acid with corticosteroids and thiamine improve morbidity/mortality in septic patients (examples include NCT03389555, NCT03509350, NCT03258684, NCT03422159, and NCT03338569) and separately, whether ascorbic acid alone improves outcomes in patients with adult respiratory distress syndrome (NCT02106975). Although many possibilities exist, mechanism(s) of ascorbic acid efficacy are uncertain. Nevertheless, these trials were launched and are proceeding because of existent poor treatment options. Use of ascorbic acid pharmacologically is the foundation of these trials.

Tissue Transport and Unexpected Consequences

Included in the NIH pharmacokinetic studies were isolation from subjects of circulating neutrophils, monocytes, lymphocytes, and platelets, utilizing apheresis or specific cell purification techniques. Evaluation of ascorbate concentrations served as a proxy for tissue transport and accumulation. These data showed that tissue transport was a function of ingested dose and that all isolates achieved near-maximal ascorbic acid concentrations at the dose of 100 mg daily (Figure 1.4). The plasma concentration at which tissues saturated was similar to or lower than that concentration at which plasma saturated. Isolated cells and platelets had ascorbic acid concentrations approximately 20- to 50-fold greater than plasma across the range of plasma concentrations. The data show that tissue transport was a second fundamental mechanism contributing to tight control of ascorbate in healthy people.

Ascorbic acid is a charged molecule at physiologic pH. Ascorbic acid pK_a is approximately 4.2, and physiologic pH is 7.4. Because of its charge, transport would be predicted as an essential requirement for ascorbic acid movement across the intestine, into cells, and, as discussed later, for renal reabsorption. Candidate transported substances are ascorbic acid as such, or its oxidized product dehydroascorbic acid [98,99]. However, only ascorbic acid is detectable intracellularly. For dehydroascorbic acid to be a candidate substrate, it would be predicted to undergo near-instantaneous

intracellular reduction to ascorbate, the observed findings [99,100]. With these points as background foundation, we can address a fundamental issue in ascorbate physiology. Which substrate is essential for transport and activity: dehydroascorbic acid, ascorbic acid, or both? Endocrinology has several analogies. Both T4 (tetraiodothyronine, or levothyroxine) and T3 (triiodothyronine) are found in blood, the former approximately 100-fold higher than the latter, but only the latter is biologically active to regulate thyroid response elements in DNA. Similarly, in blood, 25-hydroxy vitamin D (calcidiol) is found at approximately 1000-fold higher concentrations than 1,25-dihydroxy vitamin D (calcitriol), but the latter is the active hormone. In comparison to ascorbic acid, dehydroascorbic acid concentrations in plasma are estimated as nearly two orders of magnitude less, and it is unclear whether they are reliably distinguished from zero [40,101]. To determine the biological importance of ascorbic acid and dehydroascorbic acid, transport characterization is an essential prerequisite (Figure 1.5).

Ascorbic acid is transported by two sodium-dependent vitamin C transporters, SVCT1 and SVCT2 [44,102,103]. SVCT1 was originally labeled as SLC23A2, and SVCT2 as SLC23A1, but these labels were later reversed, and the current nomenclature is SVCT1 as SLC23A1, and SVCT2 as SLC23A2. Neither of these transporters function as efflux transporters that allow ascorbate to exit cells on a basolateral surface, including in the intestine, the liver, and the kidney [104]. It is likely that there is either at least one additional SLC23 that functions as an ascorbic acid efflux

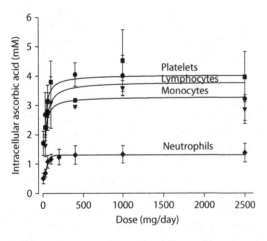

Figure 1.4. Ascorbic acid concentrations in circulating cells and platelets from healthy women as a function of dose. (From Levine, M. et al. 2001. *Proc. Natl. Acad. Sci. USA* 98, 9842–9846.)

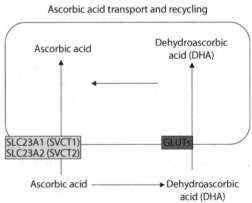

Figure 1.5. Mechanisms of ascorbic acid and dehydroascorbic acid transport and accumulation in cells.

transporter or a distinct solute carrier that is an efflux transporter, but efflux transporter activity coupled to a gene/protein has not yet been clearly demonstrated. SVCT2 knockout mice show that this transporter is essential for ascorbate uptake in many mouse tissues [105]. These mice have plasma ascorbate concentrations similar to those expected in wild-type mice, consistent with the existence of a distinct efflux transporter from the liver. SVCT1 knockout mice lose their renal threshold and ability to reabsorb ascorbic acid in the kidney [47]. However, SVCT1 knockout mice easily absorb ascorbic acid or an ascorbic acid halogen analog administered by gavage or in water. These data imply the existence of another distinct intestinal ascorbic acid transporter. It is likely that this intestinal absorption activity is distinct from dehydroascorbic acid and glucose transporters (GLUTs), because ascorbate halogen analogs and their halogenated dehydroascorbic acid products are not transported by glucose transporters [106], as discussed later. Consistent with findings for thousands of other genes and their protein products in humans, it is likely that there are mutations in SLC23A1, SLC23A2, putative efflux transporters, or an as yet unidentified intestinal transporter, any of which could produce aberrant protein function. Such cases have not yet been described but are likely to exist as yet undiagnosed. Humans with such mutations would be predicted to require massive (gram) doses of oral ascorbate or parenteral ascorbate to maintain either plasma or tissue concentrations. Predicted findings in people with such putative mutations include severely compromised bioavailability, an aberrant renal threshold, or specific cell populations with much lower ascorbate concentrations than predicted from pharmacokinetic data.

Dehydroascorbic acid in its hydrated form has a structure similar to that of glucose (Figure 1.6). Dehydroascorbic acid was predicted and found to be transported by GLUTs [98,107]. Of the 14 identified GLUTs [108], many transport dehydroascorbic acid [109]. GLUT1, expressed in many cell types, has an affinity for dehydroascorbic acid that can be estimated to be approximately three orders of magnitude more than glucose [107]. Multiple mechanisms have been proposed to explain GLUT transport activity [110]. These mechanisms are based on using glucose or glucose analogs as substrates. A confirmation model is believed to be the best explanation

the mechanism of GLUT1 transport [110], but it is unclear whether dehydroascorbic acid transport fits any of the molecular mechanistic transport models for GLUTs.

Because ascorbic acid and dehydroascorbic acid have distinct transport mechanisms, the contribution of each mechanism to ascorbic acid physiology in vivo can and has been explored. Before knockout mice were created, there was evidence supporting either a sole role for dehydroascorbic acid or a contribution by both mechanisms for cell accumulation [99,100,111]. Evidence supporting dehydroascorbic acid as the sole mechanism was based on much higher rates of transport compared to ascorbic acid, in cell systems and an animal model [112–114]. There were several flaws in this approach including use of nonphysiologic concentrations of dehydroascorbic acid, short time courses, inability to account for ascorbic acid transport under conditions where dehydroascorbic acid would not be present, and absence of explanation for sodium dependence of cell accumulation of ascorbate. A central limitation was the absence of techniques to measure dehydroascorbic acid with sensitivity, with specificity, and without interference from ascorbic acid [19,115].

Identification of distinct genes and transporters for each substrate was the necessary advance that allowed for the creation of knockout mice for ascorbate transporters. If dehydroascorbic acid transport was the primary mechanism of ascorbate accumulation, then absence of an ascorbate transporter should not affect ascorbate accumulation in knockout mice. Alternatively, if an ascorbate transporter was required for accumulation of the vitamin, knockout mice would have severe tissue deficiency, perhaps coupled to lethality. Mice lacking SVCT2 had severe generalized ascorbate deficiency and died within hours postpartum [105]. In theory, it is possible to test the converse hypothesis, for consequences of GLUT knockouts on dehydroascorbic acid transport and ascorbic acid accumulation. However, data from such knockouts would be confounded by effects of absence of glucose transport due to unintended multiple downstream consequences.

At first interpretation, data from SVCT2 knockout mice indicate the absence of a biological role for dehydroascorbic acid in cellular accumulation of ascorbate. A more balanced view is necessary. Not every mouse tissue was isolated and measured for

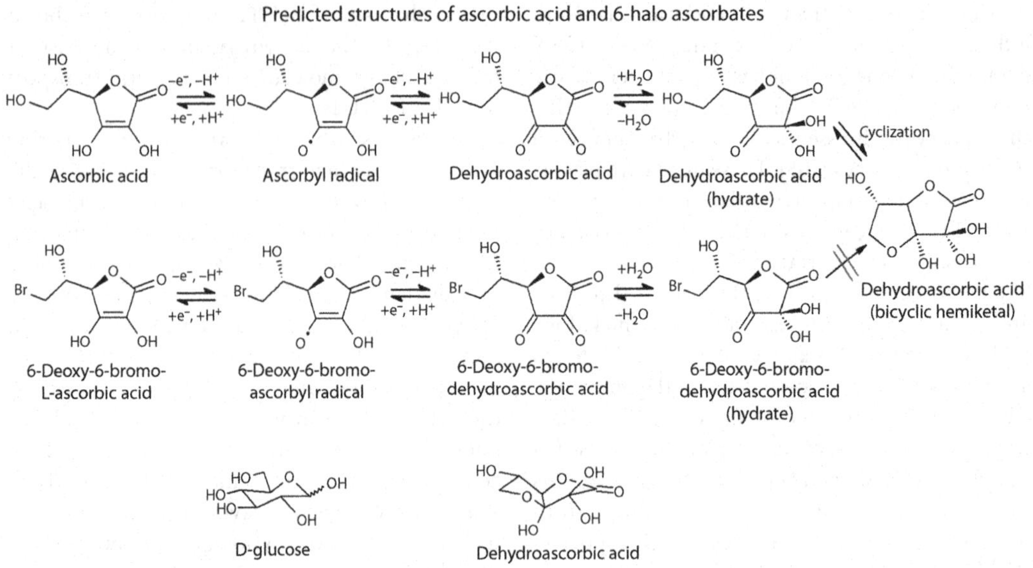

Predicted structures of ascorbic acid and 6-halo ascorbates

Figure 1.6. Structures of ascorbic acid, dehydroascorbic acid, and 6-halogen ascorbic acid. (From Corpe, C. P. et al. 2005. *J. Biol. Chem.* 280, 5211–5220.)

ascorbate accumulation, and identification could have been missed of a tissue that specifically required only dehydroascorbic acid transport. Obviously, mice and humans are different, and what is relevant for the mouse may not be relevant for humans, and vice versa. We do not have to look any further than endogenous ascorbate synthesis, found in most rodents but not in humans and nonhuman primates [116]. Because of the structural similarity of glucose and dehydroascorbic acid, and the difficulties in prevention and treatment of diabetes complications, there is an obligation to be thorough about the role of dehydroascorbic acid in ascorbate biology.

For these reasons, we pursued characterization of a possible role of dehydroascorbic acid in ascorbate accumulation. As earlier, knockout mice for GLUTs had unacceptable confounders. Therefore, we pursued a chemical knockout strategy. This path is based on the structure of dehydroascorbic acid and its oxidation and formation from ascorbic acid (Figure 1.6). Dehydroascorbic acid has multiple forms, based on whether it is hydrated. The hydrated form is the one that is structurally similar to glucose. The hydrated form requires that the sixth carbon on ascorbic acid has a hydroxyl group. The hydroxyl group is essential for formation of the cyclized hydrated structure of dehydroascorbic acid, that is, in its bicyclic hemiketal form [106]. Based on transport studies of ascorbate analogs, we hypothesized that substitution of a halogen for

the OH group on the sixth carbon of ascorbic acid (6-halo-ascorbic acid) would prevent cyclization of dehydroascorbic acid [99,106,111]. If correct, halo-ascorbic acid analogs when oxidized would be predicted to not be recognized by GLUTs and not transported. We synthesized 6-bromo-6-deoxy ascorbic acid (bromo ascorbic acid, or BromoAA) and tested transport of reduced and oxidized (BromoAA and 6-bromo-6-deoxy dehydroascorbic acid [BromoDHA]) by SVCTs and GLUTs. The findings were that BromoDHA was a complete chemical knockout for GLUTs: no BromoDHA was transported by GLUTs. Conversely, BromoAA was as good as or better than ascorbic acid as a substrate for SVCT1 and SVCT2 [106].

These findings were based on studies using expressed transporters in *Xenopus laevis* oocytes and in cells. To advance to an in vivo system, we utilized gulo$^{-/-}$ mice, those whose ability to synthesize ascorbate has been knocked out by disruption of gulonolactone oxidase, the last enzyme in the in vivo biosynthesis pathway for ascorbic acid [117]. In this regard, gulo$^{-/-}$ mice are similar to humans, where the gulonolactone oxidase gene has multiple mutations so that no enzyme is produced [116]. Both gulo$^{-/-}$ mice and humans require ascorbic acid for survival. To test whether dehydroascorbic acid had a physiologic function, gulo$^{-/-}$ mice were raised on either ascorbic acid or BromoAA for periods as long as 1 year. The findings were that red blood cells from mice raised on BromoAA visibly hemolyzed when plasma was prepared from whole blood by low speed centrifugation [118,119]. These findings were recapitulated when ascorbate repletion to mice was low. These findings pointed to an essential requirement for dehydroascorbic acid by red cells to them to maintain their ascorbic acid. Advancement of these experiments required development of a technique to measure ascorbic acid reliably in red blood cells, accomplished by HPLC with coulometric electrochemical detection [120]. Together, these experiments confirmed that only dehydroascorbic acid, and not ascorbic acid, is required for ascorbic acid in red blood cells in mice and humans.

The dependence of red cells on dehydroascorbic acid and intracellular ascorbate was characterized in some detail [118,119]. One mechanism was characterized by red cell fragility, mediated by β-spectrin. Under a threshold of approximately 10 μM in plasma and red cells, red blood cells become more rigid and osmotically fragile, producing hemolysis with centrifugation. Red cell fragility was reversed in vivo when ascorbic acid concentrations were increased above 20 μM, indicating that ascorbic acid was responsible (Figure 1.7). Similar findings were obtained in red cells from healthy humans and those with diabetes. The transporters mediating dehydroascorbic acid were identified GLUT4 in mouse red cells and GLUT1 in human red cells. Dehydroascorbic acid was competed by glucose, at concentrations similar to those in diabetes.

These findings in red cells have revealed unexpected potential roles of dehydroascorbic acid and ascorbic acid in diabetes. It has been known for more than 40 years that red cells from diabetic patients are more rigid, leading to slower flow

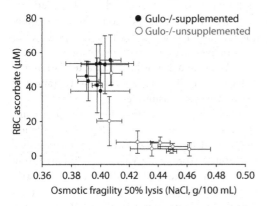

Figure 1.7. Osmotic fragility as a function of red blood cell ascorbic acid concentrations. (From Tu, H. et al. 2015. *EBioMedicine* 2, 1735–1750.)

per unit time [121,122]. Rigid red cells in diabetes would be expected to deliver less oxygen per unit time in capillaries. Clinically, this problem is all too apparent in diabetic subjects, manifested as microvascular disease. These findings with dehydroascorbic acid in red cells open new and exciting paths, with potential consequences for patients. It is possible that a contributing factor to pathogenesis of microvascular disease in diabetes is relative deficiency of ascorbic acid in red cells and perhaps plasma, deficiencies that may be reversible. Testing these possibilities will require time, great effort, and clinical experiments. Nevertheless, potential positive outcomes may be new and unanticipated means to prevent or delay microvascular disease in diabetes. What at first glance may appear to be convoluted paths are in fact open and direct. They derive linearly from pursuing findings from physiology and pharmacokinetic experiments, with concurrent studies of ascorbic acid and dehydroascorbic acid transport.

Findings with BromoAA in gulo$^{-/-}$ mice revealed that the red cell was dependent solely on dehydroascorbic acid transport for its internal ascorbate [118,119]. Mice raised on BromoAA for 1 year had minimal pathologic changes compared to controls. Nevertheless, it is possible that there are other tissues dependent on dehydroascorbic acid transport and that these were missed. The clue for red cells was visible hemolysis. Another tissue or tissues may be dependent on dehydroascorbic acid, without as clear a phenotype as found in red cells. As one example, dehydroascorbic acid has been proposed as the essential substrate to cross the blood-brain barrier [113,123]. Findings in BromoAA mice were not consistent with this possibility, as bromoAA was found in brain tissue of mice raised on BromoAA. However, dehydroascorbic acid transport remains as a possibility under some conditions, for example, with pharmacologic ascorbic acid concentrations in blood, when it is conceivable there would also be higher dehydroascorbic acid concentrations due to simple chemical equilibrium between pharmacologic ascorbic acid and its oxidation products.

Renal Reabsorption

Data from the NIH clinical physiology and pharmacokinetic studies showed that ascorbic acid was not excreted in urine at doses less than 100 mg daily in men and 60 mg in women [34,35]. When ascorbic acid was administered intravenously, so that limitations of intestinal absorption were bypassed, all of the administered doses were excreted at the highest doses of 500 and 1250 mg. From these data, it can be inferred that there is a renal threshold for ascorbic acid, and that when plasma and renal tubule ascorbic acid concentrations exceed this threshold, then ascorbic acid will be found in urine. A renal threshold is analogous to that found for phosphate and for glucose. A precise renal threshold has not been published, but it is anticipated that such a threshold can be calculated.

Ascorbic acid, as a water-soluble vitamin, would be expected to be freely filtered at renal glomeruli. Reabsorption would be predicted in proximal renal tubules, concurrent with SVCT1 localization. SVCT1 as a necessary reabsorptive ascorbic acid transporter was shown in studies from SVCT1 knockout mice [47]. These mice lose their ability to reabsorb ascorbic acid, and clear ascorbic acid similarly to inulin, which is filtered at the glomerulus but not reabsorbed. As previously mentioned, a basolateral ascorbic acid transporter should exist but has not been identified.

Once a renal threshold for ascorbic acid is described, it becomes possible to search for aberrant reabsorption. As earlier, aberrant reabsorption is predictable in people with mutations in some regions of SVCT1, or in the as yet unidentified efflux transporter. Renal tubule disease would also be predicted to result in a shift in reabsorption. One example is from a rare disease, Fabry disease, with known damage to renal tubules in humans. Another example is from a common disease, diabetes mellitus, where again renal tubular damage occurs. Similarly, drugs that have nephrotoxicity in the renal tubule may also induce inappropriate ascorbic acid appearance in urine.

Aberrant renal reabsorption of ascorbic acid might be a sign of tubular disease, as is microalbuminuria for glomerular disease in diabetes. Separately, aberrant reabsorption could produce quantitative losses in ascorbate, leading to lowered plasma and tissue concentrations. These possibilities are worth pursuing, having as a prerequisite the accurate characterization of ascorbic acid renal threshold in humans.

Utilization

The NIH clinical physiology and pharmacokinetic data show that ascorbic acid depletion occurs in approximately 30 days in healthy people. Utilization must be responsible and appears to occur at varying rates depending on plasma concentration.

Ascorbic acid concentrations have been reported to be near or at deficiency concentrations (<10 μM, or 0.2 mg/dL) in patients with critical illness of many types [83–87,124–126]. Low concentrations could have multiple causes, including absent dietary ingestion, aberrant absorption or renal excretion, and increased utilization secondary to disease-associated oxidant production leading to accelerated ascorbate oxidation. Clinical studies in sepsis suggesting improved clinical outcomes are based on replacing ascorbic acid, using pharmacologic concentrations. It is unknown whether replacement is simply correcting existent deficiency [127] or whether pharmacologic ascorbic acid has another role, analogous to that of pharmacologic ascorbic acid in cancer treatment by distinct or even similar mechanisms [63,88,128,129].

Accelerated ascorbic acid utilization in disease states may be a new frontier that can lead to discoveries for roles of ascorbic acid in disease treatment. Technologies are now available to drive this field forward. The foundation for this work is characterization of expected plasma and tissue concentrations in healthy people. With knowledge of normal ranges obtained from physiology and pharmacokinetic studies in healthy people, we can then learn when the abnormal occurs, why it occurs, and whether restoration of concentrations improves outcomes.

PARADIGM AND PROMISE

Physiology and pharmacokinetic experiments in humans for ascorbic acid are foundational for dietary recommendations, for treatment possibilities in cancer and sepsis, and perhaps for delay or even prevention of microvascular disease in diabetes. Renal threshold characterization and utilization studies may reveal new therapeutics and prevention possibilities.

Physiology and pharmacokinetic experiments with ascorbic acid utilized a depletion-repletion design, which is time consuming, labor intensive, and limited to only some vitamins. With the advent of liquid chromatography/mass spectrometry (LC/MS) and deuterated or 13C carbon molecules, advances for many vitamins are feasible with the same foundational principles. As one example, we created and utilized deuterated α-tocopherol preparations to characterize the physiology and pharmacokinetics of vitamin E in healthy people and those with hepato-steatosis [130] (manuscript submitted). Studies in ascorbic acid have shown us new ways forward. We believe that physiology and pharmacokinetic experiments in healthy people for many vitamin are not only warranted but necessary to advance nutritional principles and to give full meaning to what is optimal. The visionaries who wrote about optimal nutrition were unable to move forward because key tools were missing. Now, they are becoming available or are already here. Again, to paraphrase Alfred Harper: We have a blizzard of diseases, and we have a snowplow: let's go plow snow!

ABBREVIATIONS

BromoAA: 6-bromo-6-deoxy ascorbic acid
BromoDHA: 6-bromo-6-deoxy dehydroascorbic acid
DHA: dehydroascorbic acid
FNB: Food and Nutrition Board
GLUT: facilitated glucose transporter
RDA: recommended dietary allowance
SVCT: sodium-dependent vitamin C transporter

ACKNOWLEDGMENTS

Supported by the Intramural Research Program, National Institute of Diabetes and Digestive and Kidney Diseases (NIDDK), NIH, Bethesda, Maryland. Grants DK053211-13, DK053212-13, DK053218-13, DK54506-22.

REFERENCES

1. Harper, A. E. 1975. The recommended dietary allowances for ascorbic acid. *Ann. N.Y. Acad. Sci.* 258, 491–497.

2. Harper, A. E. 1985. Origin of recommended dietary allowances—An historic overview. *Am J Clin Nutr.* 41, 140–148.

3. Harper, A. E. 1987. Evolution of recommended dietary allowances—New directions? *Annu. Rev. Nutr.* 7, 509–537.

4. Food and Nutrition Board, Institute of Medicine. 1994. *How Should the Recommended Dietary Allowances Be Revised?* pp. 10–11, National Academy Press, Washington, DC. https://doi.org/10.17226/9194.

5. Food and Nutrition, B. 1998. Thiamin. In *Dietary Reference Intakes: Thiamin, Riboflavin, Niacin, Vitamin B6, Folate, Vitamin B12, Pantothenic Acid, Biotin, and Choline* (Food and Nutrition, B., eds.). pp. 58–86, National Academy Press, Washington, DC.

6. Food and Nutrition Board, Institute of Medicine. 2000. *Report of the Panel on Dietary Antioxidants and Related Compounds, Subcommittees on Upper Reference Levels of Nutrients and Interpretation and Uses of Dietary Reference Intakes, and the Standing Committee on the Scientific Evaluation of Dietary Reference Intakes.* pp. 95–185; 181–120; 434–435; 442–443, National Academy Press, Washington DC.

7. Tatum, E. L. and Beadle, G. W. 1942. Genetic control of biochemical reactions in Neurospora: An "Aminobenzoicless" Mutant. *Proc. Natl. Acad. Sci. USA* 28, 234–243.

8. Perla, D. and Marmortston, J. 1941. The effect of vitamin C on resistence. In *Natural Resistance and Clinical Medicine.* pp. 1038–1091, Little Brown & Co., Boston.

9. Williams, R. J. and Pelton, R. B. 1966. Individuality in nutrition: Effects of vitamin A-deficient and other deficient diets on experimental animals. *Proc. Natl. Acad. Sci. USA* 55, 126–134.

10. Williams, R. J. and Deason, G. 1967. Individuality in vitamin C needs. *Proc. Natl. Acad. Sci. USA* 57, 1638–1641.

11. Pauling, L. 1968. Orthomolecular psychiatry. Varying the concentrations of substances normally present in the human body may control mental disease. *Science* 160, 265–271.

12. Pauling, L. 1971. Vitamin C and common cold. *JAMA* 216, 332–332.

13. Cameron, E. and Pauling, L. 1978. Supplemental ascorbate in the supportive treatment of cancer: Reevaluation of prolongation of survival times in terminal human cancer. *Proc. Natl. Acad. Sci. USA* 75, 4538–4542.

14. Levine, M. 1986. New concepts in the biology and biochemistry of ascorbic acid. *N. Engl. J. Med.* 314, 892–902.

15. Levine, M. and Hartzell, W. 1987. Ascorbic acid: The concept of optimum requirements. *Ann. N.Y. Acad. Sci.* 498, 424–444.

16. Levine, M., Dhariwal, K. R., Washko, P. W., Butler, J. D., Welch, R. W., Wang, Y. H. and Bergsten, P. 1991. Ascorbic acid and in situ kinetics: A new approach to vitamin requirements. *Am. J. Clin. Nutr.* 54, 1157S–1162S.

17. Levine, M., Espey, M. G. and Padayatty, S. J. 2011. Vitamin C: A concentration-function approach yields pharmacology and therapeutic discoveries. *Adv. Nutr.* 2, 78–88.

18. Levine, M. and Eck, P. 2009. Vitamin C: Working on the x-axis. *Am. J. Clin. Nutr.* 90, 1121–1123.

19. Washko, P. W., Welch, R. W., Dhariwal, K. R., Wang, Y. and Levine, M. 1992. Ascorbic acid and dehydroascorbic acid analyses in biological samples. *Anal. Biochem.* 204, 1–14.

20. Pachla, L. A. and Kissinger, P. T. 1976. Determination of ascorbic acid in foodstuffs, pharmaceuticals, and body fluids by liquid chromatography with electrochemical detection. *Anal. Chem.* 48, 364–367.

21. Washko, P. W., Hartzell, W. O. and Levine, M. 1989. Ascorbic acid analysis using high-performance liquid chromatography with coulometric electrochemical detection. *Anal. Biochem.* 181, 276–282.

22. Dhariwal, K. R., Black, C. D. and Levine, M. 1991. Semidehydroascorbic acid as an intermediate in norepinephrine biosynthesis in chromaffin granules. *J. Biol. Chem.* 266, 12908–12914.

23. Friedman, S. and Kaufman, S. 1965. 3,4-dihydroxyphenylethylamine beta-hydroxylase. Physical properties, copper content, and role of copper in the catalytic acttivity. *J. Biol. Chem.* 240, 4763–4773.

24. Laduron, P. M. 1975. Evidence for a localization of dopamine-beta-hydroxylase within the chromaffin granules. *FEBS Lett.* 52, 132–134.

25. Dhawan, S., Duong, L. T., Ornberg, R. L. and Fleming, P. J. 1987. Subunit exchange between membranous and soluble forms of bovine dopamine beta-hydroxylase. *J. Biol. Chem.* 262, 1869–1875.

26. Kaufman, S. and Friedman, S. 1965. DOPAMINE-BETA-HYDROXYLASE. *Pharmacol. Rev.* 17, 71–100.

27. Stewart, L. C. and Klinman, J. P. 1988. Dopamine beta hydroxylase of adrenal chromaffin granules: Structure and function. *Annu. Rev. Biochem.* 57, 551–592.

28. Dhariwal, K. R., Washko, P., Hartzell, W. O. and Levine, M. 1989. Ascorbic acid within chromaffin granules. In situ kinetics of norepinephrine biosynthesis. *J. Biol. Chem.* 264, 15404–15409.

29. Fleming, P. J. and Kent, U. M. 1991. Cytochrome b561, ascorbic acid, and transmembrane electron transfer. *Am. J. Clin. Nutr.* 54, 1173S–1178S.

30. Baker, E. M., Hodges, R. E., Hood, J., Sauberlich, H. E. and March, S. C. 1969. Metabolism of ascorbic-1-14C acid in experimental human scurvy. *Am. J. Clin. Nutr.* 22, 549–558.

31. Baker, E. M., Hodges, R. E., Hood, J., Sauberlich, H. E., March, S. C. and Canham, J. E. 1971. Metabolism of 14C- and 3H-labeled L-ascorbic acid in human scurvy. *Am. J. Clin. Nutr.* 24, 444–454.

32. Kallner, A., Hartmann, D. and Hornig, D. 1979. Steady-state turnover and body pool of ascorbic acid in man. *Am. J. Clin. Nutr.* 32, 530–539.

33. Omaye, S. T., Skala, J. H. and Jacob, R. A. 1986. Plasma ascorbic acid in adult males: Effects of depletion and supplementation. *Am. J. Clin. Nutr.* 44, 257–264.

34. Levine, M., Conry-Cantilena, C., Wang, Y., Welch, R. W., Washko, P. W., Dhariwal, K. R., Park, J. B. et al. 1996. Vitamin C pharmacokinetics in healthy volunteers: Evidence for a Recommended Dietary Allowance. *Proc. Natl. Acad. Sci. USA* 93, 3704–3709.

35. Levine, M., Wang, Y., Padayatty, S. J. and Morrow, J. 2001. A new recommended dietary allowance of vitamin C for healthy young women. *Proc. Natl. Acad. Sci. USA* 98, 9842–9846.

36. King, J., Wang, Y., Welch, R. W., Dhariwal, K. R., Conry-Cantilena, C. and Levine, M. 1997. Use of a new vitamin C-deficient diet in a depletion/repletion clinical trial. *Am. J. Clin. Nutr.* 65, 1434–1440.

37. Yung, S., Mayersohn, M. and Robinson, J. B. 1982. Ascorbic acid absorption in humans: A comparison among several dosage forms. *J. Pharm. Sci.* 71, 282–285.

38. Mayersohn, M. 1992. Vitamin C bioavailability. *J. Nutr. Sci. Vitaminol. (Tokyo)* Spec No:446-9, 446–449.

39. Mangels, A. R., Block, G., Frey, C. M., Patterson, B. H., Taylor, P. R., Norkus, E. P. and Levander, O. A. 1993. The bioavailability to humans of ascorbic acid from oranges, orange juice and cooked broccoli is similar to that of synthetic ascorbic acid. *J. Nutr.* 123, 1054–1061.

40. Dhariwal, K. R., Hartzell, W. O. and Levine, M. 1991. Ascorbic acid and dehydroascorbic acid measurements in human plasma and serum. *Am. J. Clin. Nutr.* 54, 712–716.

41. German Nutrition, S. 2015. New reference values for vitamin C intake. *Ann. Nutr. Metab.* 67, 13–20.

42. Elste, V., Troesch, B., Eggersdorfer, M. and Weber, P. 2017. Emerging evidence on neutrophil motility supporting its usefulness to define vitamin C intake requirements. *Nutrients* 9.

43. Jameson, J. L., De Groot, L. J., de Kretser, D. M., Giudice, L. C., Grossman, A. B., Melmed, S., Potts, J.T., Weir, G. C. (Eds). 2016. *Endocrinology: Adult & Pediatric.* Saunders/Elsevier, Philadelphia, Pennsylvania.

44. Tsukaguchi, H., Tokui, T., Mackenzie, B., Berger, U. V., Chen, X. Z., Wang, Y., Brubaker, R. F. and Hediger, M. A. 1999. A family of mammalian Na+-dependent L-ascorbic acid transporters. *Nature* 399, 70–75.

45. Wang, Y., Mackenzie, B., Tsukaguchi, H., Weremowicz, S., Morton, C. C. and Hediger, M. A. 2000. Human vitamin C (L-ascorbic acid) transporter SVCT1. *Biochem. Biophys Res. Commun.* 267, 488–494.

46. Subramanian, V. S., Subramanya, S. B., Ghosal, A., Marchant, J. S., Harada, A. and Said, H. M. 2013. Modulation of function of sodium-dependent vitamin C transporter 1 (SVCT1) by Rab8a in intestinal epithelial cells: Studies utilizing Caco-2 cells and Rab8a knockout mice. *Dig. Dis. Sci.* 58, 641–649.

47. Corpe, C. P., Tu, H., Eck, P., Wang, J., Faulhaber-Walter, R., Schnermann, J., Margolis, S. et al. 2010. Vitamin C transporter Slc23a1 links renal reabsorption, vitamin C tissue accumulation, and perinatal survival in mice. *J. Clin. Invest.* 120, 1069–1083.

48. Frikke-Schmidt, H., Tveden-Nyborg, P. and Lykkesfeldt, J. 2016. L-dehydroascorbic acid can substitute l-ascorbic acid as dietary vitamin C source in guinea pigs. *Redox Biol.* 7, 8–13.

49. Tsujimura, M., Higasa, S., Nakayama, K., Yanagisawa, Y., Iwamoto, S. and Kagawa, Y. 2008. Vitamin C activity of dehydroascorbic acid in humans—Association between changes in the blood vitamin C concentration or urinary excretion after oral loading. *J. Nutr. Sci. Vitaminol.* (Tokyo) 54, 315–320.

50. Graumlich, J. F., Ludden, T. M., Conry-Cantilena, C., Cantilena, L. R., Jr., Wang, Y. and Levine, M. 1997. Pharmacokinetic model of ascorbic acid in healthy male volunteers during depletion and repletion. *Pharm. Res.* 14, 1133–1139.

51. Padayatty, S. J. and Levine, M. 2000. Reevaluation of ascorbate in cancer treatment: Emerging evidence, open minds and serendipity. *J. Am. Coll. Nutr.* 19, 423–425.

52. Padayatty, S. J., Sun, H., Wang, Y., Riordan, H. D., Hewitt, S. M., Katz, A., Wesley, R. A. and Levine, M. 2004. Vitamin C pharmacokinetics: Implications for oral and intravenous use. *Ann. Intern. Med.* 140, 533–537.

53. Parrow, N. L., Leshin, J. A. and Levine, M. 2013. Parenteral ascorbate as a cancer therapeutic: A reassessment based on pharmacokinetics. *Antioxid. Redox Signal.* 19(17), 2141–2156.

54. Shenoy, N., Creagan, E., Witzig, T. and Levine, M. 2018. Ascorbic acid in cancer treatment: Let the phoenix fly. *Cancer Cell.* 34, 700–706.

55. Cameron, E. and Campbell, A. 1974. The orthomolecular treatment of cancer. II. Clinical trial of high-dose ascorbic acid supplements in advanced human cancer. *Chem. Biol. Interact.* 9, 285–315.

56. Cameron, E. and Pauling, L. 1976. Supplemental ascorbate in the supportive treatment of cancer: Prolongation of survival times in terminal human cancer. *Proc. Natl. Acad. Sci. USA* 73, 3685–3689.

57. Campbell, A., Jack, T. and Cameron, E. 1991. Reticulum cell sarcoma: Two complete "spontaneous" regressions, in response to high-dose ascorbic acid therapy. A report on subsequent progress. *Oncology* 48, 495–497.

58. Creagan, E. T., Moertel, C. G., O'Fallon, J. R., Schutt, A. J., O'Connell, M. J., Rubin, J. and Frytak, S. 1979. Failure of high-dose vitamin C (ascorbic acid) therapy to benefit patients with advanced cancer. A controlled trial. *N. Engl. J. Med.* 301, 687–690.

59. Moertel, C. G., Fleming, T. R., Creagan, E. T., Rubin, J., O'Connell, M. J. and Ames, M. M. 1985. High-dose vitamin C versus placebo in the treatment of patients with advanced cancer who have had no prior chemotherapy. A randomized double-blind comparison. *N. Engl. J. Med.* 312, 137–141.

60. Wittes, R. E. 1985. Vitamin C and cancer [editorial]. *N. Engl. J. Med.* 312, 178–179.

61. Padayatty, S. J., Riordan, H. D., Hewitt, S. M., Katz, A., Hoffer, L. J. and Levine, M. 2006. Intravenously administered vitamin C as cancer therapy: Three cases. *CMAJ* 174, 937–942.

62. Chen, Q., Espey, M. G., Krishna, M. C., Mitchell, J. B., Corpe, C. P., Buettner, G. R., Shacter, E. and Levine, M. 2005. Pharmacologic ascorbic acid concentrations selectively kill cancer cells: Action as a pro-drug to deliver hydrogen peroxide to tissues. *Proc. Natl. Acad. Sci. USA* 102, 13604–13609.

63. Chen, Q., Espey, M. G., Sun, A. Y., Lee, J. H., Krishna, M. C., Shacter, E., Choyke, P. L. et al. 2007. Ascorbate in pharmacologic concentrations selectively generates ascorbate radical and hydrogen peroxide in extracellular fluid in vivo. *Proc. Natl. Acad. Sci. USA* 104, 8749–8754.

64. Chen, Q., Espey, M. G., Sun, A. Y., Pooput, C., Kirk, K. L., Krishna, M. C., Khosh, D. B., Drisko, J. and Levine, M. 2008. Pharmacologic doses

of ascorbate act as a prooxidant and decrease growth of aggressive tumor xenografts in mice. *Proc. Natl. Acad. Sci. USA* 105, 11105–11109.

65. Kuiper, C., Molenaar, I. G., Dachs, G. U., Currie, M. J., Sykes, P. H. and Vissers, M. C. 2010. Low ascorbate levels are associated with increased hypoxia-inducible factor-1 activity and an aggressive tumor phenotype in endometrial cancer. *Cancer Res.* 70, 5749–5758.

66. Vissers, M. C. M. and Das, A. B. 2018. Potential mechanisms of action for vitamin C in cancer: Reviewing the evidence. *Front. Physiol.* 9, 809.

67. Minor, E. A., Court, B. L., Young, J. I. and Wang, G. 2013. Ascorbate induces ten-eleven translocation (Tet) methylcytosine dioxygenase-mediated generation of 5-hydroxymethylcytosine. *J. Biol Chem.* 288, 13669–13674.

68. Blaschke, K., Ebata, K. T., Karimi, M. M., Zepeda-Martinez, J. A., Goyal, P., Mahapatra, S., Tam, A. et al. 2013. Vitamin C induces Tet-dependent DNA demethylation and a blastocyst-like state in ES cells. *Nature* 500, 222–226.

69. Yin, R., Mao, S. Q., Zhao, B., Chong, Z., Yang, Y., Zhao, C., Zhang, D. et al. 2013. Ascorbic acid enhances Tet-mediated 5-methylcytosine oxidation and promotes DNA demethylation in mammals. *J. Am. Chem. Soc.* 135, 10396–10403.

70. Gustafson, C. B., Yang, C., Dickson, K. M., Shao, H., Van Booven, D., Harbour, J. W., Liu, Z. J. and Wang, G. 2015. Epigenetic reprogramming of melanoma cells by vitamin C treatment. *Clin Epigenetics* 7, 51.

71. Liu, M., Ohtani, H., Zhou, W., Orskov, A. D., Charlet, J., Zhang, Y. W., Shen, H. et al. 2016. Vitamin C increases viral mimicry induced by 5-aza-2′-deoxycytidine. *Proc. Natl. Acad. Sci. USA* 113, 10238–10244.

72. Agathocleous, M., Meacham, C. E., Burgess, R. J., Piskounova, E., Zhao, Z., Crane, G. M., Cowin, B. L. et al. 2017. Ascorbate regulates haematopoietic stem cell function and leukaemogenesis. *Nature* 549, 476–481.

73. Cimmino, L., Dolgalev, I., Wang, Y., Yoshimi, A., Martin, G. H., Wang, J., Ng, V. et al. 2017. Restoration of TET2 function blocks aberrant self-renewal and leukemia progression. *Cell* 170, 1079–1095.e1020.

74. Yun, J., Mullarky, E., Lu, C., Bosch, K. N., Kavalier, A., Rivera, K., Roper, J. et al. 2015. Vitamin C selectively kills KRAS and BRAF mutant colorectal cancer cells by targeting GAPDH. *Science* 350, 1391–1396.

75. Padayatty, S. J., Sun, A. Y., Chen, Q., Espey, M. G., Drisko, J. and Levine, M. 2010. Vitamin C: Intravenous use by complementary and alternative medicine practitioners and adverse effects. *PLOS ONE* 5, e11414.

76. Monti, D. A., Mitchell, E., Bazzan, A. J., Littman, S., Zabrecky, G., Yeo, C. J., Pillai, M. V., Newberg, A. B., Deshmukh, S. and Levine, M. 2012. Phase I evaluation of intravenous ascorbic Acid in combination with gemcitabine and erlotinib in patients with metastatic pancreatic cancer. *PLOS ONE* 7, e29794.

77. Welsh, J. L., Wagner, B. A., van't Erve, T. J., Zehr, P. S., Berg, D. J., Halfdanarson, T. R., Yee, N. S. et al. 2013. Pharmacological ascorbate with gemcitabine for the control of metastatic and node-positive pancreatic cancer (PACMAN): Results from a phase I clinical trial. *Cancer Chemother. Pharmacol.* 71, 765–775.

78. Ma, Y., Chapman, J., Levine, M., Polireddy, K., Drisko, J. and Chen, Q. 2014. High-dose parenteral ascorbate enhanced chemosensitivity of ovarian cancer and reduced toxicity of chemotherapy. *Sci. Transl. Med.* 6, 222ra218.

79. Schoenfeld, J. D., Sibenaller, Z. A., Mapuskar, K. A., Wagner, B. A., Cramer-Morales, K. L., Furqan, M., Sandhu, S. et al. 2017. O2(-) and H2O2-Mediated disruption of Fe Metabolism causes the differential susceptibility of NSCLC and GBM cancer cells to pharmacological ascorbate. *Cancer Cell* 32, 268.

80. Polireddy, K., Dong, R., Reed, G., Yu, J., Chen, P., Williamson, S., Violet, P. C. et al. 2017. High dose parenteral ascorbate inhibited pancreatic cancer growth and metastasis: Mechanisms and a phase I/IIa study. *Sci. Rep.* 7, 17188.

81. Wang, F., He, M. M., Wang, Z. X., Li, S., Jin, Y., Ren, C., Shi, S. M. et al. 2019. Phase I study of high-dose ascorbic acid with mFOLFOX6 or FOLFIRI in patients with metastatic colorectal cancer or gastric cancer. *BMC Cancer* 19, 460.

82. Nauman, G., Gray, J. C., Parkinson, R., Levine, M. and Paller, C. J. 2018. Systematic review of intravenous ascorbate in cancer clinical trials. *Antioxidants (Basel)* 7.

83. Marcus, S. L., Dutcher, J. P., Paietta, E., Ciobanu, N., Strauman, J., Wiernik, P. H., Hutner, S. H., Frank, O. and Baker, H. 1987. Severe hypovitaminosis C occurring as the result of adoptive immunotherapy with high-dose interleukin 2 and lymphokine- activated killer cells. *Cancer Res.* 47, 4208–4212.

84. Schorah, C. J., Downing, C., Piripitsi, A., Gallivan, L., Al-Hazaa, A. H., Sanderson, M. J. and Bodenham, A. 1996. Total vitamin C, ascorbic acid, and dehydroascorbic acid concentrations in plasma of critically ill patients. *Am. J. Clin. Nutr.* 63, 760–765.

85. Borrelli, E., Roux-Lombard, P., Grau, G. E., Girardin, E., Ricou, B., Dayer, J. and Suter, P. M. 1996. Plasma concentrations of cytokines, their soluble receptors, and antioxidant vitamins can predict the development of multiple organ failure in patients at risk. *Crit. Care Med.* 24, 392–397.

86. Bonham, M. J., Abu-Zidan, F. M., Simovic, M. O., Sluis, K. B., Wilkinson, A., Winterbourn, C. C. and Windsor, J. A. 1999. Early ascorbic acid depletion is related to the severity of acute pancreatitis. *Br. J. Surg.* 86, 1296–1301.

87. Nathens, A. B., Neff, M. J., Jurkovich, G. J., Klotz, P., Farver, K., Ruzinski, J. T., Radella, F., Garcia, I. and Maier, R. V. 2002. Randomized, prospective trial of antioxidant supplementation in critically ill surgical patients. *Ann. Surg.* 236, 814–822.

88. Leelahavanichkul, A., Somparn, P., Bootprapan, T., Tu, H., Tangtanatakul, P., Nuengjumnong, R., Worasilchai, N. et al. 2015. High-dose ascorbate with low-dose amphotericin B attenuates severity of disease in a model of the reappearance of candidemia during sepsis in the mouse. *Am. J. Physiol. Regul. Integr. Comp. Physiol.* 309, R223–234.

89. Zhou, G., Kamenos, G., Pendem, S., Wilson, J. X. and Wu, F. 2012. Ascorbate protects against vascular leakage in cecal ligation and puncture-induced septic peritonitis. *Am. J. Physiol. Regul. Integr. Comp. Physiol.* 302, R409–416.

90. Wilson, J. X. 2013. Evaluation of vitamin C for adjuvant sepsis therapy. *Antioxid. Redox Signal.* 19, 2129–2140.

91. Zabet, M. H., Mohammadi, M., Ramezani, M. and Khalili, H. 2016. Effect of high-dose Ascorbic acid on vasopressor's requirement in septic shock. *J. Res. Pharm. Pract.* 5, 94–100.

92. Marik, P. E., Khangoora, V., Rivera, R., Hooper, M. H. and Catravas, J. 2017. Hydrocortisone, vitamin C, and thiamine for the treatment of severe sepsis and septic shock: A retrospective before-after study. *Chest* 151, 1229–1238.

93. Sadaka, F., Grady, J., Organti, N., Donepudi, B., Korobey, M., Tannehill, D. and O'Brien, J. 2019. Ascorbic acid, thiamine, and steroids in septic shock: Propensity matched analysis. *J. Intensive Care Med.* 885066619864541. doi: 10.1177/0885066619864541.

94. Rivers, E., Nguyen, B., Havstad, S., Ressler, J., Muzzin, A., Knoblich, B., Peterson, E., Tomlanovich, M. and Early Goal-Directed Therapy Collaborative, G. 2001. Early goal-directed therapy in the treatment of severe sepsis and septic shock. *N. Engl. J. Med.* 345, 1368–1377.

95. Investigators, P., Rowan, K. M., Angus, D. C., Bailey, M., Barnato, A. E., Bellomo, R., Canter, R. R. et al. 2017. Early, goal-directed therapy for septic shock—A patient-level meta-analysis. *N. Engl. J. Med.* 376, 2223–2234.

96. Pepper, D. J., Jaswal, D., Sun, J., Welsh, J., Natanson, C. and Eichacker, P. Q. 2018. Evidence underpinning the Centers for Medicare & Medicaid Services' severe sepsis and septic shock management bundle (SEP-1): A systematic review. *Ann. Intern. Med.* 168, 558–568.

97. Chow, J. H., Abuelkasem, E., Sankova, S., Henderson, R. A., Mazzeffi, M. A. and Tanaka, K. A. 2019. Reversal of vasodilatory shock: Current perspectives on conventional, rescue, and emerging vasoactive agents for the treatment of shock. *Anesth Analg.* PMID:31348056. DOI: 10.1213/ANE.0000000000004343.

98. Vera, J. C., Rivas, C. I., Fischbarg, J. and Golde, D. W. 1993. Mammalian facilitative hexose transporters mediate the transport of dehydroascorbic acid. *Nature* 364, 79–82.

99. Welch, R. W., Wang, Y., Crossman, A., Jr., Park, J. B., Kirk, K. L. and Levine, M. 1995. Accumulation of vitamin C (ascorbate) and its oxidized metabolite dehydroascorbic acid occurs by separate mechanisms. *J. Biol. Chem.* 270, 12584–12592.

100. Washko, P. W., Wang, Y. and Levine, M. 1993. Ascorbic acid recycling in human neutrophils. *J. Biol. Chem.* 268, 15531–15535.

101. Lykkesfeldt, J. 2012. Ascorbate and dehydroascorbic acid as biomarkers of oxidative stress: Validity of clinical data depends on vacutainer system used. *Nutr. Res.* 32, 66–69.

102. May, J. M. 2011. The SLC23 family of ascorbate transporters: Ensuring that you get and keep your daily dose of vitamin C. *Br. J. Pharmacol.* 164, 1793–1801.

103. Burzle, M., Suzuki, Y., Ackermann, D., Miyazaki, H., Maeda, N., Clemencon, B., Burrier, R. and Hediger, M. A. 2013. The sodium-dependent ascorbic acid transporter family SLC23. *Mol. Aspects Med.* 34, 436–454.

104. Eck, P., Kwon, O., Chen, S., Mian, O. and Levine, M. 2013. The human sodium-dependent ascorbic acid transporters SLC23A1 and SLC23A2 do not

mediate ascorbic acid release in the proximal renal epithelial cell. *Physiol. Rep.* 1, e00136.

105. Sotiriou, S., Gispert, S., Cheng, J., Wang, Y., Chen, A., Hoogstraten-Miller, S., Miller, G. F. et al. 2002. Ascorbic-acid transporter Slc23a1 is essential for vitamin C transport into the brain and for perinatal survival. *Nat. Med.* 8, 514–517.

106. Corpe, C. P., Lee, J. H., Kwon, O., Eck, P., Narayanan, J., Kirk, K. L. and Levine, M. 2005. 6-Bromo-6-deoxy-L-ascorbic acid: An ascorbate analog specific for Na+-dependent vitamin C transporter but not glucose transporter pathways. *J. Biol. Chem.* 280, 5211–5220.

107. Rumsey, S. C., Kwon, O., Xu, G. W., Burant, C. F., Simpson, I. and Levine, M. 1997. Glucose transporter isoforms GLUT1 and GLUT3 transport dehydroascorbic acid. *J. Biol. Chem.* 272, 18982–18989.

108. Yan, N. 2017. A glimpse of membrane transport through structures-advances in the structural biology of the GLUT glucose transporters. *J. Mol. Biol.* 429, 2710–2725.

109. Corpe, C. P., Eck, P., Wang, J., Al-Hasani, H. and Levine, M. 2013. Intestinal dehydroascorbic acid (DHA) transport mediated by the facilitative sugar transporters, GLUT2 and GLUT8. *J. Biol. Chem.* 288, 9092–9101.

110. Mueckler, M. and Thorens, B. 2013. The SLC2 (GLUT) family of membrane transporters. *Mol. Aspects Med.* 34, 121–138.

111. Rumsey, S. C., Welch, R. W., Garraffo, H. M., Ge, P., Lu, S. F., Crossman, A. T., Kirk, K. L. and Levine, M. 1999. Specificity of ascorbate analogs for ascorbate transport. Synthesis and detection of [(125)I]6-deoxy-6-iodo-L-ascorbic acid and characterization of its ascorbate-specific transport properties. *J. Biol. Chem.* 274, 23215–23222.

112. Vera, J. C., Rivas, C. I., Zhang, R. H., Farber, C. M. and Golde, D. W. 1994. Human HL-60 myeloid leukemia cells transport dehydroascorbic acid via the glucose transporters and accumulate reduced ascorbic acid. *Blood* 84, 1628–1634.

113. Agus, D. B., Gambhir, S. S., Pardridge, W. M., Spielholz, C., Baselga, J., Vera, J. C. and Golde, D. W. 1997. Vitamin C crosses the blood-brain barrier in the oxidized form through the glucose transporters. *J. Clin. Invest.* 100, 2842–2848.

114. Perez-Cruz, I., Carcamo, J. M. and Golde, D. W. 2003. Vitamin C inhibits FAS-induced apoptosis in monocytes and U937 cells. *Blood* 102, 336–343.

115. Levine, M., Wang, Y. and Rumsey, S. C. 1999. Analysis of ascorbic acid and dehydroascorbic acid in biological samples. *Methods Enzymol.* 299, 65–76.

116. Padayatty, S. J. and Levine, M. 2016. Vitamin C: The known and the unknown and Goldilocks. *Oral Dis.* 22, 463–493.

117. Maeda, N., Hagihara, H., Nakata, Y., Hiller, S., Wilder, J. and Reddick, R. 2000. Aortic wall damage in mice unable to synthesize ascorbic acid. *Proc. Natl. Acad. Sci. USA* 97, 841–846.

118. Tu, H., Li, H., Wang, Y., Niyyati, M., Wang, Y., Leshin, J. and Levine, M. 2015. Low red blood cell vitamin C concentrations induce red blood cell fragility: A link to diabetes via glucose, glucose transporters, and dehydroascorbic acid. *EBioMedicine* 2, 1735–1750.

119. Tu, H., Wang, Y., Li, H., Brinster, L. R. and Levine, M. 2017. Chemical transport knockout for oxidized vitamin C, dehydroascorbic acid, reveals its functions in vivo. *EBioMedicine* 23, 125–135.

120. Li, H., Tu, H., Wang, Y. and Levine, M. 2012. Vitamin C in mouse and human red blood cells: An HPLC assay. *Anal. Biochem.* 426, 109–117.

121. McMillan, D. E., Utterback, N. G. and La, P. J. 1978. Reduced erythrocyte deformability in diabetes. *Diabetes* 27, 895–901.

122. Kamada, T., McMillan, D. E., Yamashita, T. and Otsuji, S. 1992. Lowered membrane fluidity of younger erythrocytes in diabetes. *Diabetes Res. Clin.Pract.* 16, 1–6.

123. Huang, J., Agus, D. B., Winfree, C. J., Kiss, S., Mack, W. J., McTaggart, R. A., Choudhri, T. F. et al. 2001. Dehydroascorbic acid, a blood-brain barrier transportable form of vitamin C, mediates potent cerebroprotection in experimental stroke. *Proc. Natl. Acad. Sci. USA* 98, 11720–11724.

124. Story, D. A., Ronco, C. and Bellomo, R. 1999. Trace element and vitamin concentrations and losses in critically ill patients treated with continuous venovenous hemofiltration [see comments]. *Crit. Care Med.* 27, 220–223.

125. Wang, Y., Liu, X. J., Robitaille, L., Eintracht, S., Macnamara, E. and Hoffer, L. J. 2013. Effects of vitamin C and vitamin D administration on mood and distress in acutely hospitalized patients. *Am. J. Clin. Nutr.* 98, 705–711.

126. Oudemans-van Straaten, H. M., Spoelstra-de Man, A. M. and de Waard, M. C. 2014. Vitamin C revisited. *Crit. Care.* 18, 460.

127. Kawade, N., Tokuda, Y., Tsujino, S., Aoyama, H., Kobayashi, M., Murai, A. and Horio, F. 2018. Dietary intake of ascorbic acid attenuates lipopolysaccharide-induced sepsis and septic inflammation in ODS rats. *J. Nutr. Sci. Vitaminol* (*Tokyo*) 64, 404–411.

128. Mohammed, B. M., Fisher, B. J., Kraskauskas, D., Farkas, D., Brophy, D. F., Fowler, A. A., 3rd and Natarajan, R. 2013. Vitamin C: A novel regulator of neutrophil extracellular trap formation. *Nutrients* 5, 3131–3151.

129. Fisher, B. J., Kraskauskas, D., Martin, E. J., Farkas, D., Puri, P., Massey, H. D., Idowu, M. O., Brophy, D. F., Voelkel, N. F., Fowler, A. A., 3rd, and Natarajan, R. 2014. Attenuation of sepsis-induced organ injury in mice by vitamin C. *JPEN J. Parenter Enteral Nutr.* 38, 825–839.

130. Traber, M. G., Leonard, S. W., Ebenuwa, I., Violet, P.-C., Wang, Y., Niyyati, M., Padayatty, S. J. et al. 2019. Vitamin E absorption and kinetics in women as modulated by fat and fasting stuided using two deuterium labeled alpha-tocopherol in a cross over study. *Am. J. Clin. Nutr.* 110(5), 1148–1167.

131. Levine, M., Wang, Y., Katz, A., Eck, P., Kwon, O., Chen, S., Lee, J. H. and Padayatty, S. J. 2001. Ideal vitamin C intake. *Biofactors* 15, 71–74.

132. Hoffer, L. J., Levine, M., Assouline, S., Melnychuk, D., Padayatty, S. J., Rosadiuk, K., Rousseau, C., Robitaille, L. and Miller, W. H., Jr. 2008. Phase I clinical trial of i.v. ascorbic acid in advanced malignancy. *Ann. Oncol.* 19, 1969–1974.

133. Levine, M. and Violet, P. C. 2017. Data Triumph at C. *Cancer Cell* 31, 467–469.

PART TWO

Chemistry and Biology
of Vitamin C

CHAPTER TWO

Chemistry and Biochemistry of Vitamin C in Mammalian Systems

Margreet C.M. Vissers, Juliet M. Pullar, and Nicholas Smirnoff

DOI: 10.1201/9780429442025-2

CONTENTS

INTRODUCTION

Interest in the biological functions of vitamin C (ascorbate) has been fueled by the fact that humans, together with other primates and some other animal species, have lost the capacity to synthesize this compound and are dependent on dietary intake, making them prone to deficiency [1–3]. Since its discovery early in the twentieth century, there has been an ongoing discussion regarding the daily intake requirement to support the many biological functions of ascorbate in the body that is informed by knowledge and understanding of its chemistry and metabolism [4–11]. Regardless of whether ascorbate is acquired via the diet, as occurs in humans, or as a result of biosynthesis in the liver, as in most other mammals, there is essentially no difference in its means of distribution throughout the body. Ascorbate is highly water soluble and is distributed to the tissues via the circulation, with active transport into cells occurring largely via the sodium-dependent vitamin C transporters (SVCTs) [12–14]. When plasma levels are below a maximum of around 80 μM, reuptake occurs in the kidney, and ascorbate does not appear in the urine until plasma and tissue saturation is reached [15–17]. Ascorbate uptake and turnover are considered in some detail in other chapters in this book; hence, the discussion in this chapter is limited to a consideration of the biochemistry and metabolism of ascorbate in mammalian cells.

ASCORBATE CHEMISTRY

The complex chemistry of ascorbate belies the apparently simple nature of this small molecule, and the reader is referred to excellent reviews of this topic [2,18–21]. Ascorbic acid (AscH$_2$) is a highly water-soluble compound with a pK_a of 4.17, which means that at neutral pH, greater than 99% is present as the anionic ascorbate form (AscH$^-$), and this is the chemically active form of vitamin C in most biological settings [2,18,21]. Diffusion of the anion across cell membranes is limited, and active transport via the SVCTs is vital [12–14]. The pK_a suggests that there may be biological circumstances where AscH$_2$ exists at significant concentrations. This would occur in the acid environment of the stomach, at some inflammatory sites [22,23], and also in the localized environment of solid tumors [24–26], and under these conditions some passive diffusion of $AscH_2$ into cells is a possibility. However, the extent to which this means of uptake contributes to ascorbate bioavailability is unknown.

Ascorbate undergoes single-electron oxidation with biological free radicals or transition metals to generate the ascorbyl (monodehydroascorbate) radical (Figure 2.1). Two-electron oxidants such as hypochlorous acid also react with ascorbate to generate dehydroascorbate (DHA) [27–29], but it is believed that such reactions also occur in two single-electron steps [30]. The reaction rate constants for ascorbate with biologically relevant radicals such as superoxide, tocopheroxyl, alkoxyl/peroxyl, tyrosyl, and tryptophan radicals are generally $>10^5$ M^{-1} s^{-1}, and this indicates a significant capacity as an efficient free radical scavenger [18,30]. Therefore, at sufficiently high concentrations, ascorbate could play an important radical scavenging antioxidant role. However, as recently pointed out, for maximum efficiency and in competition with other targets, this would require concentrations above 10 mM [21], which is a value somewhat higher than the low millimolar levels that are reported for most intracellular locations [18,31–38]. Some radical scavenging antioxidant activity is to be expected, particularly in cellular subcompartments, and there is substantial evidence that ascorbate regenerates tocopherol from the lipid-phase tocopheroxyl radicals and thereby limits lipid peroxidation damage to cells and tissues [39–48].

Recycling of the Ascorbyl Radical

As the primary product of these radical scavenging reactions, the ascorbyl radical can dismutate, regenerating $AscH^-$ and producing DHA [49,50] (Figure 2.1). In addition, a number of means of reduction of the radical have been identified in mammalian cells. A cytochrome b_5 ascorbyl radical reductase activity that utilizes nicotinamide adenine dinucleotide (NADH) has been found associated with cell membranes, including the mitochondrial outer membrane, microsomes, and plasma membrane [51–55]. In addition, some cytosolic enzymes such as thioredoxin reductase have ascorbyl reductase activity [56]. Of interest is that the localization of these enzyme systems coincides with sites of high oxidant flux (mitochondrial membranes), the site of ascorbate synthesis in animals with functional gulonolactone oxidase activity (liver microsomes) or sites of ascorbate-dependent enzyme activity (the adrenal medulla) [57–59]. These observations suggest that there may be a functional requirement for the regeneration of reduced ascorbate at these locations.

Autoxidation

Strictly speaking, the autoxidation of ascorbate describes the uncatalyzed reaction with oxygen. This reaction occurs relatively slowly, with

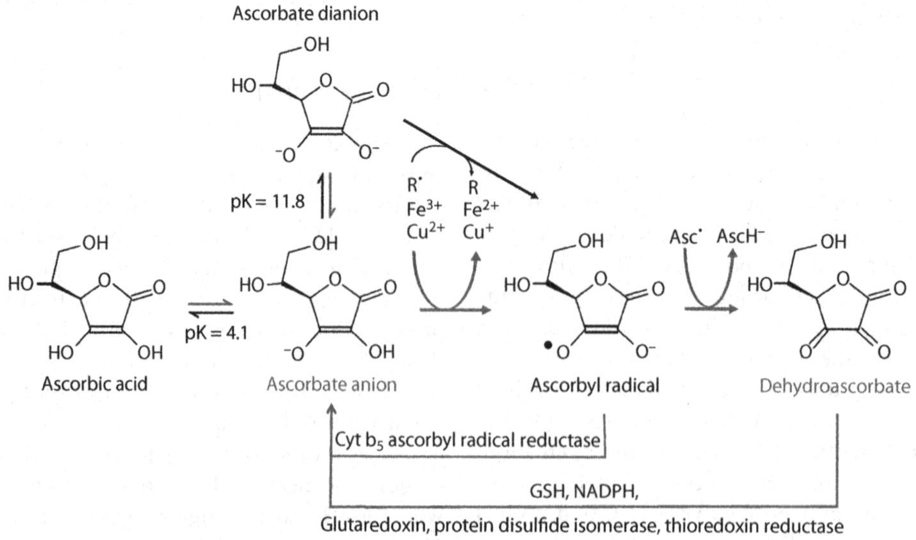

Figure 2.1. Forms of ascorbic acid present in mammalian systems. The predominant forms present at neutral pH are labeled in red, and relevant directions of reactions are indicated by red arrows.

a second-order rate constant estimated at approximately $10^{-4} \, M^{-1} \, s^{-1}$ [60], and the extent to which this oxidation contributes to the turnover of ascorbate in biological systems is unknown. The availability of oxygen in many tissues is far below levels present in ambient air solutions in vitro [61–63], and the combination of a slow reaction and a limited supply of oxygen is likely to result in low turnover of ascorbate by autoxidation in vivo. The poor availability of oxygen is particularly pertinent to the consideration of ascorbate autoxidation at sites of inflammation or in tumors, sites that are known to be hypoxic, but where the functions and availability of ascorbate are of great interest [63–71].

The reaction of ascorbate with oxygen is greatly accelerated in the presence of metal catalysts, and it has been proposed that most oxidation observed in vitro is due to the presence of contaminant metals in cell culture media or buffer solutions [3]. Ascorbate was shown to be stable over a 6-hour period in phosphate-buffered saline but was rapidly lost from tissue culture media in a metal-catalyzed reaction that was inhibited by the presence of a metal chelator [3]. The ascorbyl radical has also been detected in a number of biological fluids, which indicates oxidative turnover, but notably, detection has generally been in tissues under conditions of stress, or in the presence of a metal catalyst [72–77].

CHEMISTRY OF DEHYDROASCORBATE

In the course of carrying out its antioxidant and radical scavenging functions, ascorbate oxidation can result in the generation of DHA, either by further oxidation or by dismutation of the ascorbyl radical (Figure 2.1). Ascorbate is also oxidized when acting as a reducing agent for copper (Cu)- or iron (Fe)-containing enzymes, being required to maintain the active site transition metal in a reduced state [78–82], and this activity could contribute significantly to the daily turnover [12,13,83–86]. Increased ascorbate consumption and turnover in locations with high ascorbate-dependent enzymatic activity may explain the active accumulation of millimolar intracellular concentrations of the vitamin at these sites [35,87]; these concentrations are well above the reported K_m (Michaelis constant) [81–83,88] and may provide a buffer to ensure the continuation of optimal enzyme activity in these tissues. The cofactor activity of ascorbate is the topic of Chapter 5.

Just how much of the total body pool in vivo is present as DHA remains a point of debate. Although significant quantities may be generated under conditions of severe oxidative stress, such as at inflammatory sites, the concentration of DHA in most biological fluids is very low and represents only a very small percentage of the measured ascorbate in any biological sample [3,89–94]. It is highly likely that DHA concentrations are often overestimated when analyzing tissue or plasma samples unless scrupulous care is taken with sample handling to prevent in vitro oxidation [3].

GLUT-Mediated Uptake of DHA

DHA can compete with glucose for the glucose transporters (GLUTs), particularly GLUT1, GLUT3, and to some extent GLUT4, and can thereby be taken up into cells [95–99]. The uptake is efficient in the absence of glucose but has been shown to be inhibited by physiological glucose concentrations in a number of cell types [94,100] and, given the low concentrations of DHA predicted in vivo, it is likely that DHA uptake into cells contributes little to the overall pharmacokinetics of ascorbate distribution in the tissues. Animal models that lack SVCT1 and SVCT2 exhibit severe deficiency phenotypes implying that DHA transport is unlikely to be a major pathway for ascorbate distribution [14,101]. However, there are likely exceptions to this, and the red blood cell is a particular case in point. Mature circulating red cells lack the SVCTs, these having been lost during maturation from the bone marrow precursors, but express significant amounts of GLUT1 [102]. Interestingly, red cell ascorbate concentrations are generally equivalent to plasma levels [103], and given the large bulk of these cells in the circulation, they could form a significant reserve ascorbate pool. In support of this idea, it has been reported that GLUT1 expression in the red cell appears to be higher in those species that have lost ascorbate biosynthetic capacity [104]. Ascorbate is transported efficiently by GLUT1 [105,106], and red cells from mice that synthesize ascorbate express low levels of GLUT1 but do express GLUT4, which results in much less efficient uptake of DHA, and red cell levels are low in these animals [104,107–109].

Neutrophils are dependent on glycolysis for energy and express high levels of GLUT1, which enhances their capacity for DHA uptake [98,110–112]. DHA can be scavenged by these cells at inflammatory sites, which has been proposed to be a significant ascorbate recycling mechanism in these oxidative environments [98,110–114].

GLUT1 expression is often increased in cancer cells in response to upregulation of the hypoxia-inducible factors (HIFs) [115–117]. This expression is associated with the increased dependency of the cancer cells on glycolysis, particularly in the oxygen-deprived tumor environments, but whether GLUT1 contributes to ascorbate accumulation in cancer is unknown. A connection between the uptake of DHA and the upregulation of GLUT1 in KRAS and BRAF mutant cells was proposed as an anticancer mechanism in colorectal cancer [118]. KRAS and BRAF mutations induce a glycolytic phenotype and are common in colorectal cancer. In an in vitro setting, KRAS and BRAF mutant cancer cells accumulated ascorbate by a GLUT1-dependent mechanism. The rapid uptake of DHA by this means was associated with loss of cell viability that was attributed to the oxidative stress on the cell. Loss of GSH from the cells was noted, but the dependency of this on DHA uptake versus an oxidative stress from H_2O_2 generated by ascorbate in the medium was not determined. When KRAS and BRAF cancer cells were implanted in a nude mouse model, tumor growth was slowed with high doses of vitamin C administered daily by intraperitoneal injection. When considering the potential for DHA uptake via the GLUTs to effect an antitumor activity, it would be of interest to determine whether DHA is present following high-dose ascorbate administration at concentrations sufficient to be able to compete with glucose for GLUT-mediated uptake. Two other studies have indicated that ascorbate can reverse the glycolytic Warburg phenotype in cetuximab-treated KRAS mutant colon cancer cell lines [119], and this activity occurred following SVCT2-dependent ascorbate uptake [120].

Regeneration of DHA

DHA exists in equilibrium with a more stable hydrated bicyclic form, and turnover of ascorbate reflects the degradation of DHA [18,21], as shown in Figure 2.2. DHA has a half-life of around 6 minutes in solution at neutral pH and 37°C, undergoing hydrolysis in an irreversible reaction to form 2,3-diketogulonic acid (2,3-DKG) (Figure 2.2) [121]. This reaction represents effective loss of ascorbate from the system. The daily intake of ascorbate to maintain tissue saturation in healthy humans is around 100 mg per day [6,16,122,123], which indicates a daily loss of approximately 10% of the total body pool of 900–2000 mg [124–126]. This is likely to represent only a fraction of the ascorbate that is oxidized in redox cycling and suggests that much of the ascorbate pool is recycled [126].

In the previous section, we discussed the regeneration of ascorbate from the ascorbyl radical.

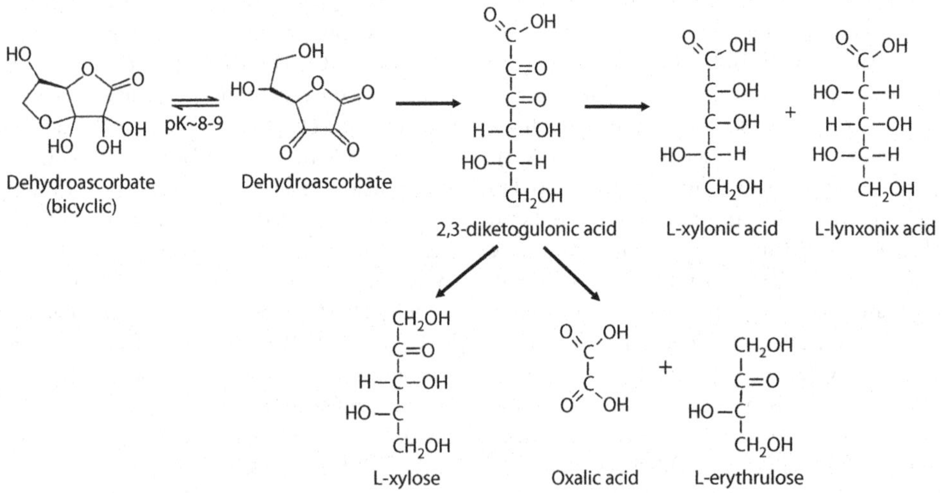

Figure 2.2. The metabolism and breakdown of DHA in mammalian systems. DHA exists predominantly in the more stable bicyclic form. It is degraded by irreversible hydrolysis to generate 2,3-DKG, which can be further hydrolyzed to generate the products shown.

Similarly, a number of cellular mechanisms are able to reduce DHA, and this occurs by two-electron reduction [10], with GSH, NADH, and NADPH serving as electron donors (Figure 2.1) [10,19,127]. GSH can react directly with DHA, but this reaction may not be highly favorable at physiologic GSH concentrations [128]. Rather, enzyme-dependent reduction of DHA mediated by glutaredoxin [129], protein disulfide isomerase [130,131], and thioredoxin reductase [33,132–134] is considered to be responsible for most of this activity in vivo. The dependency of the body on GSH to support this activity has been well demonstrated in studies with GSH-deficient animals in which tissue ascorbate levels were lowered, together with an increased DHA:AscH$^-$ ratio [135,136]. In addition, GSH was shown to be required for ascorbate recycling in cultured endothelial cells [133], and a NADPH-dependent thioredoxin reductase activity was identified in rat liver cells [132]. These enzymatic systems are considered to be the major pathways for ascorbate regeneration from DHA, but other options may also be available. For example, 3α-hydroxysteroid dehydrogenase, an NADPH-dependent oxidoreductase, could reduce DHA under pathologic conditions when its concentrations are high [137]. Therefore, there are numerous possibilities for the recycling of ascorbate that together will allow for many cycles of ascorbate oxidation and will minimize loss from the body pool.

Fate of DHA Breakdown Products

The breakdown of DHA to 2,3-DKG via hydrolysis not only results in the irreversible loss of ascorbate, it also introduces a plethora of metabolic possibilities due to the complex chemistry of 2,3-DKG breakdown (Figure 2.2). This topic was previously well reviewed [10,18,19,21,138] and is only summarized here. The 2,3-DKG readily breaks down, with decarboxylation generating L-xylonate and L-lyxonate [139]. This may be a less common reaction in mammalian systems, but if formed, these compounds can contribute to pentose phosphate pathways [10,19]. Alternatively, 2,3-DKG degradation can result in the production of L-erythrulose and oxalic acid [2,19,21], with potential metabolic consequences. The reactive compound L-erythrulose can glycate proteins and cause protein cross-linking, and these reactions have been proposed to contribute to cataract

formation in the lens of individuals with diabetes [140].

The production of oxalate through the oxidation and breakdown of ascorbate has for many years attracted clinical interest because of the potential to contribute to calcium oxalate crystals and the formation of kidney stones [141–144]. In 1999, an analysis of a random cohort of U.S. citizens from the Second National Health and Nutrition Examination Survey, 1976–1980, concluded that there is no association between serum ascorbate levels and the incidence of kidney stones [145]. Several prospective cohort analyses have suggested that regular supplement use could double the rate of kidney stone formation in men [146,147] but not in women [148,149]. However, it should be noted that in these analyses, vitamin C intake is often estimated from dietary recall or supplement use at a single time point, but regardless of potential confounders, the evidence appears sufficient to warrant more controlled intervention studies. Other studies have been more direct and have indicated low levels of urinary oxalate excretion as a proportion of intake following high daily intake of ascorbic acid [150] including by high-dose intravenous infusions [144]. At this point in time, a significant degree of uncertainty remains around the question of long-term risk of kidney stones, but it appears that there is little evidence for an increased risk with elevated intake of ascorbate over the short term.

PRO-OXIDANT ACTIVITY OF ASCORBATE

Despite the many aspects of ascorbate function that depend on its acting as a reducing agent, there is great interest in its ability to act as a pro-oxidant. As described in Figure 2.3, the autoxidation of ascorbate, that is, the reaction with oxygen, generates the superoxide radical $O_2^{\cdot-}$, which dismutates to produce H_2O_2 [151–153]. As mentioned previously, this reaction occurs spontaneously but is relatively slow in the absence of a metal chelator [3,152,153]. Metal-catalyzed oxidation is directly proportional to the concentration of ascorbate, and at millimolar concentrations, this can result in H_2O_2 generation in quantities high enough to be cytotoxic to cells in vitro [154–159]. The rate of ascorbate oxidation is slowed in the presence of iron chelators, and ascorbate-mediated cytotoxicity is inhibited by catalase, confirming dependency on H_2O_2

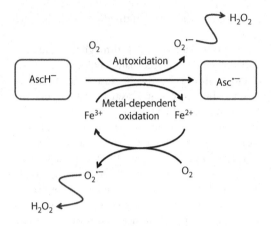

Figure 2.3. Pro-oxidant reactions of ascorbate. Either autoxidation, or the recycling of Fe^{2+} from Fe^{3+} in the presence of oxygen, allows the generation of $O_2^{\cdot-}$ and H_2O_2.

[154,156,160,161]. These data have stimulated interest in the potential for intravenous infusions of ascorbate that deliver millimolar plasma concentrations for periods up to 4–6 hours [162] to generate sufficient H_2O_2 in vivo to exert an anticancer effect [163,164]. The ascorbyl radical has been detected in biological fluids that contain traces of catalytic iron [77,156] but is not readily detected in plasma and at neutral pH [158].

A major pro-oxidant activity of ascorbate is in the recycling of Fe^{2+} (Equation 2.1), which is required for the Fenton reaction, that results in the generation of the highly reactive hydroxyl radical ($OH^{\cdot}$) (Equation 2.2) [2,3,5,152,165,166] (Figure 2.3).

$$AscH^- + Fe^{3+} \rightarrow Asc^{\cdot-} + Fe^{2+} + H^+ \qquad (2.1)$$

$$Fe^{2+} + H_2O_2 + H^+ \rightarrow Fe^{3+} + H_2O + HO^{\cdot} \qquad (2.2)$$

These pro-oxidant activities require specific conditions, such as free iron, and it is still unclear as to whether this species is present in biological fluids in sufficient quantities. In most tissues and in plasma, iron is normally sequestered into proteins such as transferrin or ferritin and is unavailable for redox cycling [152,165,167]. Hence, the pro-oxidant activities of ascorbate are likely to be significant only in certain pathological settings, such as in conditions of iron overload, when there is hemolysis, or at sites of acute inflammation. In addition, the presence of supraphysiological concentrations that are achieved by intravenous infusions could result in the generation of H_2O_2 in significant quantities, albeit for short periods of time.

SUMMARY

The chemistry of ascorbate is complex, and turnover in vivo will reflect the availability of oxygen, free transition metals, and the concentration of ascorbate. Many of the reactions of ascorbate recycling in mammalian cells are dependent on enzymes localized at different cell sites, and this indicates a need for a more sophisticated understanding of ascorbate transport and turnover. The capacity for ascorbate to affect cell metabolism, particularly glucose metabolism, remains to be explored.

REFERENCES

1. Wheeler, G., Ishikawa, T., Pornsaksit, V. and Smirnoff, N. 2015. Evolution of alternative biosynthetic pathways for vitamin C following plastid acquisition in photosynthetic eukaryotes. *eLife* 4.

2. Smirnoff, N. 2000. Ascorbic acid: Metabolism and functions of a multi-facetted molecule. Curr. Opin. Plant Biol. 3, 229–235.

3. Michels, A. J. and Frei, B. 2013. Myths, artifacts, and fatal flaws: Identifying limitations and opportunities in vitamin C research. Nutrients 5, 5161–5192.

4. Levine, M., Rumsey, S. C., Daruwala, R., Park, J. B. and Wang, Y. 1999. Criteria and recommendations for vitamin C intake. *JAMA* 281, 1415–1423.

5. Levine, M., Padayatty, S. J. and Espey, M. G. 2011. Vitamin C: A concentration-function approach yields pharmacology and therapeutic discoveries. Adv. Nutr. (Bethesda, MD) 2, 78–88.

6. Levine, M., Dhariwal, K. R., Welch, R. W., Wang, Y. and Park, J. B. 1995. Determination of optimal vitamin C requirements in humans. Am. J. Clin. Nutr. 62, 1347S–1356S.

7. Levine, M., Dhariwal, K. R., Washko, P. W., Welch, R. W. and Wang, Y. 1993. Cellular functions of ascorbic acid: A means to determine vitamin C requirements. Asia. Pacific. J. Clin. Nutr. 2 (Suppl 1), 5–13.

8. Levine, M., Dhariwal, K. R., Washko, P. W., Butler, J. D., Welch, R. W., Wang, Y. H. and Bergsten, P. 1991. Ascorbic acid and in situ kinetics: A new approach to vitamin requirements. Am. J. Clin. Nutr. 54, 1157s–1162s.

9. Frei, B., Birlouez-Aragon, I. and Lykkesfeldt, J. 2012. Authors' perspective: What is the

optimum intake of vitamin C in humans? *Critical Rev. Food Sci. Nutr.* 52, 815–829.

10. Linster, C. L. and Van Schaftingen, E. 2007. Vitamin C. Biosynthesis, recycling and degradation in mammals. *FEBS J.* 274, 1–22.

11. Lykkesfeldt, J. and Poulsen, H. E. 2010. Is vitamin C supplementation beneficial? Lessons learned from randomised controlled trials. *British J. Nutr.* 103, 1251–1259.

12. May, J. M. 2012. Vitamin C transport and its role in the central nervous system. *Subcell Biochem.* 56, 85–103.

13. May, J. M. 2011. The SLC23 family of ascorbate transporters: Ensuring that you get and keep your daily dose of vitamin C. *Br. J. Pharmacol.* 164, 1793–1801.

14. Sotiriou, S., Gispert, S., Cheng, J., Wang, Y., Chen, A., Hoogstraten-Miller, S., Miller, G. F., Kwon, O., Levine, M., Guttentag, S. H. and Nussbaum, R. L. 2002. Ascorbic-acid transporter Slc23a1 is essential for vitamin C transport into the brain and for perinatal survival. *Nat. Med.* 8, 514–517.

15. Graumlich, J. F., Ludden, T. M., Conry-Cantilena, C., Cantilena, L. R., Jr., Wang, Y. and Levine, M. 1997. Pharmacokinetic model of ascorbic acid in healthy male volunteers during depletion and repletion. *Pharmaceut. Res.* 14, 1133–1139.

16. Levine, M., Wang, Y., Padayatty, S. J. and Morrow, J. 2001. A new recommended dietary allowance of vitamin C for healthy young women. *Proc. Natl. Acad. Sci. USA* 98, 9842–9846.

17. Carr, A. C., Pullar, J. M., Moran, S. and Vissers, M. C. 2012. Bioavailability of vitamin C from kiwifruit in non-smoking males: Determination of "healthy" and "optimal" intakes. *J. Nutr. Sci.* 1, e14.

18. Du, J., Cullen, J. J. and Buettner, G. R. 2012. Ascorbic acid: Chemistry, biology and the treatment of cancer. *Biochim. Biophys. Acta* 1826, 443–457.

19. Banhegyi, G., Braun, L., Csala, M., Puskas, F. and Mandl, J. 1997. Ascorbate metabolism and its regulation in animals. *Free Radic. Biol. Med.* 23, 793–803.

20. Mandl, J., Szarka, A. and Banhegyi, G. 2009. Vitamin C: Update on physiology and pharmacology. *Br. J. Pharmacol.* 157, 1097–1110.

21. Smirnoff, N. 2018. Ascorbic acid metabolism and functions: A comparison of plants and mammals. *Free Radic. Biol. Med.* 122, 116–129.

22. Peyssonnaux, C., Datta, V., Cramer, T., Doedens, A., Theodorakis, E. A., Gallo, R. L., Hurtado-Ziola, N.,

Nizet, V. and Johnson, R. S. 2005. HIF-1alpha expression regulates the bactericidal capacity of phagocytes. *J. Clin. Invest.* 115, 1806–1815.

23. Peyssonnaux, C., Cejudo-Martin, P., Doedens, A., Zinkernagel, A. S., Johnson, R. S. and Nizet, V. 2007. Cutting edge: Essential role of hypoxia inducible factor-1alpha in development of lipopolysaccharide-induced sepsis. *J. Immunol.* 178, 7516–7519.

24. Potter, C. and Harris, A. L. 2004. Hypoxia inducible carbonic anhydrase IX, marker of tumour hypoxia, survival pathway and therapy target. *Cell Cycle* 3, 164–167.

25. Semenza, G. L. 2010. HIF-1: Upstream and downstream of cancer metabolism. *Curr. Opin. Genet. Dev.* 20, 51–56.

26. Vaupel, P., Schmidberger, H. and Mayer, A. 2019. The Warburg effect: Essential part of metabolic reprogramming and central contributor to cancer progression. *Int. J. Radiat. Biol.* 95, 912–919.

27. Carr, A. C., Tijerina, T. and Frei, B. 2000. Vitamin C protects against and reverses specific hypochlorous acid- and chloramine-dependent modifications of low-density lipoprotein. *Biochem. J.* 346 (Pt 2), 491–499.

28. Parker, A., Cuddihy, S. L., Son, T. G., Vissers, M. C. and Winterbourn, C. C. 2011. Roles of superoxide and myeloperoxidase in ascorbate oxidation in stimulated neutrophils and H_2O_2-treated HL60 cells. *Free Radic. Biol. Med.* 51, 1399–1405.

29. Pattison, D. I., Hawkins, C. L. and Davies, M. J. 2009. What are the plasma targets of the oxidant hypochlorous acid? A kinetic modeling approach. *Chem. Res. Toxicol.* 22, 807–817.

30. Buettner, G. R. and Schafer, F. Q. 2003. Ascorbate as an antioxidant. In *Vitamin C: Functions and Biochemistry in Animals and Plants* (Asard, M., May, J. M. and Smirnoff, N., eds.). pp. 173–188, BIOS Scientific Publishers, Taylor and Francis Group, London and New York.

31. May, J. M. and Qu, Z. C. 2009. Ascorbic acid efflux and re-uptake in endothelial cells: Maintenance of intracellular ascorbate. *Mol. Cell. Biochem.* 325, 79–88.

32. May, J. M. and Qu, Z. C. 2005. Transport and intracellular accumulation of vitamin C in endothelial cells: Relevance to collagen synthesis. *Arch. Biochem. Biophys.* 434, 178–186.

33. May, J. M., Mendiratta, S., Qu, Z. C. and Loggins, E. 1999. Vitamin C recycling and function in human monocytic U-937 cells. *Free Radic. Biol. Med.* 26, 1513–1523.

34. Bergsten, P., Yu, R., Kehrl, J. and Levine, M. 1995. Ascorbic acid transport and distribution in human B lymphocytes. *Arch. Biochem. Biophys.* 317, 208–214.

35. Banhegyi, G., Benedetti, A., Margittai, E., Marcolongo, P., Fulceri, R., Nemeth, C. E. and Szarka, A. 2014. Subcellular compartmentation of ascorbate and its variation in disease states. *Biochim. Biophys. Acta* 1843, 1909–1916.

36. Chaudiere, J. and Ferrari-Iliou, R. 1999. Intracellular antioxidants: From chemical to biochemical mechanisms. *Food Chem. Toxicol.* 37, 949–962.

37. Kuiper, C., Dachs, G. U., Currie, M. J. and Vissers, M. C. 2014. Intracellular ascorbate enhances hypoxia-inducible factor (HIF)-hydroxylase activity and preferentially suppresses the HIF-1 transcriptional response. *Free Radic. Biol. Med.* 69, 308–317.

38. May, J. M., Li, L., Qu, Z. C. and Huang, J. 2005. Ascorbate uptake and antioxidant function in peritoneal macrophages. *Arch. Biochem. Biophys.* 440, 165–172.

39. Frei, B., Stocker, R. and Ames, B. N. 1988. Antioxidant defenses and lipid peroxidation in human blood plasma. *Proc. Natl. Acad. Sci. USA* 85, 9748–9752.

40. Thomas, S. R. and Stocker, R. 2000. Molecular action of vitamin E in lipoprotein oxidation: Implications for atherosclerosis. *Free Radic. Biol. Med.* 28, 1795–1805.

41. Bowry, V. W., Mohr, D., Cleary, J. and Stocker, R. 1995. Prevention of tocopherol-mediated peroxidation in ubiquinol-10-free human low density lipoprotein. *J. Biol. Chem.* 270, 5756–5763.

42. Stewart, M. S., Cameron, G. S. and Pence, B. C. 1996. Antioxidant nutrients protect against UVB-induced oxidative damage to DNA of mouse keratinocytes in culture. *J. Invest. Dermatol.* 106, 1086–1089.

43. Lin, J. Y., Selim, M. A., Shea, C. R., Grichnik, J. M., Omar, M. M., Monteiro-Riviere, N. A. and Pinnell, S. R. 2003. UV photoprotection by combination topical antioxidants vitamin C and vitamin E. *J. Am. Acad. Dermatol.* 48, 866–874.

44. Darr, D., Dunston, S., Faust, H. and Pinnell, S. 1996. Effectiveness of antioxidants (vitamin C and E) with and without sunscreens as topical photoprotectants. *Acta Dermato-venereologica* 76, 264–268.

45. Dreher, F., Gabard, B., Schwindt, D. A. and Maibach, H. I. 1998. Topical melatonin in combination with vitamins E and C protects skin from ultraviolet-induced erythema: A human study in vivo. *British J. Dermatol.* 139, 332–339.

46. Mukai, K. 1989. Kinetic study of the reaction of vitamin C derivatives with tocopheroxyl (vitamin E radical) and substituted phenoxyl radicals in solution. *Biochim. Biophys. Acta* 993, 168–173.

47. Tanaka, K., Hashimoto, T., Tokumaru, S., Iguchi, H. and Kojo, S. 1997. Interactions between vitamin C and vitamin E are observed in tissues of inherently scorbutic rats. *J. Nutr.* 127, 2060–2064.

48. Li, X., Huang, J. and May, J. M. 2003. Ascorbic acid spares alpha-tocopherol and decreases lipid peroxidation in neuronal cells. *Biochem. Biophys. Res. Commun.* 305, 656–661.

49. Bielski, B. H. J. 1982. Chemistry of ascorbic acid radicals. In *Ascorbic Acid: Chemistry, Metabolism and Uses* (Seib, P. A. and Tolbert, B. M., eds.). pp. 81–100, American Chemical Society, Washington, DC.

50. Bielski, B. H. J., Allen, A. O. and Schwarz, H. A. 1981. Mechanism of disproportionation of ascorbate radicals. *J. Am. Chem. Soc.* 103, 3516–3518.

51. Diliberto, E. J., Jr., Dean, G., Carter, C. and Allen, P. L. 1982. Tissue, subcellular, and submitochondrial distributions of semidehydroascorbate reductase: Possible role of semidehydroascorbate reductase in cofactor regeneration. *J. Neurochem.* 39, 563–568.

52. Hara, T. and Minakami, S. 1971. On functional role of cytochrome b5. II. NADH-linked ascorbate radical reductase activity in microsomes. *J. Biochem.* 69, 325–330.

53. Ito, A., Hayashi, S. and Yoshida, T. 1981. Participation of a cytochrome b5-like hemoprotein of outer mitochondrial membrane (OM cytochrome b) in NADH-semidehydroascorbic acid reductase activity of rat liver. *Biochem. Biophys. Res. Commun.* 101, 591–598.

54. Nishino, H. and Ito, A. 1986. Subcellular distribution of OM cytochrome b-mediated NADH- semidehydroascorbate reductase activity in rat liver. *J. Biochem.* 100, 1523–1531.

55. Schweinzer, E. and Goldenberg, H. 1993. Monodehydroascorbate reductase activity in the surface membrane of leukemic cells. Characterization by a ferricyanide-driven redox cycle. *Eur. J. Biochem.* 218, 1057–1062.

56. May, J. M., Cobb, C. E., Mendiratta, S., Hill, K. E. and Burk, R. F. 1998. Reduction of the ascorbyl free radical to ascorbate by thioredoxin reductase. *J. Biol. Chem.* 273, 23039–23045.

57. Njus, D., Knoth, J., Cook, C. and Kelly, P. M. 1983. Electron transfer across the chromaffin granule membrane. J. Biol. Chem. 258, 27–30.

58. Njus, D., Wigle, M., Kelley, P. M., Kipp, B. H. and Schlegel, H. B. 2001. Mechanism of ascorbic acid oxidation by cytochrome b(561). Biochemistry 40, 11905–11911.

59. Wakefield, L. M., Cass, A. E. and Radda, G. K. 1986. Functional coupling between enzymes of the chromaffin granule membrane. J. Biol. Chem. 261, 9739–9745.

60. Buettner, G. R. and Jurkiewicz, B. A. 1996. Catalytic metals, ascorbate and free radicals: Combinations to avoid. Radiation Res. 145, 532–541.

61. Vaupel, P. W., Frinak, S. and Bicher, H. I. 1981. Heterogeneous oxygen partial pressure and pH distribution in C3H mouse mammary adenocarcinoma. Cancer Res. 41, 2008–2013.

62. Vogelberg, K. H. and Konig, M. 1993. Hypoxia of diabetic feet with abnormal arterial blood flow. Clin. Invest. 71, 466–470.

63. Semenza, G. L. 2016. The hypoxic tumor microenvironment: A driving force for breast cancer progression. Biochim. Biophys. Acta. 1863, 382–391.

64. Acker, T., Fandrey, J. and Acker, H. 2006. The good, the bad and the ugly in oxygen-sensing: ROS, cytochromes and prolyl-hydroxylases. Cardiovasc. Res. 71, 195–207.

65. Arena, E. T., Tinevez, J. Y., Nigro, G., Sansonetti, P. J. and Marteyn, B. S. 2017. The infectious hypoxia: Occurrence and causes during Shigella infection. Microbes. Infect. 19, 157–165.

66. Ceradini, D. J. and Gurtner, G. C. 2005. Homing to hypoxia: HIF-1 as a mediator of progenitor cell recruitment to injured tissue. Trends Cardiovasc. Med. 15, 57–63.

67. Walmsley, S. R. and Whyte, M. K. 2014. Neutrophil energetics and oxygen sensing. Blood 123, 2753–2754.

68. Colgan, S. P., Campbell, E. L. and Kominsky, D. J. 2016. Hypoxia and Mucosal Inflammation. Ann. Rev. Pathol. 11, 77–100.

69. Colgan, S. P. and Taylor, C. T. 2010. Hypoxia: An alarm signal during intestinal inflammation. Nat. Rev. Gastro Hepatol. 7, 281–287.

70. Kuiper, C. and Vissers, M. C. 2014. Ascorbate as a co-factor for Fe- and 2-oxoglutarate dependent dioxygenases: Physiological activity in tumor growth and progression. Front. Oncol. 4, 359.

71. Kuiper, C., Vissers, M. C. and Hicks, K. O. 2014. Pharmacokinetic modeling of ascorbate diffusion through normal and tumor tissue. Free Radic. Biol. Med. 77C, 340–352.

72. Buettner, G. R. and Jurkiewicz, B. A. 1993. Ascorbate free radical as a marker of oxidative stress: An EPR study. Free Radic. Biol. Med. 14, 49–55.

73. Buettner, G. R., Motten, A. G., Hall, R. D. and Chignell, C. F. 1987. ESR detection of endogenous ascorbate free radical in mouse skin: Enhancement of radical production during UV irradiation following application of chlorpromazine. Photochem. Photobiol. 46, 161–164.

74. Jurkiewicz, B. A. and Buettner, G. R. 1994. Ultraviolet light-induced free radical formation in skin: An electron paramagnetic resonance study. Photochem. Photobiol. 59, 1–4.

75. Pagan-Carlo, L. A., Garcia, L. A., Hutchison, J. L., Buettner, G. R. and Kerber, R. E. 1999. Captopril lowers coronary venous free radical concentration after direct current cardiac shocks. Chest 116, 484–487.

76. Sharma, M. K. and Buettner, G. R. 1993. Interaction of vitamin C and vitamin E during free radical stress in plasma: An ESR study. Free Radic. Biol. Med. 14, 649–653.

77. Buettner, G. R. and Chamulitrat, W. 1990. The catalytic activity of iron in synovial fluid as monitored by the ascorbate free radical. Free Radic. Biol. Med. 8, 55–56.

78. Myllyla, R., Majamaa, K., Gunzler, V., Hanauske-Abel, H. M. and Kivirikko, K. I. 1984. Ascorbate is consumed stoichiometrically in the uncoupled reactions catalyzed by prolyl 4-hydroxylase and lysyl hydroxylase. J. Biol. Chem. 259, 5403–5405.

79. Nietfeld, J. J. and Kemp, A. 1981. The function of ascorbate with respect to prolyl 4-hydroxylase activity. Biochim. Biophys. Acta. 657, 159–167.

80. Myllyharju, J. and Kivirikko, K. I. 1997. Characterization of the iron- and 2-oxoglutarate-binding sites of human prolyl 4-hydroxylase. EMBO J. 16, 1173–1180.

81. Koivunen, P., Hirsila, M., Gunzler, V., Kivirikko, K. I. and Myllyharju, J. 2004. Catalytic properties of the asparaginyl hydroxylase (FIH) in the oxygen sensing pathway are distinct from those of its prolyl 4-hydroxylases. J. Biol. Chem. 279, 9899–9904.

82. Hirsila, M., Koivunen, P., Gunzler, V., Kivirikko, K. I. and Myllyharju, J. 2003. Characterization of the human prolyl 4-hydroxylases that modify the hypoxia-inducible factor. J. Biol. Chem. 278, 30772–30780.

83. Herman, H. H., Wimalasena, K., Fowler, L. C., Beard, C. A. and May, S. W. 1988. Demonstration of the ascorbate dependence of membrane-bound dopamine beta-monooxygenase in adrenal chromaffin granule ghosts. *J. Biol. Chem.* 263, 666–672.

84. Wimalasena, K., Herman, H. H. and May, S. W. 1989. Effects of dopamine beta-monooxygenase substrate analogs on ascorbate levels and norepinephrine synthesis in adrenal chromaffin granule ghosts. *J. Biol. Chem.* 264, 124–130.

85. Wimalasena, K. and Wimalasena, D. S. 1995. The reduction of membrane-bound dopamine beta-monooxygenase in resealed chromaffin granule ghosts. Is intragranular ascorbic acid a mediator for extragranular reducing equivalents? *J. Biol. Chem.* 270, 27516–27524.

86. Harrison, F. E. and May, J. M. 2009. Vitamin C function in the brain: Vital role of the ascorbate transporter SVCT2. *Free Radic. Biol. Med.* 46, 719–730.

87. Lykkesfeldt, J., Trueba, G. P., Poulsen, H. E. and Christen, S. 2007. Vitamin C deficiency in weanling guinea pigs: Differential expression of oxidative stress and DNA repair in liver and brain. *Br. J. Nutr.* 98, 1116–1119.

88. Myllyla, R., Kuutti-Savolainen, E. R. and Kivirikko, K. I. 1978. The role of ascorbate in the prolyl hydroxylase reaction. *Biochem. Biophys. Res. Commun.* 83, 441–448.

89. Chung, W. Y., Chung, J. K., Szeto, Y. T., Tomlinson, B. and Benzie, I. F. 2001. Plasma ascorbic acid: Measurement, stability and clinical utility revisited. *Clin. Biochem.* 34, 623–627.

90. Karlsen, A., Blomhoff, R. and Gundersen, T. E. 2007. Stability of whole blood and plasma ascorbic acid. *Eur. J. Clin. Nutr.* 61, 1233–1236.

91. Leonard, S. W., Bobe, G. and Traber, M. G. 2018. Stability of antioxidant vitamins in whole human blood during overnight storage at 4 degrees C and frozen storage up to 6 months. *Int. J. Vitam. Nutr. Res.* 88, 151–157.

92. Lykkesfeldt, J. 2012. Ascorbate and dehydroascorbic acid as biomarkers of oxidative stress: Validity of clinical data depends on vacutainer system used. *Nutr. Res.* 32, 66–69.

93. Pullar, J. M., Bayer, S. and Carr, A. C. 2018. Appropriate handling, processing and analysis of blood samples is essential to avoid oxidation of vitamin C to dehydroascorbic acid. *Antioxidants (Basel, Switzerland)* 7.

94. Washko, P. W., Welch, R. W., Dhariwal, K. R., Wang, Y. and Levine, M. 1992. Ascorbic acid and dehydroascorbic acid analyses in biological samples. *Anal. Biochem.* 204, 1–14.

95. Liang, W. J., Johnson, D. and Jarvis, S. M. 2001. Vitamin C transport systems of mammalian cells. *Mol. Membr. Biol.* 18, 87–95.

96. Vera, J. C., Rivas, C. I., Velasquez, F. V., Zhang, R. H., Concha, II and Golde, D. W. 1995. Resolution of the facilitated transport of dehydroascorbic acid from its intracellular accumulation as ascorbic acid. *J. Biol. Chem.* 270, 23706–23712.

97. Vera, J. C., Rivas, C. I., Zhang, R. H., Farber, C. M. and Golde, D. W. 1994. Human HL-60 myeloid leukemia cells transport dehydroascorbic acid via the glucose transporters and accumulate reduced ascorbic acid. *Blood* 84, 1628–1634.

98. Washko, P. and Levine, M. 1992. Inhibition of ascorbic acid transport in human neutrophils by glucose. *J. Biol. Chem.* 267, 23568–23574.

99. Rumsey, S. C., Kwon, O., Xu, G. W., Burant, C. F., Simpson, I. and Levine, M. 1997. Glucose transporter isoforms GLUT1 and GLUT3 transport dehydroascorbic acid. *J. Biol. Chem.* 272, 18982–18989.

100. Wilson, J. X. 2005. Regulation of vitamin C transport. *Ann. Rev. Nutr.* 25, 105–125.

101. Corpe, C. P., Tu, H., Eck, P., Wang, J., Faulhaber-Walter, R., Schnermann, J., Margolis, S. et al. 2010. Vitamin C transporter Slc23a1 links renal reabsorption, vitamin C tissue accumulation, and perinatal survival in mice. *J. Clin. Invest.* 120, 1069–1083.

102. May, J. M., Qu, Z. C., Qiao, H. and Koury, M. J. 2007. Maturational loss of the vitamin C transporter in erythrocytes. *Biochem. Biophys. Res. Commun.* 360, 295–298.

103. Evans, R. M., Currie, L. and Campbell, A. 1982. The distribution of ascorbic acid between various cellular components of blood, in normal individuals, and its relation to the plasma concentration. *Br. J. Nutr.* 47, 473–482.

104. Montel-Hagen, A., Blanc, L., Boyer-Clavel, M., Jacquet, C., Vidal, M., Sitbon, M. and Taylor, N. 2008. The Glut1 and Glut4 glucose transporters are differentially expressed during perinatal and postnatal erythropoiesis. *Blood* 112, 4729–4738.

105. Angulo, C., Rauch, M. C., Droppelmann, A., Reyes, A. M., Slebe, J. C., Delgado-Lopez, F., Guaiquil, V. H., Vera, J. C. and Concha, II. 1998. Hexose transporter expression and function in mammalian spermatozoa: Cellular localization and transport of hexoses and vitamin C. *J. Cell. Biochem.* 71, 189–203.

106. Sage, J. M. and Carruthers, A. 2014. Human erythrocytes transport dehydroascorbic acid and sugars using the same transporter complex. *Am. J. Physiol. Cell. Physiol.* 306, C910–917.

107. Montel-Hagen, A., Kinet, S., Manel, N., Mongellaz, C., Prohaska, R., Battini, J. L., Delaunay, J., Sitbon, M. and Taylor, N. 2008. Erythrocyte Glut1 triggers dehydroascorbic acid uptake in mammals unable to synthesize vitamin C. *Cell* 132, 1039–1048.

108. Tu, H., Li, H., Wang, Y., Niyyati, M., Wang, Y., Leshin, J. and Levine, M. 2015. Low red blood cell vitamin C concentrations induce red blood cell fragility: A link to diabetes via glucose, glucose transporters, and dehydroascorbic. *Acid. EBioMedicine* 2, 1735–1750.

109. Tu, H., Wang, Y., Li, H., Brinster, L. R. and Levine, M. 2017. Chemical transport knockout for oxidized vitamin C, dehydroascorbic acid, reveals its functions in vivo. *EBioMedicine* 23, 125–135.

110. Washko, P., Rotrosen, D. and Levine, M. 1989. Ascorbic acid transport and accumulation in human neutrophils. *J. Biol. Chem.* 264, 18996–19002.

111. Washko, P., Rotrosen, D. and Levine, M. 1990. Ascorbic acid accumulation in plated human neutrophils. *FEBS Lett.* 260, 101–104.

112. Washko, P., Rotrosen, D. and Levine, M. 1991. Ascorbic acid in human neutrophils. *Am. J. Clin. Nutr.* 54, 1221s–1227s.

113. Washko, P. W., Wang, Y. and Levine, M. 1993. Ascorbic acid recycling in human neutrophils. *J. Biol. Chem.* 268, 15531–15535.

114. Winterbourn, C. C. and Vissers, M. C. 1983. Changes in ascorbate levels on stimulation of human neutrophils. *Biochim. Biophys. Acta* 763, 175–179.

115. Ratcliffe, P. J. 2013. Oxygen sensing and hypoxia signalling pathways in animals: The implications of physiology for cancer. *J. Physiol.* 591, 2027–2042.

116. Ratcliffe, P., Koivunen, P., Myllyharju, J., Ragoussis, J., Bovee, J. V., Batinic-Haberle, I., Vinatier, C. B. et al., 2017. Update on hypoxia-inducible factors and hydroxylases in oxygen regulatory pathways: From physiology to therapeutics. *Hypoxia (Auckland, N.Z.)* 5, 11–20.

117. Semenza, G. L. 2017. Hypoxia-inducible factors: Coupling glucose metabolism and redox regulation with induction of the breast cancer stem cell phenotype. *EMBO J.* 36, 252–259.

118. Yun, J., Mullarky, E., Lu, C., Bosch, K. N., Kavalier, A., Rivera, K., Roper, J. et al. 2015. Vitamin C selectively kills KRAS and BRAF mutant colorectal cancer cells by targeting GAPDH. *Science* 350, 1391–1396.

119. Aguilera, O., Munoz-Sagastibelza, M., Torrejon, B., Borrero-Palacios, A., Del Puerto-Nevado, L., Martinez-Useros, J., Rodriguez-Remirez, M. et al. 2016. Vitamin C uncouples the Warburg metabolic switch in KRAS mutant colon cancer. *Oncotarget* 7, 47954–47965.

120. Jung, S. A., Lee, D. H., Moon, J. H., Hong, S. W., Shin, J. S., Hwang, I. Y., Shin, Y. J. et al. 2016. L-Ascorbic acid can abrogate SVCT-2-dependent cetuximab resistance mediated by mutant KRAS in human colon cancer cells. *Free Radic. Biol. Med.* 95, 200–208.

121. Bode, A. M., Cunningham, L. and Rose, R. C. 1990. Spontaneous decay of oxidized ascorbic acid (dehydro-L-ascorbic acid) evaluated by high-pressure liquid chromatography. *Clin. Chem.* 36, 1807–1809.

122. Carr, A. C. and Frei, B. 1999. Toward a new recommended dietary allowance for vitamin C based on antioxidant and health effects in humans. *Am. J. Clin. Nutr.* 69, 1086–1107.

123. Levine, M., Conry-Cantilena, C., Wang, Y., Welch, R. W., Washko, P. W., Dhariwal, K. R., Park, J. B. et al. 1996. Vitamin C pharmacokinetics in healthy volunteers: Evidence for a recommended dietary allowance. *Proc. Natl. Acad. Sci. USA* 93, 3704–3709.

124. Mayland, C. R., Bennett, M. I. and Allan, K. 2005. Vitamin C deficiency in cancer patients. *Palliative Med.* 19, 17–20.

125. Kallner, A. B., Hartmann, D. and Hornig, D. H. 1981. On the requirements of ascorbic acid in man: Steady-state turnover and body pool in smokers. *Am. J. Clin. Nutr.* 34, 1347–1355.

126. Kallner, A., Hartmann, D. and Hornig, D. 1979. Steady-state turnover and body pool of ascorbic acid in man. *Am. J. Clin. Nutr.* 32, 530–539.

127. May, J. M. 2002. Recycling of vitamin C by mammalian thioredoxin reductase. *Methods Enzymol.* 347, 327–332.

128. Winkler, B. S., Orselli, S. M. and Rex, T. S. 1994. The redox couple between glutathione and ascorbic acid: A chemical and physiological perspective. *Free Radic. Biol. Med.* 17, 333–349.

129. Park, J. B. and Levine, M. 1996. Purification, cloning and expression of dehydroascorbic acid-reducing activity from human neutrophils:

Identification as glutaredoxin. *Biochem. J.* 315. (Pt 3), 931–938.

130. Wells, W. W. and Xu, D. P. 1994. Dehydroascorbate reduction. *J. Bioenerg. Biomembr.* 26, 369–377.

131. Wells, W. W., Xu, D. P., Yang, Y. F. and Rocque, P. A. 1990. Mammalian thioltransferase (glutaredoxin) and protein disulfide isomerase have dehydroascorbate reductase activity. *J. Biol. Chem.* 265, 15361–15364.

132. May, J. M., Mendiratta, S., Hill, K. E. and Burk, R. F. 1997. Reduction of dehydroascorbate to ascorbate by the selenoenzyme thioredoxin reductase. *J. Biol. Chem.* 272, 22607–22610.

133. May, J. M., Qu, Z. and Li, X. 2001. Requirement for GSH in recycling of ascorbic acid in endothelial cells. *Biochem. Pharmacol.* 62, 873–881.

134. May, J. M., Qu, Z. C., Neel, D. R. and Li, X. 2003. Recycling of vitamin C from its oxidized forms by human endothelial cells. *Biochim. Biophys. Acta* 1640, 153–161.

135. Meister, A. 1992. On the antioxidant effects of ascorbic acid and glutathione. *Biochem. Pharmacol.* 44, 1905–1915.

136. Meister, A. 1992. Biosynthesis and functions of glutathione, an essential biofactor. *J. Nutr. Sci. Vitaminol (Tokyo)* Spec No, 1–6.

137. Del Bello B., Maellaro E., Sugherini L., Santucci A., Comporti M., and Casini A. F. 1994. Purification of NADPH-dependent dehydroascorbate reductase from rat liver and its identification with 3 alpha-hydroxysteroid dehydrogenase. *Biochem. J.* 304 (Pt 2), 385–390.

138. Wilson, J. X. 2002. The physiological role of dehydroascorbic acid. *FEBS Lett.* 527, 5–9.

139. Simpson, G. L. and Ortwerth, B. J. 2000. The non-oxidative degradation of ascorbic acid at physiological conditions. *Biochim. Biophys. Acta* 1501, 12–24.

140. Nagaraj, R. H., Shamsi, F. A., Huber, B. and Pischetsrieder, M. 1999. Immunochemical detection of oxalate monoalkylamide, an ascorbate-derived Maillard reaction product in the human lens. *FEBS Lett.* 453, 327–330.

141. Urivetzky, M., Kessaris, D. and Smith, A. D. 1992. Ascorbic acid overdosing: A risk factor for calcium oxalate nephrolithiasis. *J. Urol.* 147, 1215–1218.

142. Chalmers, A. 1994. Re: Ascorbate acid overdosing: A risk factor for calcium oxalate nephrolithiasis. *J. Urol.* 152, 171.

143. Traxer, O., Huet, B., Poindexter, J., Pak, C. Y. and Pearle, M. S. 2003. Effect of ascorbic acid consumption on urinary stone risk factors. *J. Urol.* 170, 397–401.

144. Robitaille, L., Mamer, O. A., Miller, W. H., Jr., Levine, M., Assouline, S., Melnychuk, D., Rousseau, C. and Hoffer, L. J. 2009. Oxalic acid excretion after intravenous ascorbic acid administration. *Metabolism, Clin Exp.* 58, 263–269.

145. Simon, J. A. and Hudes, E. S. 1999. Relation of serum ascorbic acid to serum vitamin B12, serum ferritin, and kidney stones in US adults. *Arch. Int. Med.* 159, 619–624.

146. Curhan, G. C., Willett, W. C., Rimm, E. B. and Stampfer, M. J. 1996. A prospective study of the intake of vitamins C and B6, and the risk of kidney stones in men. *J. Urol.* 155, 1847–1851.

147. Thomas, L. D., Elinder, C. G., Tiselius, H. G., Wolk, A. and Akesson, A. 2013. Ascorbic acid supplements and kidney stone incidence among men: A prospective study. *JAMA Int. Med.* 173, 386–388.

148. Curhan, G. C., Willett, W. C., Speizer, F. E. and Stampfer, M. J. 1999. Intake of vitamins B6 and C and the risk of kidney stones in women. *J. Am. Soc. Nephr.* 10, 840–845.

149. Jiang, K., Tang, K., Liu, H., Xu, H., Ye, Z. and Chen, Z. 2019. Ascorbic acid supplements and kidney stones incidence among men and women: A systematic review and meta-analysis. *Urol. J.* 16, 115–120.

150. Auer, B. L., Auer, D. and Rodgers, A. L. 1998. The effect of ascorbic acid ingestion on the biochemical and physicochemical risk factors associated with calcium oxalate kidney stone formation. *Clin. Chem. Lab. Med.* 36, 143–147.

151. Liochev, S. I. and Fridovich, I. 1999. Superoxide and iron: Partners in crime. *IUBMB Life* 48, 157–161.

152. Koppenol, W. H. 2001. The Haber-Weiss cycle—70 years later. *Redox. Rep.* 6, 229–234.

153. Winterbourn, C. C. and Kettle, A. J. 2003. Radical-radical reactions of superoxide: A potential route to toxicity. *Biochem. Biophys. Res. Commun.* 305, 729–736.

154. Halliwell, B., Clement, M. V., Ramalingam, J. and Long, L. H. 2000. Hydrogen peroxide. Ubiquitous in cell culture and in vivo? *IUBMB Life* 50, 251–257.

155. Chen, Q., Espey, M. G., Krishna, M. C., Mitchell, J. B., Corpe, C. P., Buettner, G. R., Shacter, E. and Levine, M. 2005. Pharmacologic ascorbic acid concentrations selectively kill cancer cells: Action as a pro-drug to deliver hydrogen

peroxide to tissues. *Proc. Natl. Acad. Sci. USA* 102, 13604–13609.

156. Chen, Q., Espey, M. G., Sun, A. Y., Lee, J. H., Krishna, M. C., Shacter, E., Choyke, P. L., Pooput, C., Kirk, K. L., Buettner, G. R. and Levine, M. 2007. Ascorbate in pharmacologic concentrations selectively generates ascorbate radical and hydrogen peroxide in extracellular fluid in vivo. *Proc. Natl. Acad. Sci. USA* 104, 8749–8754.

157. Miller, D. M., Buettner, G. R. and Aust, S. D. 1990. Transition metals as catalysts of "autoxidation" reactions. *Free Radic. Biol. Med.* 8, 95–108.

158. Buettner, G. R. 1988. In the absence of catalytic metals ascorbate does not autoxidize at pH 7: Ascorbate as a test for catalytic metals. *J. Biochem. Biophys. Meth.* 16, 27–40.

159. Chen, Q., Espey, M. G., Sun, A. Y., Pooput, C., Kirk, K. L., Krishna, M. C., Khosh, D. B., Drisko, J. and Levine, M. 2008. Pharmacologic doses of ascorbate act as a prooxidant and decrease growth of aggressive tumor xenografts in mice. *Proc. Natl. Acad. Sci. USA* 105, 11105–11109.

160. Wee, L. M., Long, L. H., Whiteman, M. and Halliwell, B. 2003. Factors affecting the ascorbate- and phenolic-dependent generation of hydrogen peroxide in Dulbecco's Modified Eagles Medium. *Free Rad. Res.* 37, 1123–1130.

161. Clement, M. V., Ramalingam, J., Long, L. H. and Halliwell, B. 2001. The in vitro cytotoxicity of ascorbate depends on the culture medium used to perform the assay and involves hydrogen peroxide. *Antioxid. Redox. Signal* 3, 157–163.

162. Padayatty, S. J., Sun, H., Wang, Y., Riordan, H. D., Hewitt, S. M., Katz, A., Wesley, R. A. and Levine, M. 2004. Vitamin C pharmacokinetics: Implications for oral and intravenous use. *Ann. Intern. Med.* 140, 533–537.

163. Belin, S., Kaya, F., Duisit, G., Giacometti, S., Ciccolini, J. and Fontes, M. 2009. Antiproliferative effect of ascorbic acid is associated with the inhibition of genes necessary to cell cycle progression. *PLOS ONE* 4, e4409.

164. Carosio, R., Zuccari, G., Orienti, I., Mangraviti, S. and Montaldo, P. G. 2007. Sodium ascorbate induces apoptosis in neuroblastoma cell lines by interfering with iron uptake. *Mol. Cancer* 6, 55.

165. Winterbourn, C. C. 1981. Hydroxyl radical production in body fluids: Roles of metal ions, ascorbate and superoxide. *Biochem. J.* 198, 125–131.

166. Buettner, G. R. 1987. Activation of oxygen by metal complexes and its relevance to autoxidative processes in living systems. *Bioelectrochem. Bioenergy* 18, 29–36.

167. Koppenol, W. H. 1983. Thermodynamics of Fenton-driven Haber-Weiss and related reactions. In *Oxy Radicals and Their Scavenging Systems. Vol. I. Molecular Aspects* (Cohen, G. and Greenwald, R. A., eds.). pp. 84–88, Elsevier Biomedical, New York.

CHAPTER THREE

Vitamin C Alimentation via SLC Solute Carriers

Damian Nydegger, Gergely Gyimesi, and Matthias A. Hediger

DOI: 10.1201/9780429442025-3

CONTENTS

INTRODUCTION

L-Ascorbic acid, also known as vitamin C, is an important cofactor for a great variety of metal ion–dependent enzymes, as well as an efficient antioxidant. All plants and most animals can synthetize vitamin C from glucose, except humans, guinea pigs, and certain birds, fishes, and nonhuman primates [1,2]. In humans, severe ascorbic acid deficiency (e.g., due to long voyages during the "Age of Sail") results in scurvy, leading to impaired wound healing, anemia, fatigue, and ultimately death. To avoid ascorbic acid deficiency, humans rely on taking up ascorbic acid from the diet.

There are two ascorbic acid transporters belonging to the SLC23 family, SLC23A1/SVCT1 and SLC23A2/SVCT2. Both are Na^+-coupled cotransporters [2]. SVCT1 is expressed in the intestines, the kidneys, the liver, the lungs, and skin [2,3]. It is the major uptake pathway for ascorbic acid in the intestine and, thus, is responsible for maintaining whole-body vitamin C levels [4]. SVCT2 is expressed in a variety of tissues such as the brain, lungs, liver, skin, spleen, muscles, and adrenal glands [3,5–7]. The uptake of ascorbic acid is regulated tightly, thereby limiting intestinal uptake, renal absorption, and delivery into target tissues. Ascorbic acid is an essential cofactor of many enzymes, for example, as part of the synthesis of collagens [8]. Given the importance of vitamin C, it is not surprising that altered functions of the transporters of this essential vitamin are linked to different diseases. Thus, it is of fundamental importance to understand in detail the vitamin C uptake and dissemination pathways in our body.

INITIAL CLONING AND CHARACTERIZATION

The SLC23 family consists of four members. Of these, SVCT1 (SLC23A1) and SVCT2 (SLC23A2) are Na^+-dependent vitamin C (ascorbic acid) transporters. The third member of the family, SVCT3 (SLC23A3), is an orphan transporter. A fourth member, SLC23A4 (SVCT4 or SNBT1), found in several organisms, exists in humans only as a pseudogene [2]. SVCT1, SVCT2, and SVCT3 sequences were previously identified as yolk sac permease-like proteins, YSPL3, YSPL2, and YSPL1, respectively [9,10], but were not functionally characterized. The Na^+-coupled vitamin C transporter SVCT1 was first identified by functional expression of rat cDNA in *Xenopus oocytes* in our laboratory. The gene encoding SVCT2, SLC23A2, was identified by homologue screening of a rat brain cDNA library [6]. Several groups further cloned and characterized SVCT1 and SVCT2 from different species [2,6,11,12]. SVCT3 is an orphan transporter, since its transport substrates and physiological roles are still unknown [2].

FUNCTIONAL PROPERTIES OF THE SLC23 FAMILY

SVCT1 and SVCT2 facilitate the transport of ascorbic acid in a Na^+-dependent manner across the cell membranes. Both transporters have a high affinity for L-ascorbic acid [4,13–15]. SVCT1 exhibits a $K_{0.5}$ of 20–100 µM for L-ascorbic acid, depending on the species and the expression system. It displays a unique preference for L-ascorbic acid over the stereoisomer D-isoascorbic acid, as well as dehydroascorbic acid (DHA), various analogs, and intermediates of vitamin C metabolism [6]. The optimal pH for the electrogenic transport is around 7.5, while at pH 5.5, the transport rate is reduced by 50%–60% [2,6,14,15]. Both SVCT1 and SVCT2 cotransport ascorbic acid and Na^+, most likely in a 1:2 stoichiometry, using the Na^+ gradient as the driving force for the accumulation of ascorbic acid [3,15,16]. The binding order of the substrates has been determined, with the first Na^+ binding, then ascorbic acid, and finally the second Na^+ [15,16]. SVCT2 function is not only Na^+-dependent but is also affected by Ca^{2+} and Mg^{2+}. Even in the presence of Na^+, there is no transport in the absence of Mg^{2+}/Ca^{2+}. Without these ions, SVCT2 seems to be in an inactive state [2,16].

DHA is the oxidized form of vitamin C, which is not a substrate for SVCT1 or SVCT2. It is a substrate of SLC2A1/GLUT1, SLC2A3/GLUT3, and SLC2A4/GLUT4 [17]. DHA is toxic at high concentrations. When it enters into cells via GLUT transporters, it is reduced to ascorbic acid as part of the recycling process [18].

PHYLOGENETIC ASPECTS

The human SLC23 proteins belong to the nucleobase:cation symporter 2 (NCS2) family of membrane transporters, also called the nucleobase-ascorbate transporter (NAT) family. This family of proteins was originally described as a nucleobase:H^+ symporter 2 family [19,20], containing transporters from Gram-negative and Gram-positive bacteria, fungi, plants, and animals, transporting various purines and pyrimidines, such as uracil, xanthine, or uric acid (Figure 3.1) [20,21]. Representative members of the family include the uracil transporter UraA [22] and the xanthine transporter YgfO/XanQ [23] from *Escherichia coli*, and UapA from *Aspergillus nidulans/Emericella nidulans* [24]. At around this time, rat SVCT1 and SVCT2 were identified by our group [6] and were subsequently included in the family as ascorbic acid transporters [21]. Interestingly, the only nucleobase transporter from the NCS2 family existing in mammals is SLC23A4 (SVCT4/SNBT1), which is active only in nonprimate mammals [25] and has been characterized to be a uric acid transporter [26].

STRUCTURE-FUNCTION RELATIONSHIP OF THE SLC23 FAMILY

Overall Fold and Transmembrane Architecture

Early transmembrane topology predictions based on hydropathy plots predicted 12 transmembrane helices for SLC23 proteins [6,9,14,32,33].

SLC23 transporters show an inverted repeat architecture with 7+7 transmembrane helices (TMHs) [34–36], due to which this fold family has been termed "7-TM Inverted Repeat" (7TMIR) fold [37]. One symmetric pair of the transmembrane

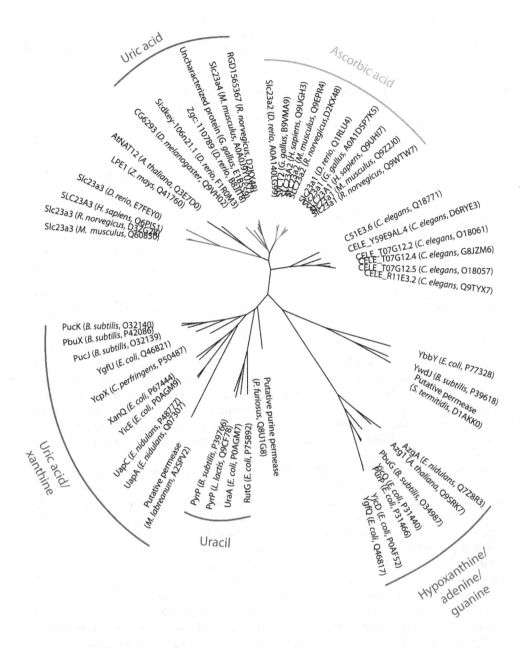

Figure 3.1. Phylogenetic tree of nucleobase:cation symporter 2 family transporters. Sequences were taken from family 2.A.40 of the Transporter Classification Database [27] and from in-house similarity searches within UniProt [28] sequences. SLC23A1-4 family branches are marked with orange, green, red, and blue, respectively. Predominant substrate groups are shown. The sequence alignment and the tree were created using ClustalO [29] and PhyML [30,31], respectively.

helices, TMH3 and TMH10, are shorter than usual and do not span the entire membrane. Instead, they are precluded by short β-strands that interact with each other to form a two-strand antiparallel β-sheet at the center of the protein, which is a hallmark element of the UraA/7TMIR fold. This structural feature is likely the reason why two TMHs were missed in the original topology predictions [38].

The UraA/7TMIR fold is represented in mammals by the SLC4/SLC23/SLC26 families, which share overall similar structural architectures [34,37,38].

Currently, there are three structures of NCS2-family transporters available at near-atomic resolution (Table 3.1), constituting a useful resource to predict the structural features of SLC23 transporters. In addition, structurally

TABLE 3.1

Currently available resolved three-dimensional structures for transporters of the 7-TM Inverted Repeat (7TMIR/UraA) fold

Protein	Protein Data Bank ID	Resolution [Å]	Cocrystallized Substrate	Conformation	References
		SLC23 Homologues			
UraA (*Escherichia coli*)	3QE7	2.781	Uracil	Inward-open	[34]
UraA (*E. coli*)	5XLS	2.5	Uracil	Occluded	[36]
UapA (*Emericella nidulans*)	5I6C	3.7	Xanthine	Inward-open	[35]
		SLC4 Homologues			
AE1 (*Homo sapiens*)	4YZF	3.5	–	Outward-open	[39]
Bor1 (*Arabidopsis thaliana*)	5L25	4.11	–	Occluded	[40]
Bor1p (*Saccharomyces cerevisiae*)	5SV9	5.9	–	Inward-facing	[41]
		SLC26 Homologues			
SLC26Dg (*Deinococcus geothermalis*)	5IOF	4.2	–	Inward-facing	[42]
SLC26Dg (*D. geothermalis*)	5DA0	3.2	–	Inward-facing	[42] (truncated construct)

similar homologues of SLC4 and SLC26 proteins can also be useful to understand the mechanism of transport (Table 3.1). Here we have used the substrate-bound occluded state of the UraA transporter (Protein Data Bank [PDB] ID: 5XLS) to generate a homology-based model of human SVCT1, SVCT2, and SVCT3 (see Figure 3.2a for hSVCT1; hSVCT2 and hSVCT3 models are not shown). Based on the structure of UraA and UapA, we also present a putative, refined transmembrane topology of SLC23 proteins (Figure 3.2c).

Interestingly, while members of the NCS1 family have been suggested to be structurally similar to APC transporters, such as LeuT (PDB ID: 2Q6H), this was initially not apparent for NAT/NCS2 family members [43]. As the structure of UraA was solved, it became clear that the overall structural fold of NCS2 and thus SLC23 transporters is distinct from APC transporters [34]. Despite the apparent structural dissimilarity, it was proposed that NCS2 transporters in fact belong to the APC superfamily, as they share a common evolutionary origin [44]. Later it was shown that the structural folds of LeuT and UraA indeed share common supersecondary structural elements [38].

Domain Architecture and Mechanism of Transport

The overall structure of the protein is often divided into the core domain comprising TMHs 1–4 and 8–11, and the gate domain that is formed by TMHs 5–7 and 12–14 [34]. While the core domain harbors most of the hydrophilic substrate-binding residues and a large number of buried hydrogen bonds, the gate domain as well as the interface between the two domains remain mostly hydrophobic [34].

Currently available structures of the 7TMIR fold show a variety of conformations (Table 3.1). Based on the superposition of these homologous structures, it has been suggested that SLC4/SLC23/SLC26 proteins function according to an elevator mechanism, as was also proposed for SLC1 transporters [35–37]. However, compared to a typical elevator model, the gate domain of 7TMIR transporters, unlike a typical scaffold domain, could undergo substantial local conformational changes. The overall mechanism is likely to be conserved among SLC23 family members [36].

Substrate-Binding Site

The structures of UraA and UapA have been crystallized in complex with uracil and xanthine, respectively (see Table 3.1) [34–36]. In all cases, the substrates have been unambiguously identified in the binding site, and residues likely to take part in substrate binding have been identified. In UraA, which is a H^+-coupled symporter, residues E241, H245, and E290 cluster at the interface between the core and the gate domains

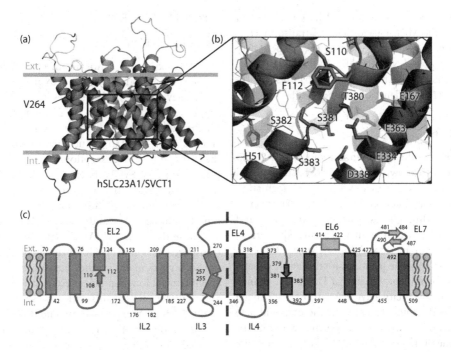

Figure 3.2. Structure and transmembrane topology of SLC23 transporters. (a) Homology-based model of human SLC23A1/SVCT1 based on the structure of the *Escherichia coli* UraA uracil:H+ symporter (Protein Data Bank [PDB] ID: 5XLS). Membrane orientation is based on the orientation of 5XLS from the Orientations of Proteins in Membranes (OPM) database [49]. The regions showing internal structural symmetry are colored orange and blue, respectively. The V264 residue corresponds to the location of the missense SNP rs33972313. The alignments for structure prediction were generated by PSI-COFFEE [50,51] and AlignMe [52] and modified manually. The structural model was generated using MODELLER 9.21 [53]. (b) Putative binding site of human SLC23A1/SVCT1 based on the UraA structure. Amino acid side chains that possibly take part in vitamin C and Na+ binding are highlighted in stick representation. (c) Predicted membrane topology based on the structure of UraA (PDB ID: 5XLS). The α-helices are shown as rectangles, β-strands as arrows, and bounding residues are numbered, based on similarity to the *E. coli* UraA structure and OPM predictions. Internally symmetric regions are colored orange and blue and separated by a red dashed line. (EL, extracellular loop; IL, intracellular loop.)

and have been implicated in proton binding [34]. Molecular dynamics simulations of the occluded UraA structure have shown that E241 needs to be deprotonated and H245 protonated to stabilize the uracil substrate in the binding site [36]. Based on these findings, the residue corresponding to H245 (D338 in hSVCT1 according to our alignment, see Figure 3.2b) in human SVCT1, SVCT2, and SNBT1 was suggested to bind the cotransported Na+ ion [34]. Interestingly, both hSVCT1 and hSVCT2 have been reported to be exclusively dependent on Na+ for transport and do not function with other relevant cations tested [33], suggesting that the functionally relevant cation binding sites in these proteins are specific for Na+.

The pH dependence of transport by hSVCT1 and hSVCT2 was early on linked with the presence of four conserved histidine residues [33]. A later study found that of these, only H51 in hSVCT1 (H109 in hSVCT2) is essential for transport activity but is

not responsible for the pH dependence of transport [45]. In line with this, the H109Q variant of hSVCT2 proved to be inactive, leading to the suggestion that H51 (hSVCT1) and H109 (hSVCT2) might be part of the substrate-binding site. In our model of hSVCT1, H51 is the only one of the four histidine residues that is buried in the protein; the other three are exposed to the solvent. Due to its orientation (Figure 3.2b), H51 is likely not in contact with the extracellular medium, which would explain why it does not affect the pH dependence of transport. Additionally, Varma et al. found that mutations of H51 affect the affinity of hSVCT1 toward ascorbic acid [45]. Based on our structural model, this is likely to be an indirect effect, as H51 is not predicted to be directly lining the substrate-binding site (Figure 3.2b). Interestingly, the analogous residue in the *E. coli* xanthine transporter YgfO/XanQ has also been reported to alter substrate selectivity [46]. Several histidine residues outside the putative substrate-binding site in hSVCT2 have been shown to affect the Michaelis-constant of transport (K_m) of transport, while not affecting the cooperativity of cotransported Na$^+$ ions [47]. In the same report, H413 in hSVCT2 has been suggested to be responsible for the pH dependence of transport [47].

Dimerization

Both UraA and UapA, two homologues of SLC23 proteins with known structure, have been suggested to function as a dimer, with dimer formation essential for function [35,36]. Dimerization is mediated through the gate domains [36], and UapA was shown to be in a dynamic equilibrium between monomeric and dimeric states [36].

Interestingly, the positions identified by genetic screens for mutant variants of UapA that change substrate specificity have been mostly localized to either the interface between the core and gate domains or the linker between the two domains [35]. One of these residues, R481 [48], seems to lie closer to the substrate-binding site of the opposite UapA protomer in a homodimeric transporter complex. Molecular dynamics simulations also suggested that R481 could approach the central binding cavity and form cation-π interactions with the xanthine substrate, thereby modulating substrate specificity. This result also corroborates the idea that the functional unit of the transporter is likely a dimeric unit [35].

PHYSIOLOGICAL ROLES

Tissue Distribution

SVCT1 is expressed primarily in epithelial cells of the small intestine and the proximal tubule of the kidney [54]. It is also expressed in the liver, the lungs, and skin [3,5,6]. SVCT1 is responsible for maintaining whole-body ascorbic acid levels (Figure 3.3), facilitating intestinal uptake from the diet and renal reabsorption of L-ascorbic acid [4]. In contrast, SVCT2 is expressed in different tissues and organs that require vitamin C, such as the brain, the lungs, the liver, skin, the spleen, muscles, the adrenal glands, the eyes, the prostate, and the testis [2,6,12,14]. SVCT2 is the predominant transporter that delivers ascorbate into tissues that are in high demand of the vitamin, to support specific metal ion–dependent enzymatic activities and to protect cells against oxidative stress [1].

SVCT2 is not expressed in endothelial cells of the blood-brain barrier (BBB) but rather in choroid plexus (Figure 3.3). Thus, ascorbic acid is not transported across the BBB. DHA, the oxidized form of vitamin C, is delivered across the BBB. The latter is facilitated by the GLUT1/SLC2A1 transporter that is expressed at the luminal and abluminal membranes of the BBB [55]. Under physiological conditions, however, primarily ascorbic acid, the reduced form of vitamin C, is present in human plasma, and only 5%–10% of vitamin C exists in the oxidized form of DHA.

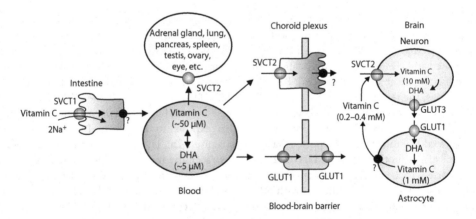

Figure 3.3. Vitamin C transport and tissue accumulation. SVCT1 absorbs vitamin C across the intestinal brush border membrane. The way it leaves the epithelial cells is yet unknown. To enter the brain, there are two possible routes. Either through the choroid plexus (blood-cerebrospinal fluid barrier) via SVCT2 or, in the form of DHA, through the blood-brain barrier via GLUT1. Since DHA is present at relatively low concentrations in the blood compared to L-ascorbic acid, the SVCT2 route across the choroid plexus may be the predominant pathway. In neurons, vitamin C is oxidized to DHA, which is released by GLUT3. Astrocytes import DHA via GLUT1 and reduce it to vitamin C. The exit pathway of vitamin C from astrocytes is currently unknown. The neurons import ascorbic acid via SVCT2, thereby closing the vitamin C recycling pathway.

Therefore, it is likely that DHA transport via GLUT1 across the BBB is not a major route for uptake of vitamin C into the brain. In contrast, in the epithelial cells of the choroid plexus, forming the blood–cerebrospinal fluid barrier, SVCT2 is highly expressed [6], indicating that ascorbic acid is mainly absorbed via this pathway [18]. In red blood cells, the main entry pathway of vitamin C is the uptake of DHA, which is immediately reduced to ascorbic acid after entering the cell. Red blood cells do not express SVCT2. GLUT1 and GLUT3 are responsible for the transport of DHA [56].

Knockout Mice

As already noted, all plants and most animals are capable of synthesizing ascorbic acid from glucose. Humans and other species, however, lost the ability to synthesize ascorbic acid. While mice are able to synthesize ascorbic acid [1,2], humans cannot synthesize it because the gene for the enzyme gulonolactone oxidase is mutated, resulting in a nonfunctional protein [1,2,57]. Gulo$^{-/-}$ mice lacking this gene were used as model organisms for research on vitamin C metabolism [2].

Studies with SVCT1 or SVCT2 knockout mice revealed that Slc23a1$^{-/-}$ mice have high ascorbic acid levels in the urine, implying that SVCT1 is involved in reabsorption of ascorbic acid from urine. However, these knockout mice were able to absorb ascorbic acid from the intestinal lumen. This indicates that another transporter makes up for the lack of SVCT1 in the intestine [1]. Interestingly, Slc23a2$^{-/-}$ mice lacking SVCT2, the predominant ascorbic acid transporter, die within minutes after birth. All fetal cells except the liver, which is the site of ascorbic acid synthesis, show very low ascorbic acid levels [58]. The immediate cause of death is respiratory failure and hemorrhage in the brain [2].

Role of SVCT in Human Health and Ascorbic Acid Recycling

Ascorbic acid is an essential nutrient involved in important processes in the whole body. It acts as a cofactor for many enzymes and is an antioxidant, which can scavenge reactive oxygen species (ROS). The two transporters of ascorbic acid, SVCT1 and SVCT2, are responsible for the uptake from food and distribution in the body [1,2,59].

The mechanisms of vitamin C delivery into the brain have already been discussed. Neurons require relatively high amounts of vitamin C, since they have a high rate of oxidative metabolism compared to other cells, leading to the oxidation of ascorbic acid to DHA. The mechanisms of ascorbic acid recycling are presented in Figure 3.3. DHA leaves the neurons, avoiding toxic effects of DHA accumulation. DHA efflux is facilitated via the GLUT3 transporter. Via the GLUT1 transporter, DHA is then imported into astrocytes. Astrocytes do not express SVCT2, but they take up DHA via GLUT1, which is then converted back into ascorbic acid [2,18]. Mechanisms for conversion into vitamin C involve glutathione and reducing enzymes. Indeed, the glutathione level of astrocytes is four times higher than that in neurons [18]. How ascorbic acid leaves astrocytes is still under investigation. The released ascorbic acid is then delivered back into neurons via SVCT2 [2,18]. In the extracellular space between the neutrons and astrocytes, the ascorbic acid concentration is between 200 and 400 μM, while in neurons it is 10 mM and in astrocytes 1 mM [2].

REGULATION OF THE EXPRESSION OF SVCT1 AND SVCT2

The substrate ascorbic acid of SVCT1 and SVCT2 is an important regulator for these transporters. In Gulo$^{-/-}$ mice, which depend on ascorbic acid uptake from the diet, ascorbic acid starvation led to increased mRNA and protein expression of SVCT1 and SVCT2 in the liver, increased expression of SVCT2 in the cerebellum, and increased expression of SVCT1 in the small intestine. Also during development in mice, expression levels and ascorbic acid levels change. In late embryonic and early natal stages, the cortex and cerebellum of these mice have high ascorbic acid levels and low SVCT2 mRNA and protein levels. During adolescence, the expression of SVCT2 mRNA and protein is increased, and the ascorbic acid levels decrease. A development in a similar direction occurs in the liver of the mice: at birth, the expression levels of SVCT1 and SVCT2 are low, increasing over time [2]. It was shown that the hepatocyte nuclear transcription factor HNF1α (hepatocyte nuclear factor 1 homeobox alpha) increases the SVCT1 promotor activity (Figure 3.4) [4,59,60]. HNF1α transcription is inhibited when the NF-κB (nuclear factor–kappa light chain enhancer of activated B cells) signaling pathway is activated. The activated NF-κB pathway leads to increased levels of inflammatory cytokines and cell death [4]. SVCT2 is also regulated in the

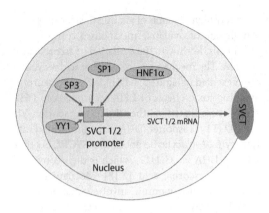

Figure 3.4. The promoter region of SVCT1 and SVCT2 is regulated by different transcription factors, Sp1, Sp3, HNF1α, and YY1. Binding of these transcription factors to the promoter of SVCT1 or SVCT2 leads to transcription of mRNA.

depends on many factors, and it is currently an active research topic because of their important role in vitamin C homeostasis.

Regulation of the cell surface distribution: SVCT1 and SVCT2 are responsible for the uptake of ascorbic acid from the extracellular space, fulfilling their task as plasma membrane transporters. For successful cell surface targeting and cell membrane incorporation of SVCT1 and SVCT2, intact C- and N-termini are required. The cell surface expression of SVCT2 also depends on microtubules and microfilaments. Nocodazole, a microtubule depolarizing agent, disturbs the transport of SVCT2 to the plasma membrane. The localization of SVCT2 is also increased by inhibitors of the myosin-II ATPase. It was furthermore shown that prostaglandin E2 increases the localization of SVCT2 in the plasma membrane [2].

same manner by HNF1α. SVCT1 and SVCT2 are both regulated on the transcriptional level by transcription factor Sp1. A decrease in Sp1 leads to decreased mRNA and protein levels of SVCT1 and SVCT2 [60]. Other transcription factors such as SP3 and YY1 are also reported to regulate the expression of these transporters [61]. The euchromatin markers H3K4me3 [62,63] and H3K9ac also increase the expression of SVCT1 and SVCT2 [62]. The expression of SVCT1 and SVCT2

PATHOLOGICAL ROLE

Inflammation, Infection

Prolonged, severe infections are associated with disturbance in the ascorbic acid homeostasis, resulting in low ascorbic acid levels (Figure 3.5a). It is not only infections that can disturb ascorbic acid homeostasis, but also inflammatory diseases such as inflammatory bowel disease are linked to low

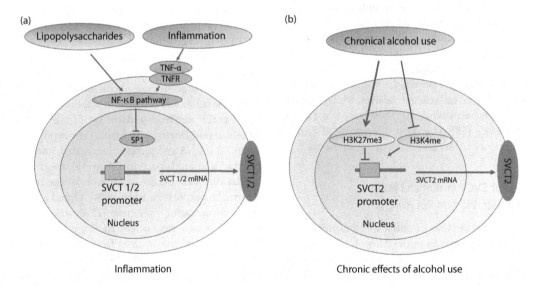

Figure 3.5. (a) Chronic inflammation or lipopolysaccharides lead to the activation of the NF-κB pathway. This pathway inhibits the transcription factor Sp1, leading to a decreased expression of SVCT1/2. (b) Chronic alcohol use leads to increased levels of H3K27me3, which leads to heterochromatin in the promoter region of the *SLC23A2* gene. At the same time, the levels of H3K4me are decreased, which leads to euchromatin in the promoter region of the *SLC23A2* gene. The activity of the promoter region is decreased, resulting in a decreased expression of SVCT2.

VITAMIN C

levels of ascorbic acid. It is known that ascorbic acid deficiency is linked to delayed immune responses by natural killer cells and suppression of cytotoxic T-cell activity [4]. This has led to the assumption that the uptake of ascorbic acid is decreased during inflammation. Lipopolysaccharides are released by various bacteria such as the pathologic *Salmonella*. These molecules bind to toll-like receptors, which leads to activation of inflammatory signaling pathways, for example, NF-κB and p38 mitogen-activated protein kinase (MAPK) (Figure 3.5a). Activation of these pathways leads to decreased ascorbic acid uptake in Caco-2 cells and mouse jejunum [60]. Increased levels of the cytokine TNF-α also lead to decreased ascorbic acid uptake in Caco-2 cells. TNF-α seems to regulate the ascorbic acid uptake the same way as lipopolysaccharides, because it also activates the NF-κB pathway. It was shown that treating cells with TNF-α leads to a decreased expression of HNF1-α, which interacts with the SVCT promotor region. In Caco-2 cells and mouse enteroids, treatment with TNF-α leads to decreased mRNA and protein levels of SVCT1 [4]. Experiments with Caco-2 cells and mouse jejunum treated with lipopolysaccharides revealed decreased levels of TNA-α and SP1, both interacting with the promoters of SLC23A1 and SLC23A2, resulting in a decreased expression of SVCT1 and SVCT2 at the mRNA and protein levels [60].

Chronic Alcohol Abuse

In developed countries, severe vitamin C deficiency leading to scurvy is relatively rare. Moderate vitamin C deficiency, however, is common and occurs especially in the elderly, smokers, and alcoholics [59,60]. As already highlighted, ascorbic acid is an important scavenger of ROS, and it is able to reduce alcohol-induced oxidative damage. Chronic alcohol abuse is linked to lower plasma ascorbic acid levels. In humans and mouse pancreatic acinar cells (PACs), SVCT2 is highly expressed (Figure 3.5b). PACs chronically exposed to alcohol exhibited reduced ascorbic acid uptake. Both SVCT2 mRNA and protein levels were decreased, indicating inhibition at the transcriptional level. Changes in chromatin modeling could be shown, with an increase of H3K27me3, leading to heterochromatin in the SLC23A2 promoter region. At the same time, H3K4me, which leads to euchromatin in the promoter region of the SLC23A2 gene, was decreased. Due to these changes, the expression level of SVCT2 was reported to be decreased [63].

Cancer

Cancer patients are frequently ascorbic acid deficient [3]. The effect of ascorbic acid on cancer is discussed controversially. In the 1970s, it was shown by Ewan Cameron and Linus Pauling that high oral doses of ascorbic acid lead to survival benefits in patients with advanced cancer [64–66]. A follow-up double-blind placebo-controlled study could not confirm the results, and ascorbic acid was dismissed as an anticancer agent [66–68]. New findings revealed that oral doses of ascorbic acid can be increased to 80-fold without great changes in the plasma levels because of limited intestinal uptake, tissue saturation, renal reabsorption, and excretion. This may explain the different outcomes of the studies: Cameron used 10 g intravenous ascorbic acid, while in the studies in which ascorbic acid failed as an anticancer agent, ascorbate was administrated orally [66].

SVCT2, as the predominant tissue transporter of ascorbic acid, has a large influence on the efficacy of ascorbic acid as an anticancer drug. A study with different human colon cancer cell lines revealed different outcomes of ascorbic acid treatment on cancer cells, depending on the expression level of SVCT2. Specifically, there is an antiproliferative effect of ascorbic acid in cancer cells, which is dependent on the expression level of SVCT2. High expression resulted in high intracellular levels of ascorbic acid, slowing down proliferation. The cells with low expression of SVCT2 exhibited a "hormetic" proliferation response to ascorbic acid, meaning that low concentrations of ascorbic acid increased proliferation of colon cancer cells, whereas high concentrations still had an anticancer effect due to the accumulation of vitamin C–altering proliferation pathways. Specifically, in low SVCT2-expressing cells, ascorbic acid at a moderate dose of 10 μM increased the expression of c-myc and cyclin D and, due to this, cell proliferation. In contrast, high doses of ascorbic acid >1 mM decreased the expression of c-myc and cyclin D [69]. This likely explains the different outcomes of the studies described earlier. The plasma level of ascorbic acid by oral administration was probably not high enough to trigger the anticancer effect of ascorbic acid in cancer cells with low SVCT2 expression. The expression level of SVCT2, however, was not measured in this study, highlighting the need for additional experiments to verify this hypothesis.

There are other studies underlining the importance of SVCT2 in anticancer treatment. Colon cancer cell lines with KRAS mutations were shown to be resistant to the anticancer drug cetuximab. Cotreatment with ascorbic acid was able to overcome cetuximab resistance but only in colon cancer cells expressing SVCT2. In contrast, cotreatment in colon cancer cells not expressing SVCT2 had no effect. Expression of SVCT2 in these cells restored the anticancer effect. Furthermore, in SVCT2-expressing colon cancer cells, knock down of SVCT2 with siRNA led to resistance against the cotreatment. The cotreatment caused cetuximab-triggered apoptotic and necrotic cell death only in colon cancer cells expressing SVCT2 [70]. Whether vitamin C alters signaling pathways or affects enzymes needed for the action of cetuximab remains to be determined.

Generating vitamin C in cancer cells by a "bystander effect" has been proposed to be a mechanism of how tumor cells in an oxidative state accumulate ascorbic acid without the expression of SVCT1 or SVCT2. According to this mechanism, superoxide generated in these cells is released to the extracellular space in an oxidative burst. Extracellular superoxide then converts ascorbic acid to DHA, which is imported into the cells via GLUTs, followed by the reduction of DHA to ascorbic acid [18].

Huntington Disease

Huntington disease is a genetic disorder caused by a mutation of the gene coding for Huntingtin. Mutation of Huntingtin leads to involuntary movements, cognitive deterioration, dementia, and weight loss. This protein is responsible for the intracellular trafficking of vesicles, organelles, and proteins to the cell surface. In immortalized striatal neurons expressing mutated Huntingtin, it was shown that SVCT2 is no longer able to translocate to the plasma membrane in response to increased extracellular ascorbic acid levels. Huntingtin is known to be associated to vesicles and microtubules, suggesting a role of this protein in the transport of SVCT2-containing vesicles, in order to deliver SVCT2 to the plasma membrane of neurons. Huntington disease is connected to increased oxidative damage in lipids, proteins, and

DNA, highlighting the protecting role of vitamin C as an antioxidant in neurons. These results also highlight the possible role of SVCT2 in other neurodegenerative diseases such as Alzheimer and Parkinson diseases [71].

Single Nucleotide Polymorphisms

In human SVCT1 and SVCT2, there are several single nucleotide polymorphisms (SNPs) linked to diseases.

SVCT1 SNPs: Several studies revealed SNPs in SLC23A1 that are associated with a specific phenotype. The intronic SNP rs6596473-C is linked to increased risk of follicular lymphoma but showed no link to gastric cancer and advanced colorectal adenoma [3]. Another intronic SNP, rs10063949-G, is associated with increased risk of Crohn disease [72]. The SNP rs33972313 is the only known polymorphism that affects the coding region (see Figure 3.2a). It causes missense variations at position V264, with substitution by the bulkier side chains Leu or Met. These may cause steric clashes that compromise transporter structure and function. These SNPs were associated with low circulating concentrations of ascorbic acid [73].

SVCT2 SNPs: SNP rs6139591 in the SLC23A2 gene is connected to a higher risk for spontaneous preterm delivery [74]. Two SNPs localized in the intron of the gene SLC23A2 were associated with increased risk for chronic lymphocytic leukemia. Different studies in different cohorts of patients showed SNPs linked with increased risk for gastric cancer and chemoradiotherapy-induced toxicity [3]. The proposed impact of SNPs on diseases and ascorbic acid homeostasis has been discussed in detail in recent reviews [75,76].

Pharmacological Relevance

SVCT1 and SVCT2 are promising drug transporters. SVCT2 may be exploited to facilitate the delivery of certain drugs to the brain via the choroid plexus. For example, nipecotic acid, a γ-aminobutyric acid (GABA) uptake inhibitor, was reported to be transported by SVCT2 when conjugated to 6-Br-ascorbic acid [2]. Saquinavir is a protease inhibitor used in the treatment of HIV. It has a high anti-HIV potency but a low bioavailability. A new prodrug, ascorbyl-succinic-saquinavir, was shown to increase absorptive permeability and metabolic stability. This prodrug was transported by SVCT2 in Caco-2 cells, which could be a potent drug delivery mechanism for anti-HIV protease inhibitors [77]. Ascorbic acid in combination with SVCT2 can also be used as a targeting agent for nanocarriers to transport drugs [2]. Overall, especially SVCT2 with its broad tissue distribution is a promising drug transporter for drugs linked to ascorbic acid or 6-Br-ascorbic acid.

SVCT2 is not well established as a direct drug target. In one clinical trial, sepsis was treated with a combination of hydrocortisone, vitamin C, and thiamine. There was no change in mortality, but there were significant reductions in the requirement for vasopressin [78]. Hydrocortisone belongs to the glucocorticoids, a class of steroid hormones, which upregulate the expression of SVCT2 [79,80]. As already discussed, SVCT2 expression is reduced in sepsis, leading to reduced ascorbic acid levels.

SVCT2 is more interesting as a drug transporter than SVCT1, because it is expressed in many different tissues and organs. SVCT1 is only expressed in epithelial cells. The expression of SVCT2 in the choroid plexus is especially interesting for drug delivery to the brain.

REFERENCES

1. Padayatty, S. J. and Levine, M. 2016. Vitamin C: The known and the unknown and Goldilocks. *Oral Dis.* 22, 463–93.
2. Bürzle, M., Suzuki, Y., Ackermann, D., Miyazaki, H., Maeda, N., Clémençon, B., Burrier, R. and Hediger, M. A. 2013. The sodium-dependent ascorbic acid transporter family SLC23. *Mol. Aspects Med.* 34, 436–54.
3. Wohlrab, C., Phillips, E. and Dachs, G. U. 2017. Vitamin C transporters in cancer: Current understanding and gaps in knowledge. *Front. Oncol.* 7, 5–10.
4. Subramanian, V. S., Sabui, S., Subramenium, G. A., Marchant, J. S. and Said, H. M. 2018. Tumor necrosis factor alpha reduces intestinal vitamin C uptake: A role for NF-κB-mediated signaling. *Am. J. Physiol. Liver Physiol.* 315, G241–G248.
5. Lee, J. H., Oh, C. S., Mun, G. H., Kim, J. H., Chung, Y. H., Hwang, Y. Il, Shin, D. H. and Lee, W. J. 2006. Immunohistochemical localization of sodium-dependent l-ascorbic acid transporter 1 protein in rat kidney. *Histochem. Cell Biol.* 126, 491–4.
6. Tsukaguchi, H., Tokui, T., Mackenzie, B., Berger, U. V., Chen, X.-Z., Wang, Y., Brubaker, R. F. and Hediger, M. A. 1999. A family of mammalian Na⁺-dependent L-ascorbic acid transporters. *Nature* 399, 70–75.
7. Bornstein, S. R., Yoshida-Hiroi, M., Sotiriou, S., Levine, M., Hartwig, H.-G., Nussbaum, R. L. and Eisenhofer, G. 2003. Impaired adrenal catecholamine system function in mice with deficiency of the ascorbic acid transporter (SVCT2). *FASEB J.* 17, 1928–30.
8. Kishimoto, Y., Saito, N., Kurita, K., Shimokado, K., Maruyama, N. and Ishigami, A. 2013. Ascorbic acid enhances the expression of type 1 and type 4 collagen and SVCT2 in cultured human skin fibroblasts. *Biochem. Biophys. Res. Commun., Elsevier Inc.* 430, 579–84.
9. Faaland, C. A., Race, J. E., Ricken, G., Warner, F. J., Williams, W. J. and Holtzman, E. J. 1998. Molecular characterization of two novel transporters from human and mouse kidney and from LLC-PK1 cells reveals a novel conserved family that is homologous to bacterial and Aspergillus nucleobase transporters. *Biochim. Biophys. Acta—Gene Struct. Expr., Elsevier* 1442, 353–60.
10. Nagase, T., Seki, N., Ishikawa, K., Ohira, M., Kawarabayasi, Y., Ohara, O., Tanaka, A., Kotani, H., Miyajima, N. and Nomura, N. 1996. Prediction of the coding sequences of unidentified human genes. VI. The coding sequences of 80 new genes (KIAA0201-KIAA0280) deduced by analysis of cDNA clones from cell line KG-1 and brain. *DNA Res.* 3, 321–9, 341–54.
11. Daruwala, R., Song, J., Koh, W. S., Rumsey, S. C. and Levine, M. 1999. Cloning and functional characterization of the human sodium-dependent vitamin C transporters hSVCT1 and hSVCT2. *FEBS Lett.* 460, 480–4.
12. Rajan, D. P., Huang, W., Dutta, B., Devoe, L. D., Leibach, F. H., Ganapathy, V. and Prasad, P. D. 1999. Human Placental Sodium-Dependent Vitamin C Transporter (SVCT2): Molecular Cloning and Transport Function. *Biochem. Biophys. Res. Commun.* 262, 762–8.
13. Luo, S., Wang, Z., Kansara, V., Pal, D. and Mitra, A. K. 2008. Activity of a sodium-dependent vitamin C transporter (SVCT) in MDCK-MDR1 cells and mechanism of ascorbate uptake. *Int. J. Pharm., NIH Public Access.* 358, 168–76.
14. Wang, Y., Mackenzie, B., Tsukaguchi, H., Weremowicz, S., Morton, C. C. and Hediger, M. A.

2000. Human Vitamin C (l-Ascorbic Acid) Transporter SVCT1. *Biochem. Biophys. Res. Commun.* 267, 488–94.

15. Mackenzie, B., Illing, A. C. and Hediger, M. A. 2008. Transport model of the human Na+-coupled L-ascorbic acid (vitamin C) transporter SVCT1. *Am. J. Physiol. Cell Physiol.* 294, C451–9.

16. Godoy, A., Ormazabal, V., Moraga-Cid, G., Zúñiga, F. A., Sotomayor, P., Barra, V., Vasquez, O., et al. 2007. Mechanistic insights and functional determinants of the transport cycle of the ascorbic acid transporter SVCT2. Activation by sodium and absolute dependence on bivalent cations. J. Biol. Chem., *Am. Soc. Biochem. Mol. Biol.* 282, 615–24.

17. Blaszczak, W., Barczak, W., Masternak, J., Kopczyński, P., Zhitkovich, A. and Rubiś, B. 2019. Vitamin C as a Modulator of the Response to Cancer Therapy. *Molecules*, Multidisciplinary Digital Publishing Institute 24, 453.

18. Nualart, F. 2014. Vitamin C Transporters, Recycling and the Bystander Effect in the Nervous System: SVCT2 versus Gluts. *J. Stem Cell Res. Ther.* 04, 209.

19. Saier, M. H. 1998. Molecular phylogeny as a basis for the classification of transport proteins from bacteria, archaea and eukarya. *Adv. Microb. Physiol.* 40, 81–136.

20. Saier, M. H., Eng, B. H., Fard, S., Garg, J., Haggerty, D. A., Hutchinson, W. J., Jack, D. L. et al. 1999. Phylogenetic characterization of novel transport protein families revealed by genome analyses. *Biochim. Biophys. Acta* 1422, 1–56.

21. de Koning, H. and Diallinas, G. 2000. Nucleobase transporters (review). *Mol. Membr. Biol.* 17, 75–94.

22. Andersen, P. S., Frees, D., Fast, R. and Mygind, B. 1995. Uracil uptake in *Escherichia coli* K-12: Isolation of uraA mutants and cloning of the gene. *J. Bacteriol.* 177, 2008–13.

23. Karatza, P. and Frillingos, S. 2005. Cloning and functional characterization of two bacterial members of the NAT/NCS2 family in *Escherichia coli*. *Mol. Membr. Biol.* 22, 251–61.

24. Diallinas, G. and Scazzocchio, C. 1989. A gene coding for the uric acid-xanthine permease of *Aspergillus nidulans*: Inactivational cloning, characterization, and sequence of a cis-acting mutation. *Genetics* 122, 341–50.

25. Frillingos, S. 2012. Insights to the evolution of nucleobase-ascorbate transporters (NAT/NCS2 family) from the Cys-scanning analysis of xanthine permease XanQ. *Int. J. Biochem. Mol. Biol.* 3, 250–72.

26. Yamamoto, S., Inoue, K., Murata, T., Kamigaso, S., Yasujima, T., Maeda, J., Yoshida, Y., Ohta, K. and Yuasa, H. 2010. Identification and functional characterization of the first nucleobase transporter in mammals: Implication in the species difference in the intestinal absorption mechanism of nucleobases and their analogs between higher primates and other mammals. *J. Biol. Chem.* 285, 6522–31.

27. Saier, M. H., Reddy, V. S., Tsu, B. V., Ahmed, M. S., Li, C. and Moreno-Hagelsieb, G. 2016. The Transporter Classification Database (TCDB): Recent advances. *Nucleic Acids Res.* 44, D372–9.

28. UniProt Consortium. 2019 UniProt: A worldwide hub of protein knowledge. *Nucleic Acids Res.* 47, D506–D515.

29. Sievers, F., Wilm, A., Dineen, D., Gibson, T. J., Karplus, K., Li, W., Lopez, R. et al. 2011. Fast, scalable generation of high-quality protein multiple sequence alignments using Clustal Omega. *Mol. Syst. Biol.* 7, 539.

30. Guindon, S. and Gascuel, O. 2003. A simple, fast, and accurate algorithm to estimate large phylogenies by maximum likelihood. *Syst. Biol.* (Rannala, B., ed.) 52, 696–704.

31. Guindon, S., Dufayard, J.-F., Lefort, V., Anisimova, M., Hordijk, W. and Gascuel, O. 2010. New algorithms and methods to estimate maximum-likelihood phylogenies: Assessing the performance of PhyML 3.0. *Syst. Biol.* 59, 307–21.

32. Wang, H., Dutta, B., Huang, W., Devoe, L. D., Leibach, F. H., Ganapathy, V. and Prasad, P. D. 1999. Human Na+-dependent vitamin C transporter 1 (hSVCT1): Primary structure, functional characteristics and evidence for a non-functional splice variant. *Biochim. Biophys. Acta—Biomembr.* 1461, 1–9.

33. Liang, W. J., Johnson, D. and Jarvis, S. M. 2001. Vitamin C transport systems of mammalian cells. *Mol. Membr. Biol.* 18, 87–95.

34. Lu, F., Li, S., Jiang, Y., Jiang, J., Fan, H., Lu, G., Deng, D. et al. 2011. Structure and mechanism of the uracil transporter UraA. *Nature* 472, 243–6.

35. Alguel, Y., Amillis, S., Leung, J., Lambrinidis, G., Capaldi, S., Scull, N. J., Craven, G. et al. 2016. Structure of eukaryotic purine/H+ symporter UapA suggests a role for homodimerization in transport activity. *Nat. Commun.* 7, 11336.

36. Yu, X., Yang, G., Yan, C., Baylon, J. L., Jiang, J., Fan, H., Lu, G. et al. 2017. Dimeric structure

of the uracil:proton symporter UraA provides mechanistic insights into the SLC4/23/26 transporters. *Cell Res.* 27, 1020–33.

37. Chang, Y.-N. and Geertsma, E. R. 2017. The novel class of seven transmembrane segment inverted repeat carriers. *Biol. Chem.* 398, 165–74.

38. Vastermark, A., Wollwage, S., Houle, M. E., Rio, R. and Saier, M. H. 2014. Expansion of the APC superfamily of secondary carriers. *Proteins Struct. Funct. Bioinforma.* 82, 2797–811.

39. Arakawa, T., Kobayashi-Yurugi, T., Alguel, Y., Iwanari, H., Hatae, H., Iwata, M. et al. 2015. Crystal structure of the anion exchanger domain of human erythrocyte band 3. *Science* 350, 680–4.

40. Thurtle-Schmidt, B. H. and Stroud, R. M. 2016. Structure of Bor1 supports an elevator transport mechanism for SLC4 anion exchangers. *Proc. Natl. Acad. Sci. USA* 113, 10542–6.

41. Coudray, N. L., Seyler, S., Lasala, R., Zhang, Z., Clark, K. M., Dumont, M. E., Rohou, A., Beckstein, O. and Stokes, D. L. 2017. Structure of the SLC4 transporter Bor1p in an inward-facing conformation. *Protein Sci.* 26, 130–45.

42. Geertsma, E. R., Chang, Y.-N., Shaik, F. R., Neldner, Y., Pardon, E., Steyaert, J. and Dutzler, R. 2015. Structure of a prokaryotic fumarate transporter reveals the architecture of the SLC26 family. *Nat. Struct. Mol. Biol.* 22, 803–8.

43. Diallinas, G. and Gournas, C. 2008. Structure-function relationships in the nucleobase-ascorbate transporter (NAT) family: Lessons from model microbial genetic systems. *Channels (Austin).* 2, 363–72.

44. Wong, F. H., Chen, J. S., Reddy, V., Day, J. L., Shlykov, M. A., Wakabayashi, S. T. and Saier, Jr., M. H. 2012. The Amino Acid-Polyamine-Organocation Superfamily. *J. Mol. Microbiol. Biotechnol.* 22, 105–13.

45. Varma, S., Campbell, C. E. and Kuo, S.-M. 2008. Functional role of conserved transmembrane segment 1 residues in human sodium-dependent vitamin C transporters. *Biochemistry* 47, 2952–60.

46. Karena, E. and Frillingos, S. 2009. Role of intramembrane polar residues in the YgfO xanthine permease: HIS-31 and ASN-93 are crucial for affinity and specificity, and ASP-304 and GLU-272 are irreplaceable. *J. Biol. Chem.* 284, 24257–68.

47. Ormazabal, V., Zuñiga, F. A., Escobar, E., Aylwin, C., Salas-Burgos, A., Godoy, A., Reyes, A. M., Vera, J. C. and Rivas, C. I. 2010. Histidine residues in the Na+-coupled ascorbic acid transporter-2

(SVCT2) are central regulators of SVCT2 function, modulating pH sensitivity, transporter kinetics, Na+ cooperativity, conformational stability, and subcellular localization. *J. Biol. Chem.* 285, 36471–85.

48. Kosti, V., Papageorgiou, I. and Diallinas, G. 2010. Dynamic elements at both cytoplasmically and extracellularly facing sides of the UapA transporter selectively control the accessibility of substrates to their translocation pathway. *J. Mol. Biol.* 397, 1132–43.

49. Lomize, M. A., Lomize, A. L., Pogozheva, I. D. and Mosberg, H. I. 2006. OPM: Orientations of proteins in membranes database. *Bioinformatics* 22, 623–5.

50. Floden, E. W., Tommaso, P. D., Chatzou, M., Magis, C., Notredame, C. and Chang, J.-M. 2016. PSI/TM-Coffee: A web server for fast and accurate multiple sequence alignments of regular and transmembrane proteins using homology extension on reduced databases. *Nucleic Acids Res.* 44, W339–43.

51. Chang, J.-M., Di Tommaso, P., Taly, J.-F. and Notredame, C. 2012. Accurate multiple sequence alignment of transmembrane proteins with PSI-Coffee. *BMC Bioinformatics* 13(Suppl 4), S1.

52. Stamm, M., Staritzbichler, R., Khafizov, K. and Forrest, L. R. 2014. AlignMe—a membrane protein sequence alignment web server. *Nucleic Acids Res.* 42, W246–W251.

53. Webb, B. and Sali, A. 2016. Comparative protein structure modeling using MODELLER. *Curr. Protoc. Bioinforma.* 54, 5.6.1–5.6.37.

54. May, J. M. 2011. The SLC23 family of ascorbate transporters: Ensuring that you get and keep your daily dose of vitamin C. *Br. J. Pharmacol.* 164, 1793–801.

55. Patching, S. G. 2017. Glucose transporters at the blood-brain barrier: Function, regulation and gateways for drug delivery. *Mol. Neurobiol., Springer US* 54, 1046–77.

56. Tu, H., Li, H., Wang, Y., Niyyati, M., Wang, Y., Leshin, J. and Levine, M. 2015. Low red blood cell vitamin C concentrations induce red blood cell fragility: A link to diabetes via glucose, glucose transporters, and dehydroascorbic acid. *EBioMedicine, Elsevier B.V.* 2, 1735–50.

57. Boggavarapu, R., Jeckelmann, J. M., Harder, D., Schneider, P., Ucurum, Z., Hediger, M. and Fotiadis, D. 2013. Expression, purification and low-resolution structure of human vitamin C transporter SVCT1 (SLC23A1). *PLOS ONE* 8, 1–5.

58. Harrison, F. E., Dawes, S. M., Meredith, M. E., Babaev, V. R., Li, L. and May, J. M. 2010. Low vitamin C and increased oxidative stress and cell death in mice that lack the sodium-dependent vitamin C transporter SVCT2. *Free Radic. Biol. Med.*, Pergamon 49, 821–9.

59. Subramanian, V. S., Srinivasan, P., Wildman, A. J., Marchant, J. S. and Said, H. M. 2016. Molecular mechanism(s) involved in differential expression of vitamin C transporters along the intestinal tract. *Am. J. Physiol. Liver Physiol.* 312, G340–G347.

60. Subramanian, V. S., Sabui, S., Moradi, H., Marchant, J. S. and Said, H. M. 2018. Inhibition of intestinal ascorbic acid uptake by lipopolysaccharide is mediated via transcriptional mechanisms. *Biochim. Biophys. Acta—Biomembr.* 1860, 556–65.

61. Qiao, H. and May, J. M. 2011. Regulation of the human ascorbate transporter SVCT2 exon 1b gene by zinc-finger transcription factors. *Free Radic. Biol. Med.* 50, 1196–209.

62. Subramanian, V. S., Srinivasan, P., Wildman, A. J., Marchant, J. S. and Said, H. M. 2017. Molecular mechanism(s) involved in differential expression of vitamin C transporters along the intestinal tract. *Am. J. Physiol. Liver Physiol.* 312, G340–G347.

63. Subramanian, V. S., Srinivasan, P. and Said, H. M. 2016. Uptake of ascorbic acid by pancreatic acinar cells is negatively impacted by chronic alcohol exposure. *Am. J. Physiol. Cell Physiol.* 311, C129–35.

64. Cameron, E. and Pauling, L. 1974. The orthomolecular treatment of cancer I. The role of ascorbic acid in host resistance. *Chem. Biol. Interact.*, Elsevier 9, 273–83.

65. Cameron, E. and Campbell, A. 1974. The orthomolecular treatment of cancer II. Clinical trial of high-dose ascorbic acid supplements in advanced human cancer. *Chem. Biol. Interact.*, Elsevier 9, 285–315.

66. Shenoy, N., Creagan, E., Witzig, T. and Levine, M. 2018. Ascorbic acid in cancer treatment: Let the phoenix fly. *Cancer Cell*, Cell Press 34, 700–6.

67. Moertel, C. G., Fleming, T. R., Creagan, E. T., Rubin, J., O'Connell, M. J. and Ames, M. M. 1985. High-dose vitamin C versus placebo in the treatment of patients with advanced cancer who have had no prior chemotherapy. *N. Engl. J. Med.* 312, 137–41.

68. Creagan, E. T., Moertel, C. G., O'Fallon, J. R., Schutt, A. J., O'Connell, M. J., Rubin, J. and Frytak, S. 1979. Failure of high-dose vitamin C (ascorbic acid) therapy to benefit patients with advanced cancer. *N. Engl. J. Med.* 301, 687–90.

69. Cho, S., Chae, J. S., Shin, H., Shin, Y., Song, H., Kim, Y., Yoo, B. C. et al. 2018. Hormetic dose response to L-ascorbic acid as an anti-cancer drug in colorectal cancer cell lines according to SVCT-2 expression. *Sci. Rep.*, Springer US 8, 1–9.

70. Jung, S. A., Lee, D. H., Moon, J. H., Hong, S. W., Shin, J. S., Hwang, I. Y., Shin, Y. J. et al. 2016. L-Ascorbic acid can abrogate SVCT-2-dependent cetuximab resistance mediated by mutant KRAS in human colon cancer cells. *Free Radic. Biol. Med.*, Elsevier 95, 200–8.

71. Covarrubias-Pinto, A., Acuña, A. I., Beltrán, F. A., Torres-Díaz, L. and Castro, M. A. 2015. Old things new view: Ascorbic acid protects the brain in neurodegenerative disorders. *Int. J. Mol. Sci.*, Multidisciplinary Digital Publishing Institute (MDPI) 16, 28194–217.

72. Serrano León, A., Amir Shaghaghi, M., Yurkova, N., Bernstein, C. N., El-Gabalawy, H. and Eck, P. 2014. Single-nucleotide polymorphisms in SLC22A23 are associated with ulcerative colitis in a Canadian white cohort. *Am. J. Clin. Nutr.*, Narnia 100, 289–94.

73. Timpson, N. J., Forouhi, N. G., Brion, M.-J., Harbord, R. M., Cook, D. G., Johnson, P., McConnachie, A. et al. 2010. Genetic variation at the SLC23A1 locus is associated with circulating concentrations of L-ascorbic acid (vitamin C): Evidence from 5 independent studies with >15,000 participants. *Am. J. Clin. Nutr.*, Europe PMC Funders 92, 375–82.

74. Erichsen, H. C., Engel, S. A. M., Eck, P. K., Welch, R., Yeager, M., Levine, M., Siega-Riz, A. M., Olshan, A. F. and Chanock, S. J. 2006. Genetic variation in the sodium-dependent vitamin C transporters, SLC23A1, and SLC23A2 and risk for preterm delivery. *Am. J. Epidemiol.* 163, 245–54.

75. Shaghaghi, M. A., Kloss, O. and Eck, P. 2016. Genetic variation in human vitamin C transporter genes in common complex diseases. *Adv. Nutr.*, Narnia 7, 287–98.

76. Michels, A. J., Hagen, T. M. and Frei, B. 2013. Human genetic variation influences vitamin C homeostasis by altering vitamin C transport and antioxidant enzyme function. *Annu. Rev. Nutr.*, Annual Reviews 33, 45–70.

77. Luo, S., Wang, Z., Patel, M., Khurana, V., Zhu, X., Pal, D. and Mitra, A. K. 2011. Targeting SVCT for enhanced drug absorption: Synthesis and in vitro evaluation of a novel vitamin C conjugated prodrug of saquinavir. *Int. J. Pharm.* 414, 77–85.

78. Balakrishnan, M., Gandhi, H., Shah, K., Pandya, H., Patel, R., Keshwani, S. and Yadav, N. 2018. Hydrocortisone, vitamin C and thiamine for the treatment of sepsis and septic shock following cardiac surgery. *Indian J. Anaesth.* 62, 934–9.

79. Pandipati, S., Driscoll, J. E. and Franceschi, R. T. 1998. Glucocorticoid stimulation of Na$^+$-dependent ascorbic acid transport in osteoblast-like cells. *J. Cell. Physiol.* 176, 85–91.

80. Chothe, P. P., Chutkan, N., Sangani, R., Wenger, K. H., Prasad, P. D., Thangaraju, M., Hamrick, M. W., Isales, C. M., Ganapathy, V. and Fulzele, S. 2013. Sodium-coupled vitamin C transporter (SVCT2): Expression, function, and regulation in intervertebral disc cells. *Spine J.* 13, 549–57.

CHAPTER FOUR

Vitamin C Pharmacokinetics*

Jens Lykkesfeldt

DOI: 10.1201/9780429442025-4

CONTENTS

INTRODUCTION

The enzyme L-gulonolactone oxidase catalyzes the final step in the biosynthesis of ascorbic acid in almost all vertebrates [1]. However, evolutionarily conserved deletions have rendered the corresponding gene inactive or lost in primates and a few other species, making us dependent on ingestion of vitamin C [1]. Although not investigated in full detail, some evidence suggests that the same evolutionary process has also adapted humans and other vitamin C–dependent species to this dependency by altering their pharmacokinetic handling of vitamin C, that is, improving the ability to take up and recycle vitamin C while limiting its excretion for optimal health [2,3].

The absorption, distribution, and retention of vitamin C are primarily governed by a group of transporter proteins—the sodium-dependent vitamin C transporters (SVCTs) 1 and 2 [4]—that actively transport ascorbate between the various bodily compartments including intestinal absorption and renal reuptake [5,6]. Collectively, the diverse expression and concentration dependency of these transporters throughout the body has resulted in the highly complex, compartmentalized, and nonlinear pharmacokinetics of vitamin C at physiologic levels [7].

The essential role of vitamin C in scurvy prevention has been known for nearly a century. However, much more recently the importance of

---

* This chapter is based in part on Lykkesfeldt, J. and Tveden-Nyborg, P. 2019. Nutrients 11. pii: E2412. doi: 10.3390/nu11102412, 1–20.

"active transport" of vitamin C for sustainability of life was illustrated by Sotiriou et al., showing that SVCT2 knockout mice die immediately after birth from respiratory failure and display severe brain hemorrhage [8]. Studies of human SVCTs have more recently identified a number of polymorphisms and suggested that these may have significant impact on the pharmacokinetics of vitamin C. Thus, modeling studies have proposed that the functionally poorest SVCT allele identified so far results in a plasma saturation level of only one-fourth of that of the background population [9] corresponding to a condition of permanent vitamin C deficiency based on the plasma concentration definition of hypovitaminosis C of $<23\ \mu M$ [10,11].

In addition to its many biological functions, the putative effect of vitamin C in cancer treatment has been the subject of much debate for decades. In particular, interest was spiked by Pauling's and Cameron's findings [12–15] that were quickly dismissed by randomized controlled trials [16,17]. However, following the identification of serious discrepancies between the studies including the critical realization that oral administration of vitamin C profoundly limits the maximum achievable plasma concentration [18], vitamin C is currently being reinvestigated for its specific toxicity to cancer cells at high concentrations—typically referred to as pharmacologic concentrations—that are only achievable by intravenous infusion [19]. Here, at supraphysiologic concentrations, the pharmacokinetics of vitamin C appears to change from zero to first order displaying a constant and dose-independent half-life [20].

Given its primary pK_a value of about 4.2, the reduced form of vitamin C predominantly exists as its anion ascorbate at physiologic pH. It is a powerful antioxidant and acts as an electron donor in all physiologic reactions to which it is known to contribute [21]. Participating both as a cofactor of various enzymes and as a chain-breaking antioxidant, ascorbate is oxidized to ascorbyl radical, which subsequently may undergo dismutation to form ascorbate and dehydroascorbic acid [22]. In healthy individuals, dehydroascorbic acid is efficiently and quantitatively recycled to ascorbate intracellularly, but studies have shown that increased turnover resulting from, for example, smoking or disease may exceed the cellular recycling capacity, rendering part of the ascorbate pool inactive [23,24]. As the oxidized

form of vitamin C has a short half-life physiologic pH [25], inadequate recycling of vitamin C per se will presumably increase the turnover of the vitamin C pool and thus increase the intake necessary to achieve homeostasis.

This chapter discusses the present knowledge on pharmacokinetics of vitamin C under various conditions in vivo and its implications for vitamin C homeostasis.

PHARMACOKINETICS OF VITAMIN C

Pharmacokinetics comprises the descriptions of absorption, distribution, metabolism, and excretion of drugs. The use of pharmacokinetics to describe these processes is based on various mathematical models, all of which have a set of assumptions that needs to be fulfilled for their validity. In Table 4.1, vitamin C is compared to a typical orally administered low molecular weight drug with respect to general pharmacokinetic properties [26]. As can be deduced from the table, there are more differences than similarities between vitamin C and the typical pharmaceutical drug when it comes to pharmacokinetics. Unfortunately, lack of appropriate attention to the unusual behavior of vitamin C has hampered the proper evaluation of its therapeutic potential in the majority of the clinical literature as reviewed elsewhere [11,26,27]. In the following, these similarities and differences are explored in more detail.

ORAL ROUTE OF ADMINISTRATION

Except for pharmacologic doses of vitamin C and the treatment of severe deficiency at hospitals where vitamin C is administered by intravenous infusion, we all get our vitamin C from ingestion of food and supplements, that is, through oral ingestion.

Absorption

As mentioned in the introduction, vitamin C exists primarily in two forms in vivo, ascorbate (reduced form) and dehydroascorbic acid (oxidized form). Due to the efficient intracellular conversion of dehydroascorbic acid to ascorbate by many cell types, the vitamin C activities of both species are considered identical. However, their abilities to cross biological membranes are highly different [6]. From a vitamin C point of view, there are

TABLE 4.1

Pharmacokinetic properties of a typical orally administered pharmaceutical drug versus those of vitamin C from food sources and supplements

Pharmacokinetic Property	Typical Orally Administered Pharmaceutical Drug	Vitamin C from Food Sources or Supplements
Absorption	Linear absorption kinetics within the therapeutic range resulting in peak drug plasma concentrations in nano- to micromolar range.	Nonlinear absorption kinetics within the therapeutic range via saturable active transport resulting in micromolar plasma concentrations and millimolar tissue concentrations of vitamin C.
Distribution	Generally distributed through passive transport. Immediate distribution primarily determined by blood flow and perfusion. Homeostasis largely based on physical–chemical properties of the drugs including lipophilicity, pK_a, and protein binding.	Primarily distributed through active transport. Immediate distribution based on tissue priority. Distribution largely regulated by tissue-specific active transporter expression and saturation kinetics. Homeostasis depends on adequacy of dose and present vitamin C status of bodily compartments.
Metabolism	Unspecifically metabolized by phase I and II enzymes potentially generating a range of metabolites with increased water solubility.	Specifically and unspecifically oxidized in vivo through electron donor and antioxidant properties, respectively, but highly efficiently regenerated intracellularly to its reduced form by various cell types.
Excretion	First-order elimination kinetics. Generally rapid excretion of parent compound and metabolites through urine and bile.	Nonlinear concentration dependent elimination kinetics resulting in 0–100% active renal reabsorption depending on availability and saturation of bodily compartments.
Pharmacokinetic modeling	Can usually be modeled well by both compartmental and noncompartmental models.	Does not comply with the basic assumption of terminal first-order kinetics used in both compartmental and noncompartmental analysis.

SOURCE: Modified from Tveden-Nyborg, P. and Lykkesfeldt, J. 2013. *Antioxid. Redox. Signal* 19, 2084–2104.

three possible modes of membrane transport: passive diffusion, facilitated diffusion, and active transport.

Simple diffusion takes place when uncharged molecules with a certain degree of lipophilicity cross the lipid bilayer. This process is strictly driven by a concentration gradient, and transport ceases when the concentration equilibrium between the undissociated forms on either side of the membrane is reached. For most low molecular weight drugs, simple diffusion is the primary means of membrane transport. However, as previously indicated, reduced vitamin C predominantly exists (>99.9%) in its anionic form at neutral pH and is highly water soluble. As such, it will only be able to diffuse across the plasma membrane at a relatively slow rate even in the presence of a considerable concentration gradient. However, in the milieu of the stomach (pH 1) or small intestine (pH 5), the proportion of unionized ascorbic acid is increased to 99.9% and 15%, respectively, and under these special conditions, passive diffusion could perhaps play a more significant role in vitamin C uptake. Studies in individuals with normal vitamin C status have reported similar times to maximal plasma concentration following oral administration of ascorbic and erythorbic acid, a stereoisomer of ascorbic acid, respectively [28,29], even though erythorbic acid—an isoform of ascorbate with low vitamin C activity—is poorly transported by epithelial SVCT1 [30].

Facilitated diffusion across membranes occurs through carrier proteins but is, like passive diffusion, dependent on an electrochemical gradient. Dehydroascorbic acid has been shown to compete with glucose for transport through several glucose transporters [31,32]. While only present in negligible amounts in the blood of healthy individuals [23,33], intestinal concentrations are presumably much higher, most likely due to the absence of intracellular recycling and relatively higher concentration in foodstuffs. This may explain the repeated finding of similar bioavailability of ascorbate and dehydroascorbic acid as vitamin C sources [2,34–36]. Dehydroascorbic acid uptake is expectedly inhibited by excess glucose, while the maximal rates of uptake for ascorbate and dehydroascorbic acid are similar when glucose is absent [37].

Finally, concentration gradient–independent active transport plays a significant role in vitamin C absorption. As early as the 1970s, it was observed that the bioavailability of ascorbate is highly dose dependent [38]. Increasing oral doses were shown to lead to decreasing absorption fractions, and it was concluded by several authors that intestinal ascorbate absorption is subject to saturable active transport [38,39]. Malo and Wilson discovered that dehydroascorbic acid and ascorbate are taken up by separate mechanisms in the intestine and that uptake of ascorbate is sodium dependent [37]. This coincided with the discovery and characterization of the SVCT family of transporters by Tsukaguchi et al. [4]. They subsequently showed that the intestine contains the low-affinity/high-capacity active transporter SVCT1 [30]. Thus, ascorbate is efficiently transported across the apical membrane of the intestinal epithelial cells via active transport, but its release into the bloodstream is less well understood. As intracellular vitamin C is effectively kept reduced, facilitating further uptake of dehydroascorbic acid, efflux to the blood through glucose transporters is unlikely to provide a significant contribution. Moreover, the intracellular pH of 7.0 renders the anionic ascorbate predominant (99.9%), and given its hydrophilic nature, passive efflux of ascorbic acid via simple diffusion will be relatively slow. However, as the cellular release of vitamin C to the bloodstream is vital for the absorption process and must occur to a high extent considering the rapid uptake of vitamin C (plasma T_{max} of about 3 hours [29]), it strongly implies the existence of yet undiscovered channels or transporters facilitating vitamin C efflux. It has been proposed that ascorbate efflux may occur through volume-sensitive anion channels in the basolateral membranes of epithelial cells [6]. In the brain, however, studies in human microvascular pericytes have shown that volume-sensitive anion channels are apparently not involved in the ascorbate efflux from these cells and may therefore not represent a general mechanism of basal ascorbate efflux [40]. A schematic overview of intestinal vitamin C absorption is shown in Figure 4.1.

Distribution

The distribution of vitamin C throughout the body is highly compartmentalized (Figure 4.2). As previously discussed, simple passive diffusion is unlikely to play a major role in vitamin C transport across membranes, at least in the further

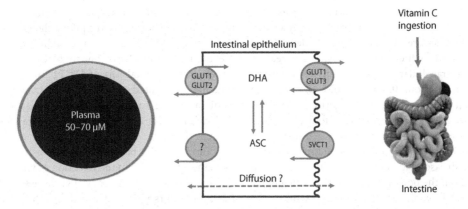

Figure 4.1. Absorption of vitamin C. Ingested vitamin C is absorbed across the intestinal epithelium primarily by membrane transporters in the apical brush border membrane, either as ascorbate by sodium-coupled active transport via the SVCT1 transporter or as dehydroascorbic acid (DHA) through facilitated diffusion via GLUT1 or GLUT3 transporters. Once inside the cell, DHA is efficiently converted to ASC or transported to the bloodstream by GLUT1 and GLUT2 in the basolateral membrane, hereby maintaining a low intracellular concentration and facilitating further DHA uptake. ASC is conveyed to plasma by diffusion, possibly also by facilitated diffusion through volume-sensitive anion channels, although the precise mechanisms remain unknown. (Modified from Lindblad, M. M., Tveden-Nyborg, P. and Lykkesfeldt, J. 2013. Nutrients. 5, 2860–2879.)

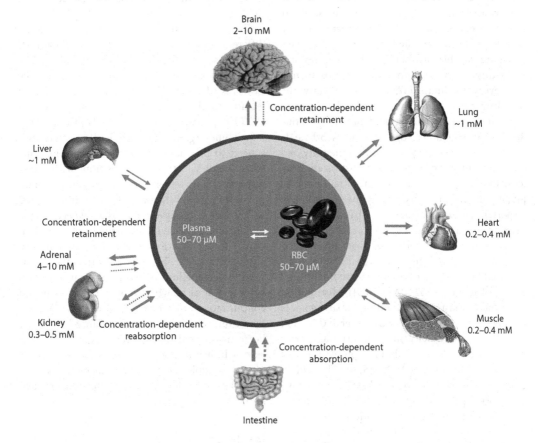

Figure 4.2. Highly differential distribution of vitamin C in the body. Several organs have concentration-dependent mechanisms for the retention of vitamin C, maintaining high levels during times of inadequate supply at the expense of other organs. Particularly protected is the brain. Also, the concentration-dependent absorption and reabsorption mechanisms contribute to the homeostatic control of vitamin C in the body. (Modified from Lindblad, M. M., Tveden-Nyborg, P. and Lykkesfeldt, J. 2013. Nutrients. 5, 2860–2879.)

distribution from the bloodstream. Moreover, ascorbate would favor plasma over tissue by a factor of 2.5 as calculated by a dissociation-determined equilibrium, which grossly contrasts reality, where intracellular concentrations of ascorbate range from about 0.5 to 10 mM compared to 50–80 μM in the blood of healthy individuals [7], that is, confirming a many-fold preference for tissue. Although the glucose transporters (GLUTs 1–4 and 8) capable of facilitating diffusion of dehydroascorbic acid are widely represented throughout the body [31,32,41–43], the negligible amount of oxidized vitamin C present in the plasma of healthy individuals precludes that GLUT-mediated transport per se is of major importance in the diverse distribution of vitamin C. One apparent exception is erythrocytes that do not contain SVCTs but are able to take up vitamin C through facilitated diffusion [44–46]. Human erythrocytes are able to recycle dehydroascorbic acid to ascorbate and maintain an intracellular vitamin C concentration similar to that of plasma [24]. It has been estimated that the erythrocytes alone are capable of reducing the total amount of vitamin C present in blood approximately once every 3 minutes [47,48]. Consequently, the recycling capacity of the red blood cells may constitute a substantial antioxidant reserve in vivo. Recent investigations suggest that ascorbate is necessary for erythrocyte structural integrity and that intracellular erythrocyte ascorbate is essential to maintain ascorbate plasma concentrations in vivo [49,50]. However, considering the quantitative importance of mechanisms, ascorbate is primarily distributed via active transport.

In contrast to epithelial ascorbate uptake and reuptake mediated by the high-capacity/low-affinity SVCT1 (V_{max} of about 15 pmol/min/cell and K_m of about 65–252 μM [5,30,51]), distribution from the bloodstream to body compartments is mainly governed by the slightly larger SVCT2 [52]. SVCT2 is a low-capacity/high-affinity transporter of vitamin C (V_{max} of about 1 pmol/min/cell and K_m of about 8–69 μM [5,30,51]) and is widely expressed in all organs [4]. The respective transport capacities and affinities for vitamin C fit well with the accepted notion that SVCT1 mediates the systemic vitamin C homeostasis, while SVCT2 secures local demands [53]. This is particularly evident for the brain, which upholds one of the higher concentrations of vitamin C in the body [7,54]. Transport of vitamin C into the brain is believed to take place through SVCT2s located in the choroid plexus [55], although it has been suggested that other—yet undiscovered—mechanisms may also be involved [56,57]. However, the pivotal role of SVCT2 in the brain remains undisputed as supported by convincing studies in Slc23a2 knockout mice that display severe brain hemorrhage and high perinatal mortality [8].

Apart from its remarkably high steady-state concentration, the brain also distinguishes itself by being exceptional in the retention of vitamin C during states of deficiency [54,58–64]. This retention occurs at the expense of the other organs and has been proposed to be essential for the maintenance of proper brain function [63,65,66]. Also, during repletion, the brain as well as the adrenal glands have a remarkable affinity for ascorbate, and detailed in vivo studies in guinea pigs—like humans unable to synthesize vitamin C—have revealed that these tissues in particular are the fastest to reestablish homeostasis [7].

The mechanism(s) underlying the highly differential steady-state concentrations of vitamin C in various tissues remain largely unknown. The potential existence of multiple tissue-specific isoforms of the SVCT2 has not been confirmed leading to the assumption that the individual SVCT2 expression level of the cells of the tissues may define organ steady-state levels of vitamin C subject to plasma availability. This implies that tissue- and cell-type composition are mainly responsible. However, for example, in the brain of guinea pigs, substantial differences in vitamin C steady-state levels have been observed between the individual regions, with the highest concentrations being found in the cerebellum, which also saturates first [7]. This does not directly coincide with the cerebellum being the most neuron-rich brain region, although neurons contain the highest concentrations of vitamin C of the brain's cells. Moreover, regional SVCT2 abundance has mostly been investigated through RNA expression levels leaving little information on the possible influence of, for example, posttranslational modifications, activation, and relocation of the functional protein to the cell membrane.

Metabolism

In contrast to plants where a number of ascorbate derivatives and analogs, including several glucosides, have been identified, only ascorbate

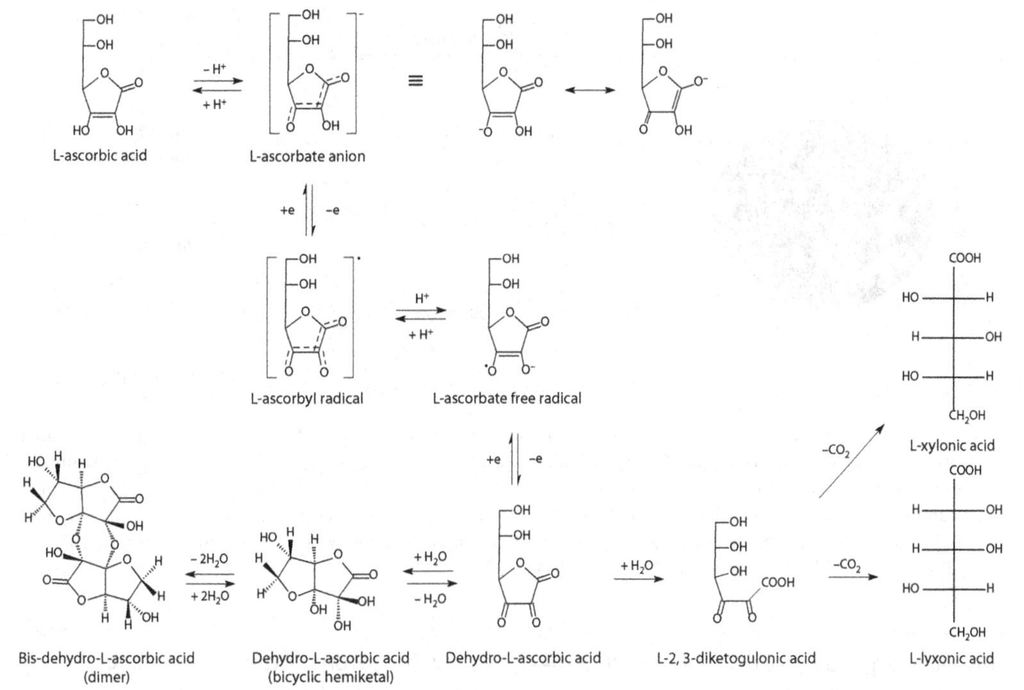

Figure 4.3. Metabolism of vitamin C. Schematic outline of ascorbate metabolism. Abstraction of a hydrogen atom from ascorbic acid leads to the formation of the ascorbate anion; the preferred state at the pH of plasma and tissue. A one-electron oxidation results in ascorbyl radical. This highly resonance-stabilized conformation renders it a comparatively long-lived intermediate. Abstraction of a hydrogen atom forms the ascorbate free radical, and further oxidation results in the formation of dehydroascorbic acid. In aqueous solution, the preferred conformation is as its bicyclic hemiketal. Dehydration of an aqueous solution forms a dimer. Hydrolysis and further oxidation irreversibly convert dehydroascorbic acid to 2,3-diketogulonic acid and other compounds, all of which have no vitamin C activity. (Modified from Frikke-Schmidt, H., Tveden-Nyborg, P. and Lykkesfeldt, J. 2011. Vitamin C in human nutrition. In *Vitamins in the Prevention of Human Disease* [Herrmann, W. and Obeid, R., eds.]. pp. 323–347, De Gruyter, Berlin.)

exists in mammals [67]. The metabolism of ascorbate is intimately linked to its antioxidant function. Through its enediol structure (Figure 4.3) that is highly resonance stabilized and influenced by the acidity of the molecule, ascorbate serves as an efficient electron donor in biological reactions. In supplying reducing equivalents as either cofactor or free radical quencher, ascorbate is oxidized to the comparatively stable radical intermediate, ascorbyl free radical, two molecules of which may disproportionate at physiologic pH to one molecule of ascorbate and one of dehydroascorbic acid [68,69]. As mentioned earlier, dehydroascorbic acid is efficiently reduced intracellularly by a number of cell types, thereby preserving the ascorbate pool. Turnover of vitamin C is therefore particularly linked to the catabolism of dehydroascorbic acid, which occurs through hydrolysis to 2,3-diketogulonic acid and decarboxylation to L-xylonate and L-lyxonate,

both of which can enter the pentose phosphate pathway for further degradation (Figure 4.3) [70].

Excretion and Reuptake

As a highly hydrophilic low molecular weight compound, ascorbate would be expected to be efficiently excreted through the kidneys. It is filtered through the glomerulus by means of the hydrostatic pressure gradient and concentrated in the pre-urine subsequent to the resorption of water (Figure 4.4). Here the pH drops to about 5, resulting in an increased proportion of unionized ascorbic acid to that of ascorbate. The ascorbic acid increase from <0.01% in plasma to about 15% in the pre-urine would for most molecules result in substantial passive reabsorption, but this does not seem to occur for ascorbic acid presumably due to its low lipid solubility. However, for individuals with saturated plasma vitamin C

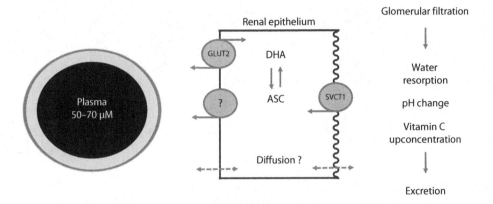

Figure 4.4. Excretion of vitamin C. In the kidney, vitamin C is excreted by glomerular filtration to the renal tubule lumen. Reabsorption is primarily achieved by SVCT1 transporters in the apical membrane, although diffusion from the luminal surface may also contribute to the overall uptake. As in the intestinal epithelium, ascorbate is presumably released to the bloodstream through diffusion, but the extent and mechanisms of this are not known in detail. GLUT2 transporters are located in the basolateral membrane enabling transport of DHA to plasma. (Modified from Lindblad, M. M., Tveden-Nyborg, P. and Lykkesfeldt, J. 2013. *Nutrients.* 5, 2860–2879.)

status—corresponding to about 70–80 μM—excretion of surplus vitamin C is quantitative [18,72]. Instead, reuptake of ascorbate in the proximal renal tubules is controlled by saturable active transport through SVCT1.

The importance of SVCT1 for intestinal vitamin C uptake and, in particular, for renal reuptake has been illustrated by Corpe et al., who showed that $Slc23a1^{-/-}$ mice display an 18-fold increased excretion of ascorbate, lower body pool and vitamin C homeostasis, and increased mortality [32]. They also modeled the effect of known human polymorphisms in the SVCT1 on the plasma saturation level and came to the astonishing conclusion that the most severely affected SNP (A772G rs35817838) would result in a maximal plasma concentration of less than 20 μM [32], that is, a lifelong state of vitamin C deficiency regardless of intake. The renal reuptake of ascorbate is highly concentration dependent. Levine and coworkers have shown in detail that the renal excretion coefficient of ascorbate ranges from 0 to 1 depending on the individual's vitamin C status, that is, corresponding to quantitative reuptake in individuals with poor vitamin C status and quantitative excretion in individuals with saturated status [18,72]. The fact that the excretion ratio is about 1 for intakes higher than about 500 mg/day in healthy individuals supports that passive reabsorption of vitamin C does not play a significant role in the kidneys.

Steady-State Homeostasis of Vitamin C Following Oral Administration/Intake

The role of the saturable active transport mechanisms that dominate the absorption, distribution, and excretion of ascorbate results in nonlinear pharmacokinetics. In case of first-order kinetics, a doubling of the dose would result in the corresponding doubling of the steady-state plasma concentration. For vitamin C, this is clearly not the case. With increasing vitamin C intake, the plasma steady-state concentration reaches a maximal level of about 70–80 μM [18,72]. From the available literature, it appears that a daily intake of about 200–400 mg of vitamin C ensures saturation of the blood in healthy individuals [27]. During periods of altered compartmentalization due to temporary physiologic needs such as pregnancy or increased turnover during disease, higher intakes are needed to maintain saturation. It may be possible to exceed the homeostatic saturation level of 70–80 μM by several-fold through multiple daily gram doses of vitamin C. However, the possible health benefits from such supraphysiologic levels have yet to be documented.

Effect of Dosing Forms and Formulations

Several attempts have been made to partially bypass the maximum plasma concentration and total exposure normally achievable through oral administration. A slow-release formulation would theoretically extend the uptake period resulting in a prolonged and thus increased accumulated uptake, thereby increasing the overall exposure. However, Viscovich et al. did not find any significant differences in exposure or other pharmacokinetic variables between plain and slow-release vitamin C supplements given to smokers—neither at study start nor after 4 weeks of supplementation [29]. Another approach to increase the maximum achievable plasma concentration through oral administration has been liposomes. This closed bilayer phospholipid was the first nano-sized drug delivery system to be successfully applied clinically [73]. Liposomes are capable of protecting the encapsulated compound from degradation and modification, protecting the host from potential unintentional localized toxicity, and improving passive diffusion of polar and hydrophilic molecules. The pharmacokinetic properties of a bolus of 4 grams of liposome-encapsulated vitamin C were compared to those of plain vitamin C and placebo in 11 volunteers in a crossover trial [74]. The authors found a 35% increase in exposure (AUC_{0-4hrs}) with a plasma C_{max} of about 200 μM after 3 hours. Unfortunately, plasma concentrations were not measured beyond the 4-hour time point. In an attempt to show a potential biological significance of increased plasma vitamin C status, the participants were subjected to a 20-minute partial ischemia induced by a blood pressure cuff at 200 mm Hg. However, no beneficial effect on ischemia reperfusion–induced oxidative stress was observed on lipid peroxidation over that of the nonencapsulated dose of vitamin C [74]. Regardless, this technology has shown some promise and continues to be explored in anticancer therapy, where chemotherapeutics can be delivered together with vitamin C for a potentially synergistic effect [75]. In another sophisticated approach, the particular ability of the brain to take up vitamin C has been used by linking ascorbate to the surface of liposomes containing chemotherapeutics, thereby making a brain-specific drug delivery system by using the endogenous vitamin C transport mechanisms [76].

INTRAVENOUS ROUTE OF ADMINISTRATION

Intravenous administration of drugs produces a predictable plasma concentration by avoiding absorption limitation and resulting in 100% bioavailability. For vitamin C specifically, intravenous administration bypasses the saturable absorption mechanisms. This basically removes the upper limit of the maximum achievable plasma concentration. Vitamin C is typically administered intravenously by infusion. Depending on the dose rate, this approach results in a predictable plasma steady-state concentration that will remain constant until infusion is discontinued.

Distribution

As for all compounds in circulation, the distribution of vitamin C following infusion depends at least initially on the vascularization of the various tissues. Whereas the millimolar plasma concentrations do not seem to affect normal tissue distribution beyond saturation, particular interest has been devoted to the poorly vascularized tumors, as ascorbate has been shown to be cytotoxic to cancer cells but not normal cells at high concentrations in in vitro and in vivo studies, possibly through a pro-oxidant function [77–79]. Campell et al. measured ascorbate concentrations in tumor tissue following high-dose vitamin C administration in a mouse model and found that daily injections were necessary to delay tumor growth and suppress the transcription factor hypoxia-inducible factor 1 [80]. Interestingly, it was also found that elimination was significantly delayed in tumor compared to normal tissue [80], which may help in preserving the effect of ascorbate in tumors between infusions. In an attempt to mimic tissue diffusion rates and availability in both normal and tumor tissue, Kuiper and coworkers used a multi-cell-layered, three-dimensional pharmacokinetic model to measure ascorbate diffusion and transport parameters through dense tissue in vitro [81]. Using the obtained data, they were able to simulate diffusion under a number of conditions, including tumors. The authors concluded that supraphysiologic concentrations of ascorbate—achievable only by intravenous infusion—are necessary for effective delivery of ascorbate into poorly vascularized tumors [81]. Using the same data, it has recently been rationalized that normal body saturation obtained by adequate oral dosing will be able to cover the distance between vessels in normal well-perfused tissue and thus provide sufficient vitamin C for the entire body. In contrast, oral intake is insufficient to increase the vitamin C content of tumors with poor vascularization, which requires millimolar plasma concentrations for effective diffusion [82].

Metabolism and Excretion

In normal tissue, metabolism of ascorbate has not been shown to deviate from the general pattern illustrated in Figure 4.3. However, in poorly vascularized tumor tissues, high-dose vitamin C combined with the hypoxic tumor environment has been proposed to promote the formation of cytotoxic levels of hydrogen peroxide, thus providing a putative mode of action and a potential role of ascorbate in cancer treatment [83–85].

Following high-dose intravenous administration of vitamin C, the dose dependency of the elimination phase—as evident at levels below saturation as previously described—is surpassed [86]. Vitamin C is primarily eliminated through glomerular filtration, and reuptake mechanisms through SVCT1 are shut off. This renders the half-life constant and the elimination kinetics first order [20]. Several pharmacokinetic studies of high-dose vitamin C have estimated a constant elimination half-life of about 2 hours following the discontinuation of infusion [20,87,88]. This suggests that the millimolar plasma concentrations achieved by intravenous infusion are normalized to physiologic levels in about 16 hours. In this perspective, the observation that tumor tissue may maintain an elevated level for as much as 48 hours is interesting [80] and may be mediated by increased stability in the hypoxic tumor environment but also likely the delayed clearance due to poor vascularization.

VITAMIN C HOMEOSTASIS

As described in detail earlier, vitamin C homeostasis is tightly controlled in healthy individuals, giving rise to a complex relationship between the steady-state levels of the various bodily organs and tissues. This interrelationship depends primarily on the availability of vitamin C in the diet and the specific "configurations" of SVCTs of the tissues. However, a number of other factors may interfere with the body's attempt to control vitamin C homeostasis, and some major contributors are suggested in the following sections.

Influence of Polymorphisms

With the acknowledgement of the importance of SVCTs for regulation of vitamin C homeostasis and the evolution of genomic sequencing techniques, it has become clear that a large number of polymorphisms exist that influence the steady-state level of vitamin C. This has been reviewed in detail elsewhere [3], but little is known about the potential clinical impact of these. A Mendelian randomization study in 83,256 individuals from the Copenhagen General Population Study used a genetic variant rs33972313 in Slc23a1 resulting in higher than average vitamin C status to test if improved vitamin C status is associated with a low risk of ischemic heart disease and all-cause mortality [89]. The authors found that a high intake of fruits and vegetables was associated with a low risk of ischemic heart disease and all-cause mortality. Effect sizes were comparable for vitamin C, albeit not significant. In view of the extensive clinical literature that persistently has shown an increased risk of heart disease in individuals with poor vitamin C status but no effect of vitamin C supplementation to individuals who already get sufficient amounts of vitamin C [11,26,27], it would probably be more relevant to test changes in disease risk resulting from genetic variants of Slc23a1 that show lower than average vitamin C status. As mentioned earlier, modeling studies have proposed that the functionally poorest SVCT allele identified so far (A772G, rs35817838) results in a plasma saturation level of only one-fourth that of the background population corresponding to a condition of lifelong vitamin C deficiency [9].

Smoking

Smoking is a major source of oxidants, and estimates have suggested that every puff of a cigarette equals the inhalation of about 10^{14} tar phase radicals and 10^{15} gas phase radicals [90]. Not surprisingly, this draws a major toll on the antioxidant defense of the body as demonstrated by a strong association between tobacco smoke and poor antioxidant status in general and poor vitamin C status in particular [23,91]. Active smoking typically depletes the vitamin C pool by 25%–50% compared to never-smokers [92], while passive smoking/environmental tobacco smoke exposure results in a drop of about half that size [93,94]. The direct cause of the smoking-induced vitamin C depletion has been investigated, and smoking cessation has been shown to immediately restore about half of the vitamin C depletion observed as a result of smoking [95]. This immediate albeit partial recovery has pointed toward an oxidative stress–mediated depletion of vitamin C caused by smoking. Moreover, both oxidative stress and ascorbate recycling are induced by smoking regardless of antioxidant intake [24,96]. However, the lack of full recovery suggests that other factors also contribute to the lower vitamin C status among smokers. Studies have suggested that the difference in vitamin C status between smokers and nonsmokers is not related to altered pharmacokinetics of vitamin C [28,29]. However, as smokers in general have a lower intake of fruits and vegetables and a larger intake of fat compared to nonsmokers [97], this may account for the difference in vitamin C levels observed between ex-smokers and never-smokers [26].

Disease

A plethora of disease conditions, including, for example, infectious diseases, cancer, cardiovascular disease, stroke, diabetes, and sepsis have been associated with poor vitamin C status, some of which are included in this book. Considerable epidemiological evidence has shown poor vitamin C status to be an independent risk factor for disease development. However, causal linkage between disease etiology and vitamin C status remains scarce. The decreased vitamin C status in disease is often explained by a combination of an—sometimes massively—increased turnover due to oxidative stress and inflammation and a decreased dietary intake of vitamin C associated with the disease.

An obvious display of increased vitamin C turnover in critical illness is the large doses often needed to replete the individual to the level of a healthy control. These doses exceed those necessary to saturate a healthy individual by many-fold [98]. One current example is sepsis patients where systemic inflammation and oxidative stress presumably increase the expenditure of vitamin C [99,100]. Whether reestablishing normal vitamin C status in critically ill patients has a significant clinical impact on disease prognosis remains to be established, but promising results are emerging [71,101], and controlled trials are under way.

CONCLUDING REMARKS

The pharmacokinetics of vitamin C are complex, strongly dose dependent, and compartmentalized at physiologic levels, while they are dose independent and first order at pharmacologic levels. The lack of this fundamental knowledge has left deep traces of design flaws, misconceptions, misinterpretations, and erroneous conclusions in the scientific literature. Unfortunately, these inherited problems continue to hamper our ability to properly evaluate the role of vitamin C in human health and its potential relevance in disease prevention and treatment. There is an important balance between the overtly exaggerated optimistic view that enough vitamin C can cure everything and the dismissive negligence that refuses to reexamine the literature based on new evidence. This balance needs to be identified for us to optimize the potential benefit of vitamin C in both health and disease in the future.

REFERENCES

1. Yang, H. 2013. Conserved or lost: Molecular evolution of the key gene GULO in vertebrate vitamin C biosynthesis. *Biochem. Genet.* 51, 413–425.
2. Frikke-Schmidt, H., Tveden-Nyborg, P. and Lykkesfeldt, J. 2016. L-dehydroascorbic acid can substitute l-ascorbic acid as dietary vitamin C source in guinea pigs. *Redox. Biol.* 7, 8–13.
3. Michels, A. J., Hagen, T. M. and Frei, B. 2013. Human genetic variation influences vitamin C homeostasis by altering vitamin C transport and antioxidant enzyme function. *Annu. Rev. Nutr.* 33, 45–70.
4. Tsukaguchi, H., Tokui, T., Mackenzie, B., Berger, U. V., Chen, X. Z., Wang, Y., Brubaker, R. F. and Hediger, M. A. 1999. A family of mammalian Na$^+$-dependent L-ascorbic acid transporters. *Nature* 399, 70–75.
5. Lindblad, M. M., Tveden-Nyborg, P. and Lykkesfeldt, J. 2013. Regulation of vitamin C homeostasis during deficiency. *Nutrients* 5, 2860–2879.
6. Wilson, J. X. 2005. Regulation of vitamin C transport. *Annu. Rev. Nutr.* 25, 105–125.
7. Hasselholt, S., Tveden-Nyborg, P. and Lykkesfeldt, J. 2015. Distribution of vitamin C is tissue specific with early saturation of the brain and adrenal glands following differential oral dose regimens in guinea pigs. *Br. J. Nutr.* 113, 1539–1549.
8. Sotiriou, S., Gispert, S., Cheng, J., Wang, Y., Chen, A., Hoogstraten-Miller, S., Miller, G. F. et al. 2002. Ascorbic-acid transporter Slc23a1 is essential for vitamin C transport into the brain and for perinatal survival. *Nat. Med.* 8, 514–517.

9. Corpe, C. P., Tu, H., Eck, P., Wang, J., Faulhaber-Walter, R., Schnermann, J., Margolis, S. et al. 2010. Vitamin C transporter Slc23a1 links renal reabsorption, vitamin C tissue accumulation, and perinatal survival in mice. *J. Clin. Invest.* 120, 1069–1083.

10. Smith, J. L. and Hodges, R. E. 1987. Serum levels of vitamin C in relation to dietary and supplemental intake of vitamin C in smokers and nonsmokers. *Ann. N Y. Acad. Sci.* 498, 144–152.

11. Lykkesfeldt, J. and Poulsen, H. E. 2010. Is vitamin C supplementation beneficial? Lessons learned from randomised controlled trials. *Br. J. Nutr.* 103, 1251–1259.

12. Cameron, E., Pauling, L. and Leibovitz, B. 1979. Ascorbic acid and cancer: A review. *Cancer Res.* 39, 663–681.

13. Cameron, E. and Pauling, L. 1978. Experimental studies designed to evaluate the management of patients with incurable cancer. *Proc. Natl. Acad. Sci. USA* 75, 6252.

14. Cameron, E. and Pauling, L. 1978. Supplemental ascorbate in the supportive treatment of cancer: Reevaluation of prolongation of survival times in terminal human cancer. *Proc. Natl. Acad. Sci. USA* 75, 4538–4542.

15. Cameron, E. and Pauling, L. 1976. Supplemental ascorbate in the supportive treatment of cancer: Prolongation of survival times in terminal human cancer. *Proc. Natl. Acad. Sci. USA* 73, 3685–3689.

16. Moertel, C. G., Fleming, T. R., Creagan, E. T., Rubin, J., O'Connell, M. J. and Ames, M. M. 1985. High-dose vitamin C versus placebo in the treatment of patients with advanced cancer who have had no prior chemotherapy. A randomized double-blind comparison. *N. Engl. J. Med.* 312, 137–141.

17. Creagan, E. T., Moertel, C. G., O'Fallon, J. R., Schutt, A. J., O'Connell, M. J., Rubin, J. and Frytak, S. 1979. Failure of high-dose vitamin C (ascorbic acid) therapy to benefit patients with advanced cancer. A controlled trial. *N. Engl. J. Med.* 301, 687–690.

18. Levine, M., Conry-Cantilena, C., Wang, Y., Welch, R. W., Washko, P. W., Dhariwal, K. R., Park, J. B. et al. 1996. Vitamin C pharmacokinetics in healthy volunteers: Evidence for a recommended dietary allowance. *Proc. Natl. Acad. Sci. USA* 93, 3704–3709.

19. Chen, Q., Espey, M. G., Krishna, M. C., Mitchell, J. B., Corpe, C. P., Buettner, G. R., Shacter, E. and Levine, M. 2005. Pharmacologic ascorbic acid concentrations selectively kill cancer cells: Action as a pro-drug to deliver hydrogen peroxide to tissues. *Proc. Natl. Acad. Sci. USA* 102, 13604–13609.

20. Nielsen, T. K., Hojgaard, M., Andersen, J. T., Poulsen, H. E., Lykkesfeldt, J. and Mikines, K. J. 2015. Elimination of ascorbic acid after high-dose infusion in prostate cancer patients: A pharmacokinetic evaluation. *Basic Clin. Pharmacol. Toxicol.* 116, 343–348.

21. Carr, A. and Frei, B. 1999. Does vitamin C act as a pro-oxidant under physiological conditions? [Review] [148 refs]. *FASEB J.* 13, 1007–1024.

22. Buettner, G. R. 1993. The pecking order of free radicals and antioxidants: Lipid peroxidation, alpha-tocopherol, and ascorbate. *Arch. Biochem. Biophys.* 300, 535–543.

23. Lykkesfeldt, J., Loft, S., Nielsen, J. B. and Poulsen, H. E. 1997. Ascorbic acid and dehydroascorbic acid as biomarkers of oxidative stress caused by smoking. *Am. J. Clin. Nutr.* 65, 959–963.

24. Lykkesfeldt, J., Viscovich, M. and Poulsen, H. E. 2003. Ascorbic acid recycling in human erythrocytes is induced by smoking in vivo. *Free Radic. Biol. Med.* 35, 1439–1447.

25. Bode, A. M., Cunningham, L. and Rose, R. C. 1990. Spontaneous decay of oxidized ascorbic acid (dehydro-L-ascorbic acid) evaluated by high-pressure liquid chromatography. *Clin. Chem.* 36, 1807–1809.

26. Tveden-Nyborg, P. and Lykkesfeldt, J. 2013. Does vitamin C deficiency increase lifestyle-associated vascular disease progression? Evidence based on experimental and clinical studies. *Antioxid. Redox. Signal* 19, 2084–2104.

27. Frei, B., Birlouez-Aragon, I. and Lykkesfeldt, J. 2012. Authors' perspective: What is the optimum intake of vitamin C in humans? *Crit. Rev. Food Sci. Nutr.* 52, 815–829.

28. Lykkesfeldt, J., Bolbjerg, M. L. and Poulsen, H. E. 2003. Effect of smoking on erythorbic acid pharmacokinetics. *Br. J. Nutr.* 89, 667–671.

29. Viscovich, M., Lykkesfeldt, J. and Poulsen, H. E. 2004. Vitamin C pharmacokinetics of plain and slow release formulations in smokers. *Clin Nutr.* 23, 1043–1050.

30. Wang, Y., Mackenzie, B., Tsukaguchi, H., Weremowicz, S., Morton, C. C. and Hediger, M. A. 2000. Human vitamin C (L-ascorbic acid) transporter SVCT1. *Biochem. Biophys. Res. Commun.* 267, 488–494.

31. Vera, J. C., Rivas, C. I., Fischbarg, J. and Golde, D. W. 1993. Mammalian facilitative hexose transporters mediate the transport of dehydroascorbic acid. *Nature* 364, 79–82.

32. Corpe, C. P., Eck, P., Wang, J., Al-Hasani, H. and Levine, M. 2013. Intestinal dehydroascorbic acid (DHA) transport mediated by the facilitative sugar transporters, GLUT2 and GLUT8. *J. Biol. Chem.* 288, 9092–9101.

33. Lykkesfeldt, J. 2007. Ascorbate and dehydroascorbic acid as reliable biomarkers of oxidative stress: Analytical reproducibility and long-term stability of plasma samples subjected to acidic deproteinization. *Cancer Epidemiol. Biomarkers Prev.* 16, 2513–2516.

34. Todhunter, E. N., Mc, M. T. and Ehmke, D. A. 1950. Utilization of dehydroascorbic acid by human subjects. *J. Nutr.* 42, 297–308.

35. Linkswiler, H. 1958. The effect of the ingestion of ascorbic acid and dehydroascorbic acid upon the blood levels of these two components in human subjects. *J. Nutr.* 64, 43–54.

36. Sabry, J. H., Fisher, K. H. and Dodds, M. L. 1958. Human utilization of dehydroascorbic acid. *J. Nutr.* 64, 457–466.

37. Malo, C. and Wilson, J. X. 2000. Glucose modulates vitamin C transport in adult human small intestinal brush border membrane vesicles. *J. Nutr.* 130, 63–69.

38. Kubler, W. and Gehler, J. 1970. Kinetics of intestinal absorption of ascorbic acid. Calculation of non-dosage-dependent absorption processes. *Int. Z. Vitaminforsch.* 40, 442–453.

39. Mayersohn, M. 1972. Ascorbic acid absorption in man—Pharmacokinetic implications. *Eur. J. Pharmacol.* 19, 140–142.

40. May, J. M. and Qu, Z. C. 2015. Ascorbic acid efflux from human brain microvascular pericytes: Role of re-uptake. *Biofactors* 41, 330–338.

41. Rumsey, S. C., Daruwala, R., Al-Hasani, H., Zarnowski, M. J., Simpson, I. A. and Levine, M. 2000. Dehydroascorbic acid transport by GLUT4 in Xenopus oocytes and isolated rat adipocytes. *J. Biol. Chem.* 275, 28246–28253.

42. Rumsey, S. C., Kwon, O., Xu, G. W., Burant, C. F., Simpson, I. and Levine, M. 1997. Glucose transporter isoforms GLUT1 and GLUT3 transport dehydroascorbic acid. *J. Biol. Chem.* 272, 18982–18989.

43. Mardones, L., Ormazabal, V., Romo, X., Jana, C., Binder, P., Pena, E., Vergara, M. and Zuniga, F. A. 2011. The glucose transporter-2 (GLUT2) is a low affinity dehydroascorbic acid transporter. *Biochem. Biophys. Res. Commun.* 410, 7–12.

44. May, J. M., Qu, Z. and Morrow, J. D. 2001 Mechanisms of ascorbic acid recycling in human erythrocytes. *Biochim Biophys Acta* 1528, 159–166.

45. Mendiratta, S., Qu, Z. C. and May, J. M. 1998 Enzyme-dependent ascorbate recycling in human erythrocytes: Role of thioredoxin reductase. *Free Radic Biol Med.* 25, 221–228.

46. Mendiratta, S., Qu, Z. C. and May, J. M. 1998. Erythrocyte ascorbate recycling: Antioxidant effects in blood. *Free Radic Biol Med.* 24, 789–797.

47. Lykkesfeldt, J. 2002. Increased oxidative damage in vitamin C deficiency is accompanied by induction of ascorbic acid recycling capacity in young but not mature guinea pigs. *Free Radic. Res.* 36, 567–574.

48. May, J. M., Qu, Z. C. and Whitesell, R. R. 1995. Ascorbic acid recycling enhances the antioxidant reserve of human erythrocytes. *Biochemistry* 34, 12721–12728.

49. Tu, H., Li, H., Wang, Y., Niyyati, M., Wang, Y., Leshin, J. and Levine, M. 2015. Low red blood cell vitamin C concentrations induce red blood cell fragility: A link to diabetes via glucose, glucose transporters, and dehydroascorbic acid. *EBioMedicine* 2, 1735–1750.

50. Tu, H., Wang, Y., Li, H., Brinster, L. R. and Levine, M. 2017. Chemical transport knockout for oxidized vitamin C, dehydroascorbic acid, reveals its functions in vivo. *EBioMedicine* 23, 125–135.

51. Daruwala, R., Song, J., Koh, W. S., Rumsey, S. C. and Levine, M. 1999. Cloning and functional characterization of the human sodium-dependent vitamin C transporters hSVCT1 and hSVCT2. *FEBS Lett.* 460, 480–484.

52. Savini, I., Rossi, A., Pierro, C., Avigliano, L. and Catani, M. V. 2008. SVCT1 and SVCT2: Key proteins for vitamin C uptake. *Amino Acids* 34, 347–355.

53. Eck, P., Kwon, O., Chen, S., Mian, O. and Levine, M. 2013. The human sodium-dependent ascorbic acid transporters SLC23A1 and SLC23A2 do not mediate ascorbic acid release in the proximal renal epithelial cell. *Physiol. Rep.* 1, e00136.

54. Lykkesfeldt, J., Trueba, G. P., Poulsen, H. E. and Christen, S. 2007. Vitamin C deficiency in weanling guinea pigs: Differential expression of oxidative stress and DNA repair in liver and brain. *Br. J. Nutr.* 98, 1116–1119.

55. Harrison, F. E. and May, J. M. 2009. Vitamin C function in the brain: Vital role of the ascorbate transporter SVCT2. *Free Radic. Biol. Med.* 46, 719–730.

56. Meredith, M. E., Harrison, F. E. and May, J. M. 2011. Differential regulation of the ascorbic acid transporter SVCT2 during development and in response to ascorbic acid depletion. *Biochem. Biophys. Res. Commun.* 414, 737–742.

57. Sogaard, D., Lindblad, M. M., Paidi, M. D., Hasselholt, S., Lykkesfeldt, J. and Tveden-Nyborg, P. 2014. In vivo vitamin C deficiency in guinea pigs increases ascorbate transporters in liver but not kidney and brain. *Nutr. Res.* 34, 639–645.

58. Frikke-Schmidt, H., Tveden-Nyborg, P., Birck, M. M. and Lykkesfeldt, J. 2011. High dietary fat and cholesterol exacerbates chronic vitamin C deficiency in guinea pigs. *Br. J. Nutr.* 105, 54–61.

59. Paidi, M. D., Schjoldager, J. G., Lykkesfeldt, J. and Tveden-Nyborg, P. 2014. Prenatal vitamin C deficiency results in differential levels of oxidative stress during late gestation in foetal guinea pig brains. *Redox. Biol.* 2, 361–367.

60. Schjoldager, J. G., Paidi, M. D., Lindblad, M. M., Birck, M. M., Kjaergaard, A. B., Dantzer, V., Lykkesfeldt, J. and Tveden-Nyborg, P. 2015. Maternal vitamin C deficiency during pregnancy results in transient fetal and placental growth retardation in guinea pigs. *Eur J. Nutr.* 54, 667–676.

61. Schjoldager, J. G., Tveden-Nyborg, P. and Lykkesfeldt, J. 2013. Prolonged maternal vitamin C deficiency overrides preferential fetal ascorbate transport but does not influence perinatal survival in guinea pigs. *Br. J. Nutr.* 110, 1573–1579.

62. Tveden-Nyborg, P., Hasselholt, S., Miyashita, N., Moos, T., Poulsen, H. E. and Lykkesfeldt, J. 2012. Chronic vitamin C deficiency does not accelerate oxidative stress in ageing brains of guinea pigs. *Basic Clin. Pharmacol. Toxicol.* 110, 524–529.

63. Tveden-Nyborg, P., Johansen, L. K., Raida, Z., Villumsen, C. K., Larsen, J. O. and Lykkesfeldt, J. 2009. Vitamin C deficiency in early postnatal life impairs spatial memory and reduces the number of hippocampal neurons in guinea pigs. *Am. J. Clin. Nutr.* 90, 540–546.

64. Tveden-Nyborg, P. and Lykkesfeldt, J. 2009. Does vitamin C deficiency result in impaired brain development in infants? *Redox. Rep.* 14, 2–6.

65. Tveden-Nyborg, P., Vogt, L., Schjoldager, J. G., Jeannet, N., Hasselholt, S., Paidi, M. D., Christen, S. and Lykkesfeldt, J. 2012. Maternal vitamin C deficiency during pregnancy persistently impairs hippocampal neurogenesis in offspring of guinea pigs. *PLOS ONE* 7, e48488.

66. May, J. M. 2012. Vitamin C transport and its role in the central nervous system. *Subcell Biochem.* 56, 85–103.

67. Smirnoff, N. 2018. Ascorbic acid metabolism and functions: A comparison of plants and mammals. *Free Radic. Biol. Med.* 122, 116–129.

68. Frikke-Schmidt, H., Tveden-Nyborg, P. and Lykkesfeldt, J. 2011. Vitamin C in human nutrition. In *Vitamins in the Prevention of Human Disease* (Herrmann, W. and Obeid, R., eds.). pp. 323–347, De Gruyter, Berlin.

69. Buettner, G. R. and Schafer, F. Q. 2004. Ascorbate as an antioxidant. In *Vitamin C: Its Functions and Biochemistry in Animals and Plants* (Asard, H., May, J. M. and Smirnoff, N., eds.). pp. 173–188, BIOS Scientific Publishers Limited, Oxford.

70. Banhegyi, G., Braun, L., Csala, M., Puskas, F. and Mandl, J. 1997. Ascorbate metabolism and its regulation in animals. [Review] [83 refs]. *Free Radic. Biol. Med.* 23, 793–803.

71. Marik, P. E., Khangoora, V., Rivera, R., Hooper, M. H. and Catravas, J. 2017. Hydrocortisone, vitamin C, and thiamine for the treatment of severe sepsis and septic shock: A retrospective before-after study. *Chest* 151, 1229–1238.

72. Levine, M., Wang, Y., Padayatty, S. J. and Morrow, J. 2001. A new recommended dietary allowance of vitamin C for healthy young women. *Proc. Natl. Acad. Sci. USA* 98, 9842–9846.

73. Bulbake, U., Doppalapudi, S., Kommineni, N. and Khan, W. 2017. Liposomal formulations in clinical use: An updated review. *Pharmaceutics* 9.

74. Davis, J. L., Paris, H. L., Beals, J. W., Binns, S. E., Giordano, G. R., Scalzo, R. L., Schweder, M. M., Blair, E. and Bell, C. 2016. Liposomal-encapsulated ascorbic acid: Influence on vitamin C bioavailability and capacity to protect against ischemia-reperfusion injury. *Nutr. Metab. Insights* 9, 25–30.

75. Miura, Y., Fuchigami, Y., Hagimori, M., Sato, H., Ogawa, K., Munakata, C., Wada, M., Maruyama, K. and Kawakami, S. 2018. Evaluation of the targeted delivery of 5-fluorouracil and ascorbic acid into the brain with ultrasound-responsive nanobubbles. *J. Drug Target* 26, 684–691.

76. Peng, Y., Zhao, Y., Chen, Y., Yang, Z., Zhang, L., Xiao, W., Yang, J., Guo, L. and Wu, Y. 2018. Dual-targeting for brain-specific liposomes drug delivery system: Synthesis and preliminary evaluation. *Bioorg. Med. Chem.* 26, 4677–4686.

77. Chen, P., Yu, J., Chalmers, B., Drisko, J., Yang, J., Li, B. and Chen, Q. 2012. Pharmacological

ascorbate induces cytotoxicity in prostate cancer cells through ATP depletion and induction of autophagy. *Anticancer Drugs* 23, 437–444.

78. Chen, Q., Espey, M. G., Sun, A. Y., Pooput, C., Kirk, K. L., Krishna, M. C., Khosh, D. B., Drisko, J. and Levine, M. 2008. Pharmacologic doses of ascorbate act as a prooxidant and decrease growth of aggressive tumor xenografts in mice. *Proc. Natl. Acad. Sci. USA* 105, 11105–11109.

79. Chen, Q., Espey, M. G., Sun, A. Y., Lee, J. H., Krishna, M. C., Shacter, E., Choyke, P. L. et al. 2007. Ascorbate in pharmacologic concentrations selectively generates ascorbate radical and hydrogen peroxide in extracellular fluid in vivo. *Proc. Natl. Acad. Sci. USA* 104, 8749–8754.

80. Campbell, E. J., Vissers, M. C. M., Wohlrab, C., Hicks, K. O., Strother, R. M., Bozonet, S. M., Robinson, B. A. and Dachs, G. U. 2016. Pharmacokinetic and anti-cancer properties of high dose ascorbate in solid tumours of ascorbate-dependent mice. *Free Radic. Biol. Med.* 99, 451–462.

81. Kuiper, C., Vissers, M. C. and Hicks, K. O. 2014. Pharmacokinetic modeling of ascorbate diffusion through normal and tumor tissue. *Free Radic. Biol. Med.* 77, 340–352.

82. Vissers, M. C. M. and Das, A. B. 2018. Potential mechanisms of action for vitamin C in cancer: Reviewing the evidence. *Front. Physiol.* 9, 809.

83. Schoenfeld, J. D., Alexander, M. S., Waldron, T. J., Sibenaller, Z. A., Spitz, D. R., Buettner, G. R., Allen, B. G. and Cullen, J. J. 2019. Pharmacological ascorbate as a means of sensitizing cancer cells to radio-chemotherapy while protecting normal Tissue. *Semin. Radiat. Oncol.* 29, 25–32.

84. Doskey, C. M., Buranasudja, V., Wagner, B. A., Wilkes, J. G., Du, J., Cullen, J. J. and Buettner, G. R. 2016. Tumor cells have decreased ability to metabolize H2O2: Implications for pharmacological ascorbate in cancer therapy. *Redox. Biol.* 10, 274–284.

85. Du, J., Cullen, J. J. and Buettner, G. R. 2012. Ascorbic acid: Chemistry, biology and the treatment of cancer. *Biochim. Biophys. Acta* 1826, 443–457.

86. Padayatty, S. J., Sun, H., Wang, Y., Riordan, H. D., Hewitt, S. M., Katz, A., Wesley, R. A. and Levine, M. 2004. Vitamin C pharmacokinetics: Implications for oral and intravenous use. *Ann. Intern. Med.* 140, 533–537.

87. Ou, J., Zhu, X., Lu, Y., Zhao, C., Zhang, H., Wang, X., Gui, X. et al. 2017. The safety and pharmacokinetics of high dose intravenous ascorbic acid synergy with modulated electrohyperthermia in Chinese patients with stage III–IV non-small cell lung cancer. *Eur. J. Pharm. Sci.* 109, 412–418.

88. Stephenson, C. M., Levin, R. D., Spector, T. and Lis, C. G. 2013. Phase I clinical trial to evaluate the safety, tolerability, and pharmacokinetics of high-dose intravenous ascorbic acid in patients with advanced cancer. *Cancer Chemother. Pharmacol.* 72, 139–146.

89. Kobylecki, C. J., Afzal, S., Davey Smith, G. and Nordestgaard, B. G. 2015. Genetically high plasma vitamin C, intake of fruit and vegetables, and risk of ischemic heart disease and all-cause mortality: A Mendelian randomization study. *Am. J. Clin. Nutr.* 101, 1135–1143.

90. Pryor, W. A. and Stone, K. 1993. Oxidants in cigarette smoke. Radicals, hydrogen peroxide, peroxynitrate, and peroxynitrite. *Ann. N Y. Acad. Sci.* 686, 12–27; discussion 27-18.

91. Lykkesfeldt, J., Christen, S., Wallock, L. M., Chang, H. H., Jacob, R. A. and Ames, B. N. 2000. Ascorbate is depleted by smoking and repleted by moderate supplementation: A study in male smokers and nonsmokers with matched dietary antioxidant intakes. *Am. J. Clin. Nutr.* 71, 530–536.

92. Lykkesfeldt, J. 2006. Smoking depletes vitamin C: Should smokers be recommended to take supplements? In *Cigarette Smoke and Oxidative Stress* (Halliwell, B. and Poulsen, H. E., eds.). pp. 237–260, Springer Verlag.

93. Preston, A. M., Rodriguez, C. and Rivera, C. E. 2006. Plasma ascorbate in a population of children: Influence of age, gender, vitamin C intake, BMI and smoke exposure. *P R. Health Sci. J.* 25, 137–142.

94. Preston, A. M., Rodriguez, C., Rivera, C. E. and Sahai, H. 2003. Influence of environmental tobacco smoke on vitamin C status in children. *Am. J. Clin. Nutr.* 77, 167–172.

95. Lykkesfeldt, J., Prieme, H., Loft, S. and Poulsen, H. E. 1996. Effect of smoking cessation on plasma ascorbic acid concentration. *BMJ* 313, 91.

96. Lykkesfeldt, J., Viscovich, M. and Poulsen, H. E. 2004. Plasma malondialdehyde is induced by smoking: A study with balanced antioxidant profiles. *Br. J. Nutr.* 92, 203–206.

97. Canoy, D., Wareham, N., Welch, A., Bingham, S., Luben, R., Day, N. and Khaw, K. T. 2005. Plasma ascorbic acid concentrations and fat distribution in 19,068 British men and women in the European Prospective Investigation into

cancer and nutrition Norfolk cohort study. *Am. J. Clin. Nutr.* 82, 1203–1209.

98. Carr, A. C., Rosengrave, P. C., Bayer, S., Chambers, S., Mehrtens, J. and Shaw, G. M. 2017. Hypovitaminosis C and vitamin C deficiency in critically ill patients despite recommended enteral and parenteral intakes. *Crit. Care* 21, 300.

99. Marik, P. E. 2018. Vitamin C for the treatment of sepsis: The scientific rationale. *Pharmacol. Ther.* 189, 63–70.

100. Carr, A. C., Shaw, G. M., Fowler, A. A. and Natarajan, R. 2015. Ascorbate-dependent vasopressor synthesis: A rationale for vitamin C administration in severe sepsis and septic shock? *Crit. Care.* 19, 418.

101. Fowler, A. A.3rd, Syed, A. A., Knowlson, S., Sculthorpe, R., Farthing, D., DeWilde, C., Farthing, C. A. et al. 2014. Phase I safety trial of intravenous ascorbic acid in patients with severe sepsis. *J. Transl. Med.* 12, 32.

CHAPTER FIVE

Ascorbate as an Enzyme Cofactor

Margreet C.M. Vissers and Andrew B. Das

DOI: 10.1201/9780429442025-5

CONTENTS

INTRODUCTION

The discovery that the compound responsible for the antiscurvy properties of fruit and vegetables, already named vitamin C [1,2], was in fact a simple hexuronic acid with reducing capability, initiated a conversation that has lasted for almost a century. The antiscorbutic properties of the compound inspired its being named *ascorbic acid* [3,4]. At neutral pH, ascorbic acid exists in the anionic form and is therefore correctly referred to as ascorbate in discussions relevant to biological settings [5–7]. Humans and other primates, guinea pigs, fruit bats, and some bird and fish species, cannot synthesize ascorbate, and consequently, the majority of the discussion over the past 80 or so years has concerned the determination of the daily intake requirement of this vitamin [8–14]. Whereas previously it was considered that ascorbate intake needed to be sufficient only to avoid the development of scurvy, this has been countered by epidemiological studies that have reported associations between vitamin C status and numerous health outcomes. Low vitamin C status has been found to be associated with an increased risk of myocardial infarction [15,16], stroke [17,18], diabetes [19–21], cancer [22–31], and all-cause mortality [32–35]. That lack of a single micronutrient could have such wide-ranging health benefits was not explained by its known biological functions, and many questions remained unanswered.

In the past two decades, however, our understanding of the many biological roles of ascorbate has expanded with improved knowledge of the uptake, transport, and pharmacokinetics of ascorbate in humans, and the discovery of new enzymes with an ascorbate cofactor requirement. Other chapters in this volume are dedicated to the uptake of vitamin C, to the facilitated transport via the dedicated sodium-dependent vitamin C transporters (SVCTs), and to the pharmacokinetics of supraphysiologic concentrations that have the potential to contribute to the pro-oxidant activity. In this chapter, we provide an overview of the cofactor activity of ascorbate and the potential biological impact when the associated enzyme activity may be compromised.

TISSUE DISTRIBUTION OF ASCORBATE IN HUMANS

Once ingested, ascorbate is absorbed through the gut in humans and distributed throughout the body via the circulation to supply the various organs and tissues. This pattern of distribution is similar in all animal species, regardless of whether ascorbate is synthesized in the liver or kidney or is sourced from the diet. Whereas plasma levels are restricted to concentrations below 100 μM by filtration and selective reuptake in the kidneys, intracellular concentrations are much higher due to SVCT-mediated active transport and are generally greater than 1 millimolar [36–48].

There is considerable variation in tissue ascorbate content. This has been investigated in humans, guinea pigs, and some other animals, with similar distribution noted in all species tested. Very high ascorbate levels are found in the adrenal and pituitary glands, and in the lens, with 30–50 mg/100 g wet weight of tissue being reported in these tissues, corresponding to around 10 mM [36,37,49]. Even higher levels of 126 mg/100 g tissue were measured in bovine corpus luteum [50]. Intracellular concentrations in neurons have also been estimated

at 10 mM [51], with other brain cell types, liver, spleen, and kidney containing approximately 1 mM [36]. Circulating neutrophils, monocytes, lymphocytes, and platelets contain approximately 1.5, 3, 3.5, and 4 mM intracellular levels, respectively [38,52]. Recently, bone marrow hematopoietic stem cells were found to have up to 20-fold more ascorbate than mature white blood cells [53]. This distribution profile of ascorbate is widely considered to reflect the functional requirements for ascorbate in the respective tissues.

ASCORBATE AS A COFACTOR FOR IRON-CONTAINING ENZYMES

Ascorbate is capable of one-electron reduction and readily restores Fe^{2+} from Fe^{3+} and Cu^+ from Cu^{2+} [54–56]. This property aligns with its capacity to act as an essential cofactor for numerous copper (Cu)- or iron (Fe)-containing enzymes that are involved in many biological processes and that are collectively found in most cells throughout the body. Most enzymes utilizing the reducing capacity of ascorbate belong to the family of Fe-containing 2-oxoglutarate dioxygenases (2-OGDDs), a large and diverse group found throughout biology. It is estimated that there are as many as 80 members of this family in animals, including humans, where they catalyze the oxidative hydroxylation of proteins, nucleic acids, and small molecules [57–59].

The posttranslational modification of collagen by hydroxylation was first described in the 1940s and 1950s [60–64], and the prolyl hydroxylase enzymes responsible were the first identified 2-OGDDs [65–68]. Since this time, the prevalence of protein hydroxylation has become apparent, and this modification is now recognized as being critical to the regulation of the hypoxic response [69–71] and to protein synthesis by modification of ribosomal proteins [57,72]. In addition, demethylation of histones, DNA, and RNA is initiated by oxidative hydroxylation of methyllysine [73–79], methylarginine [80–83], methylcytosine [84,85], and methyladenosine [86–88]. The number of enzymes identified as 2-OGDDs active in mammalian cells has increased rapidly in this century, and their widespread influence on many fundamental aspects of biology is becoming well recognized. Table 5.1 contains a summary of the more than 50 2-OGDD enzymes identified to date in mammalian cells.

TABLE 5.1

Enzymes with ascorbate cofactor requirement in humans and their biological functions

Enzyme Family	Enzymes	Target Substrate(s)	Biological Functions	References
2-Oxoglutarate–dependent dioxygenases (2-OGDDs)	Collagen hydroxylases			
	• C-P4H	Collagen—C4 proline	Stabilizes collagen triple helix	[93]
	• C-P3H	Procollagen—C3 proline	Uncertain; modification of collagen IV structure	[67]
	• Lysyl hydroxylase (PLOD)	Procollagen C5 lysine	Stabilizes collagen triple helix	[68,97,288]
	HIF hydroxylases			
	• PHD 1-3	Prolines on HIF-α	Hydroxylation targets HIF-α to the proteasome, preventing HIF transcription factor activation	[289,290]
	• FIH	Asparagine on HIF-α, asparagine in ankyrin-repeat domains	Prevents binding of HIF-α to CBP/p300 and formation of transcription complex	[291–293]
		Histidine in ankyrin-repeat domains	Stabilizes Ankyrin fold	[72,294]
			Unknown	
	Ribosomal protein hydroxylases			
	• myc-Induced nuclear antigen 53	Histidine in ribosomal proteins	Regulation of protein translation	[57]
	• Nucleolar protein 66	Histidine in ribosomal proteins	Regulation of protein translation	[295]
	• OGFOD1	Proline in ribosomal protein s23	Regulation of protein translation	[296]
	• JMJD4	Lysine in eukaryotic release factor 1	Enhances efficiency of translation termination	[57]
	• JMJD7	Lysine in developmentally regulated GTP-binding protein 1/2	Unknown	[57]
	Protein demethylases			
	• Lysine demethylases (KDMs), members of the JmjC family (>16 enzymes)	Nε-methylated lysines in histones. Mono-, di-, and trimethyl-lysines targeted via initial hydroxylation in the demethylation pathway	Extensive involvement in the regulation of gene expression, with impact on epigenetics and development and cancer	[73–79]

(Continued)

TABLE 5.1 (Continued)

Enzymes with ascorbate cofactor requirement in humans and their biological functions

Enzyme Family	Enzymes	Target Substrate(s)	Biological Functions	References
	• Arginine demethylases (JMJD6, some KDMs)	Mono- and dimethyl arginines, demethylation initiated via hydroxylation	Demethylation of histone arginines, with impact on protein expression and potential involvement in some cancers	[80–83]
	2-OGDDs targeting DNA or RNA modifications			
	• TET1-3	5mC in DNA	Generation of 5hmC, 5caC, 5fC. Demethylation and epigenetic regulation of gene transcription	[84,85]
	• tRNA hydroxylase TYW5	Wybutosine, modified nucleoside in tRNA(Phe)	Promotion of protein translation via formation of hydroxywybutosine	[297–299]
	• AlkBH8	tRNA uridine	Uridine modifications affecting protein translation efficiency	[86–88]
	• ALKB homologues 1-9	N-methyl nucleotides in DNA/RNA	DNA repair. Removal of 1-mA and 3-mC with regeneration of unsubstituted bases	[300–304]
	• ALKB5	Methyl-adenosine on RNA	N-methyl RNA demethylation, regulating mRNA stability/levels, may have role in regulation of circadian clock	
	• FTO (fat-, mass-, and obesity-associated protein)	N6,2′-O-dimethyladenosine and N6-methyladenosine in mRNA	Affects stability of mRNA; variants in FTO are linked to metabolic syndrome and obesity	
	Hydroxylases targeting small molecules			
	• Trimethyl-lysine hydroxylase	Nε-trimethyl-lysine	Catalysis of first step in the synthesis of carnitine	[135]
	• γ-Butyrobetaine hydroxylase	γ-butyrobetaine	Carnitine synthesis—catalysis of last step. Carnitine transports fatty acids into mitochondria and is obtained by synthesis and diet in humans	[134,136 305–307]
	Other			
	• ASPH1-2	Asp or Asn residues in EGF domains	Unknown, mutations in ASPH result in developmental abnormalities	[308]
Cu(+)-dependent monooxygenases	Dopamine β-monooxygenase	Dopamine	Synthesis of noradrenalin	[282,309]
	Peptidylglycine α-amidating monooxygenase	Peptide pro-hormones	Generation of peptide amide hormones including vasopressin, oxytocin, substance P, thyrotropin-releasing hormone, via hydroxyglycine intermediate	[241,255,257, 258,264,310]

Reaction Mechanism of the 2-Oxoglutarate–Dependent Dioxygenases

The active site of the 2-OGDDs is highly conserved and contains a nonheme Fe bound through three-point coordination with two histidine residues and an aspartic acid or glutamic acid, with the remaining three coordination sites occupied by water ligands [89,90]. The enzymes use O_2 and 2-oxoglutarate (2-OG) as substrates to catalyze the hydroxylation of the target substrate [57,58,90]. During the reactive cycle, the water ligands are displaced by 2-OG at two positions and by O_2 at the final available site (see Figure 5.2). The hydroxylation of the target substrate occurs by radical reactions that involve the formation of a superoxide radical that decarboxylates 2-OG to produce succinate, CO_2, and a ferryl (Fe^{IV})-oxo intermediate. This intermediate is able to hydroxylate the substrate by radical transfer, finally regenerating the Fe^{2+} active center and releasing the hydroxylated product, succinate, and CO_2 [57,58,89] (Figure 5.1).

Role of Ascorbate in the Reaction Cycle of 2-OGDDs

There is a substantial literature indicating that ascorbate is a required cofactor necessary to maintain full 2-OGDD enzyme activity (for examples see [7,66,91–98]). This is presumed to be due to its capacity to reduce Fe^{3+} and maintain iron in the ferrous state [59,99,100]. Of the mammalian enzymes, prolyl-4-hydroxylase has been the most intensively studied with respect to ascorbate dependency. The purified enzyme catalyzed the hydroxylation of proline peptide substrate in the absence of ascorbate at close to maximal rate for approximately 30 reaction cycles over the first 8 seconds [93]. Thereafter, the reaction rate rapidly declined, and further hydroxylation was not measurable after 1 minute [93]. These data suggest that repeated cycling of the enzyme leads to oxidation of the active site Fe and supports a role for ascorbate in restoring and maintaining full enzyme activity.

The specific requirement for ascorbate has been compared with other reducing agents. None were found to effectively substitute for ascorbate, with the reaction rates in the presence of millimolar concentrations of glutathione, cysteine, homocysteine, $NADH_2$, and $NADPH_2$ being less than 10% of that measured with ascorbate [93]. Similarly, the activity of the prolyl hydroxylase targeting the hypoxia-inducible factor (HIF) α subunit was found to be optimal in the presence of ascorbate, with glutathione only partially able

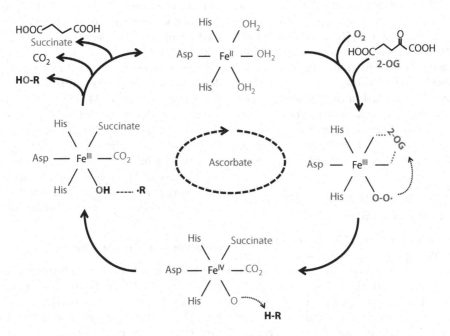

Figure 5.1. Reaction mechanism common to the family of 2-oxoglutarate–dependent dioxygenases (2-OGDDs). During the reaction cycle, the active site iron (Fe) is sequentially oxidized to enable the radical-mediated transfer of one oxygen atom from O_2 to the target molecule, while the second O is transferred to 2-OG, generating succinate with the release of CO_2. Ascorbate functions to maintain the active site Fe in the reduced state, allowing the enzyme to continue cycling.

to compensate [95]. This information supports the many biological observations where these enzymes have decreased activity under conditions of ascorbate deficiency when the concentrations of other reducing agents such as glutathione would be unaffected.

Furthermore, the 2-OGDDs have been reported to undergo "uncoupled" reaction cycles, which results in oxidation of 2-OG to succinate and CO_2 but without the formation of a hydroxylated substrate [59,101–107]. This reaction results in rapid inactivation of the enzyme that is fully reversible by ascorbate, with the generation of dehydroascorbate providing evidence for ascorbate-mediated regeneration of Fe^{2+} and maintenance of enzyme activity [93,100].

Estimation of Ascorbate Concentrations Required to Support 2-OGDDs

The ability of ascorbate to support the activity of the 2-OGDDs is widely acknowledged, and it is uniformly included in buffers for in vitro analyses of these enzymes, usually at concentrations greater than 1 mM (for examples see [66,93,95,100]). The intracellular concentration required for ascorbate to support 2-OGDD activity is unknown. The reported K_m is 300 µM for prolyl-4-hydroxylase [108], 270 µM for the asparagine hydroxylase factor-inhibiting HIF (FIH) [109], and 140–180 µM for the three HIF prolyl hydroxylase enzymes [110]. Based on these values, the millimolar intracellular levels should suffice to ensure optimal activity of the 2-OGDDs. As noted in the previous sections and in Figure 5.1, intracellular concentrations in excess of 1 mM are found in cells and organs with dedicated 2-OGDD activity, such as the bone marrow stem cells [53] and skin fibroblasts [111].

EFFECT OF ASCORBATE AVAILABILITY ON 2-OGDD–MEDIATED PROCESSES

The dependency of the 2-OGDDs on ascorbate availability suggests that deficiency, or dietary insufficiency, could have an impact on the associated biological functions. Interestingly, animals that synthesize ascorbate in the liver or kidney appear to increase synthesis in response to turnover to ensure that plasma levels are maintained at saturation [112,113]. Humans and other mammals with dysfunctional gulonolactone oxidase, however, are prone to low plasma status

when dietary intake is insufficient to compensate for ascorbate turnover, and this correlates closely with decreased ascorbate levels in organs throughout the body [38,40,114–116]. The effect of ascorbate on the 2-OGDDs and evidence for ascorbate-dependent modulation of these activities in vivo is considered in the next section.

Ascorbate and Collagen Synthesis

Three enzymes have been shown to be involved in the posttranslational modification of collagens (Table 5.1). The best known is prolyl-4-hydroxylase, which was the first identified member of the 2-OGDD family [65–67]. This enzyme is responsible for the hydroxylation of up to 10% of the abundant proline residues in collagen and stabilizes the molecule's tertiary structure via the formation of hydrogen bonds [117]. In addition, there is a less well-characterized prolyl-3-hydroxylase that specifically targets type IV collagen [68,117]. Hydroxylation of lysyl residues by lysine hydroxylase is proposed to signal the addition of galactosyl side chains, and mutations in the gene for this enzyme have been found in individuals with Ehlers-Danlos syndrome with defective connective tissue formation [118]. Patients with this condition have some symptoms in common with vitamin C deficiency, notably, the formation of skin petechiae and poor wound healing [118–123].

There is direct evidence linking vitamin C deficiency with decreased hydroxyproline content in collagen in animals [124,125], and this has been associated with impaired wound healing in some clinical studies [126] and in animal studies [127–133]. However, wound healing is complex, and ascorbate can have an impact on many aspects of this process in addition to collagen deposition. Surprisingly little clinical research has been carried out on this topic as detailed in Chapter 9.

Ascorbate and Carnitine Synthesis

The biosynthesis of carnitine involves two steps catalyzed by 2-OGDDs (Figure 5.2 and Table 5.1) [134–136]. In humans, carnitine can be acquired through the diet or synthesized in the kidney, liver, or brain, the primary organs expressing the terminal enzyme in the biosynthetic pathway, γ-butyrobetaine hydroxylase [137]. Carnitine functions to shuttle fatty acids into the mitochondria

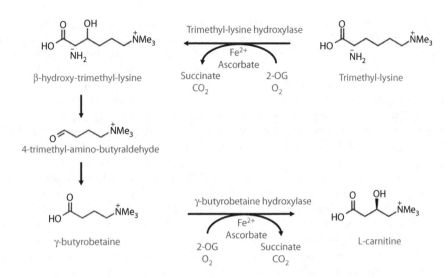

Figure 5.2. Ascorbate-dependent reactions in the biosynthesis of carnitine. Two steps in the biosynthetic pathway are dependent on ascorbate-dependent hydroxylation by 2-oxoglutarate–dependent dioxygenases.

for energy metabolism; this is a rate-limiting step in the catabolism of fatty acids, and carnitine availability is therefore closely linked to metabolism and energy production [137].

Evidence that ascorbate availability affects carnitine metabolism comes mostly from studies with guinea pigs. A complex relationship is suggested by the available data. Dietary vitamin C depletion was shown to decrease carnitine concentrations in muscle, liver, and kidney tissue in the guinea pig [135,138], corresponding to decreased activity of γ-butyrobetaine hydroxylase [135]. Plasma carnitine levels, however, do not appear to show a positive correlation with ascorbate levels, and in fact, a reverse relationship was indicated in guinea pigs [138] and also in two human studies [139,140].

There is also evidence that ascorbate status in animals and humans affects carnitine-associated energy metabolism. In one study, slow weight gain was observed in scorbutic guinea pigs, and this was reversed when the animals were supplemented with carnitine, supporting a functional impact of low ascorbate status on carnitine-mediated fatty acid metabolism [141]. Lowered carnitine levels in muscle were associated with altered mitochondrial metabolism in obese rats, together with lowered complete fatty acid oxidation [142]. In humans, low ascorbate status was shown to correlate with markers of metabolic syndrome [21,143] and obesity [144]. In a placebo-controlled intervention study with a group of young adults,

energy metabolism was monitored under conditions of mild exercise (1 hour on a treadmill at 50% of VO_2max) [140]. Physiologic measurements of respired gases and urinary nitrogen were used to estimate protein and nonprotein metabolism during the exercise period. At baseline, the group of individuals with low plasma ascorbate status (average 18 μM) were found to rely mostly on protein metabolism for energy expenditure, whereas the control group with higher plasma ascorbate (average 42 μM) expended fat energy during this time. Supplementation of the vitamin C–depleted group with 500 mg vitamin C daily or a placebo tablet for 4 weeks restored the capacity to utilize fat energy during exercise in the vitamin C–supplemented group but not in the placebo group [140]. Plasma carnitine levels were also found to be higher in the low vitamin C group than in the group with healthy levels (12 ± 0.8 versus 6.5 ± 0.9 ng/mL, $p = .000$), and marginal vitamin C status was associated with increased fatigue scores [140]. An association between low vitamin C status and fatigue has also been noted in other studies [145]. Taken together, these studies suggest that low body vitamin C status can affect carnitine-dependent fatty acid transport and energy metabolism.

Ascorbate and Regulation of the Hypoxic Response

The existence of an oxygen-sensing pathway was first suggested by observations that cells in the kidney increase production of erythropoietin, the hematopoietic growth factor capable of stimulating red blood cell production, in response to reduced oxygen supply (reviewed in [146]). The identification of the hypoxia-inducible factors (HIFs) followed, together with the realization that these transcription factors are highly conserved, ubiquitously expressed, and regulate the expression of hundreds of genes [147–151]. HIF activation results in a rapid adaptive response when oxygen supply becomes limited and enables cells to cope with the associated stress by upregulation of genes controlling glycolysis, angiogenesis, cell survival and autophagy, vascular control, iron transport, erythropoiesis, and stem cell phenotype (for recent reviews, see [71,151–157]). In addition, there is cross talk between HIF and other pathways such as NF-κB in inflammation [158,159] and microRNA networks [160,161].

The hypoxic response is particularly relevant to cancer as oxygen delivery to tumors is regularly limited due to the rapid expansion in cell numbers and their outgrowing the available blood supply [146,147,162]. Activation of the HIFs supports tumor growth through the upregulation of glycolysis, angiogenesis, and cell life and death pathways as indicated earlier [146,147,162]. Accordingly, HIF-1 activation is associated with increased tumor growth, resistance to chemotherapy and radiation, initiation of metastasis and poor prognosis in many cancers. Reduced expression of HIF-1α is associated with improved patient survival [162–172]. Taken together, the previous information illustrates the many clinical situations that involve HIF activation and highlights the importance of regulatory control of this pathway.

The active HIFs are αβ heterodimers. Their activation is controlled by posttranslational hydroxylation of the regulatory HIF-α chains that are constitutively expressed. Generation of hydroxyproline at P402 and P564 signals recruitment to the von Hippel-Lindau tumor suppressor protein and an E3 ubiquitin ligase complex, initiating ubiquitination and proteasomal degradation (Figure 5.3). Hydroxylation of asparagine N803 prevents binding to CREB binding protein and p300 and the formation of an active transcription complex (Figure 5.3) [70,173,174]. These reactions are catalyzed by four 2-OGDDs; three prolyl hydroxylases (PHD 1,2,3); and one asparagine hydroxylase, factor-inhibiting HIF (FIH) (Table 5.1 and Figure 5.3) [175–177]. When the HIF hydroxylases are inhibited, HIF-α escapes hydroxylation, cannot bind to the von Hippel-Lindau protein, and accumulates in the cytosol, from where it is recruited to the nucleus to form active transcription complexes [156,178]. The HIF hydroxylases were reported to have K_m for O_2 around 100–250 uM, making them ideal oxygen sensors, with some variability for PHD2 depending on chain length of the target peptide [109,110,146,179,180]. In addition, the absence of 2-oxoglutarate, Fe, or ascorbate all compromise hydroxylase activity and limit degradation of HIF-α, resulting in HIF activation [95,177,181–183].

In standard tissue culture conditions, there is generally a complete absence of ascorbate that results in HIF upregulation in the presence of a normal oxygen supply, indicating that the HIF hydroxylases are unable to function effectively. This basal HIF activation can be completely blocked by the addition of ascorbate to the medium [181,182,184]. Uptake of ascorbate from culture medium can suppress HIF-α protein stabilization and HIF transcriptional activity due to hydroxylase inhibition with Ni^{2+}, $CoCl_2$, the iron chelator Desferal, and insulin-like growth factor or insulin [182,184–188]. Ascorbate can also mitigate HIF activation in response to hypoxia under some conditions; prevention is possible at moderate hypoxia (1%–5% O_2) but generally not when oxygen is severely limited (≤1% O_2), reflecting the absolute hydroxylase requirement for O_2 [182,184,187–190].

In addition to the in vitro experiments reported previously, there is a substantial body of data that supports the requirement for intracellular ascorbate to regulate the HIF hydroxylases in vivo. Much of this information has come from cancer studies. HIF is often activated in tumors, and analysis of human tumor tissue has shown an inverse correlation between tumor ascorbate and levels of HIF-1 in colorectal, endometrial, breast, and thyroid cancers, and papillary renal carcinoma [191–195]. Higher tumor ascorbate was associated with increased disease-free survival for patients with colorectal cancer [192] and breast cancer [194]. In contrast with these findings, there was no correlation between ascorbate and HIF activation in clear cell renal carcinoma, a cancer that lacks a functional von Hippel-Lindau protein [193]. HIF is constitutively active in clear cell carcinoma due to loss of targeting of the hydroxylated HIF-α to the proteasome, and the protein accumulates independently of the availability of oxygen

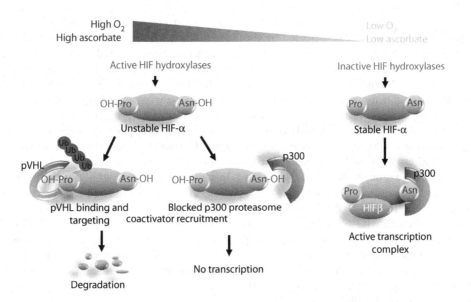

Figure 5.3. Regulation of the hypoxic response via hydroxylation of the hypoxia-inducible factor (HIF)-α subunit. Hydroxylation of critical proline residues and an asparagine on the HIF-α regulatory subunit target the protein for degradation in the proteasome and prevent the formation of an active HIF transcription complex by blocking binding to p300. Inactivity of the HIF hydroxylases under low oxygen conditions, or in the absence of 2-OG or ascorbate, allows the formation of an active transcriptional complex and the expression of many downstream genes associated with glycolysis, angiogenesis, erythropoiesis, and cell survival.

and other hydroxylase substrates or cofactors, including ascorbate. The clear cell renal carcinoma data therefore provides a causative link for the relationship between intracellular ascorbate and the HIF hydroxylase-mediated regulation of HIF transcriptional activity [193].

In vivo investigations with tumor implants in the vitamin C–dependent Gulo−/− mouse have similarly shown that high tumor ascorbate content was inversely correlated with HIF activation and was associated with slowed tumor growth [190,196,197]. Oral ascorbate supplementation of severe combined immunodeficient (SCID) mice implanted with wild-type P493 cells or P493 cells expressing a mutant, stabilized form of HIF-1α resulted in significant inhibition of wild-type tumor growth but no effect on mutant HIF-1α tumors [198]. These studies indicate that higher ascorbate concentrations can inhibit HIF-1 activation and consequent tumor growth in mice, with the likely mechanism being due to increased hydroxylation of the HIF-α chains by the ascorbate-dependent HIF hydroxylases. The data are consistent with there being a relationship between intracellular ascorbate and HIF activation in vivo, but this remains to be tested in a human clinical intervention study.

Ascorbate and Epigenetics

Recent definitions of epigenetics describe "the study of changes in gene function that are mitotically and/or meiotically heritable and that do not entail a change in DNA sequence" [199]. Processes underpinning these changes in gene function include posttranslational modifications to histones and methylation of DNA bases. In mammals, epigenetic programming directs cellular differentiation during early development and is responsible for phenotypic differences [200–202]. Epigenetics is also closely linked with the development of cancer, with global hypomethylation of DNA resulting in genomic instability and increased chromosomal fragility [203–206]. An understanding of the role of ascorbate in epigenetics had expanded rapidly in the past decade due to the direct involvement of the 2-OGDDs in epigenetic regulation.

Ten-Eleven Translocase Enzymes and Ascorbate in Epigenetics and Development

The methylation of DNA cytosine (5mC) at gene promoters is usually associated with transcriptional silencing, and the dynamic regulation of gene expression requires the controlled removal of 5mC [201,206]. This process is mediated by the ten-eleven translocase (TET) enzymes, members of the 2-OGDD family which successively oxidize 5mC to 5-hydroxymethylcytosine (5hmC), 5-formylcytosine (5fC), and 5-carboxylcytosine (5caC), leading to excision repair and effective demethylation (Table 5.1) [84,207,208]. In addition to its role as an intermediate in the demethylation process, 5hmC can be a stable epigenetic marker and is relatively abundant at ~0.032% of the genome in embryonic tissues and up to 0.6% in Purkinje cells in the cerebellum [209]. Its presence is associated with altered levels of gene transcription [85]. It can engage its own binding proteins and is most abundant at enhancers, gene bodies, promoters and CpG islands with lower GC content, and accumulates on euchromatin marked by H3K4me2/3 [210–213]. These data are indicative of the TET enzymes having complex regulatory and functional epigenetic roles via generation of 5hmC.

There are three TET proteins that are differentially expressed with distinct functions. Tet1 and Tet3 are highly expressed in development with different temporal dynamics [214,215]. Tet2 is more broadly expressed, and TET2 loss of function mutations are associated with the development of human myeloid malignancies [216,217]. The differentiation of murine embryonic stem cells (mESCs) is accompanied by decreased levels of 5hmC and downregulation of Tet1 and Tet2 [207]. Knockout of individual TET proteins results in viable mice, indicating some redundancy in function [218] but with evidence for developmental dysregulation. TET1-dependent effects on neurogenesis were noted [219], as was TET2-dependent susceptibility to atherosclerosis [220], combined effects of TET1/TET2 on differentiation [221,222], and TET3-dependent effects on neuroectodermal and mesodermal fate [215].

Like other 2-OGDDs, the TET hydroxylases utilize O_2 and 2-OG as substrates and depend on Fe and ascorbate as cofactors. The dependency of the TETs on ascorbate has been demonstrated in vitro, with ascorbate insufficiency dramatically affecting TET targets, patterns of DNA methylation, and the phenotype of cultured embryonic stem cells [223–226]. In humans, low maternal plasma ascorbate in pregnancy has been shown to correspond to low ascorbate status in the neonate [227] and to affect offspring development [228]. A close positive association between 2-OG levels and ascorbate ($P < .0001$) was reported in the placentae of a human obesity cohort, with widespread alterations in the fetal methylome, including associations with 5hmC levels [229]. Evidence that low ascorbate status during pregnancy can affect the development of offspring comes from studies with guinea pigs where phenotypic changes including growth retardation and effects on hippocampal and cerebellar neurogenesis have been reported in pups born to sows with low ascorbate status [230–232]. The irreversibility of the phenotype changes is suggestive of an effect of vitamin C on epigenetics in development [233]. It is likely that these effects are mediated via decreased TET activity, given the phenotypic overlap of ascorbate deficiency with the various TET knockout models.

Ascorbate and TET Enzymes in Cancer

Further evidence for the role of ascorbate in maintaining TET enzyme activity has come from the investigation of acute myeloid leukemia (AML, reviewed in [234]). Mutations in TET2 are relatively common in human AML, where ~10% of patients have lost function in one allele [235,236]. This results in decreased global TET activity in leukemic cells, as evidenced by loss of cytosine hydroxymethylation in DNA along with increased promoter methylation [237]. Mouse models of AML, using TET2 knockout or knockdown in the hematopoietic compartment, have demonstrated similar changes in methylation and hydroxymethylation [53,216]. Importantly, ascorbate was able to reverse these changes by upregulating residual TET. Furthermore, ascorbate deficiency was interchangeable with heterozygous loss of TET2, with either change cooperating with mutations in FLT3 to promote leukemogenesis.

A number of other mechanisms can contribute to decreased TET activity (Figure 5.4), for example, mutations in IDH1, IDH2, and WT1 seen in AML [237,238], as well as decreased expression of l-2-hydroxyglutarate dehydrogenase in clear cell renal cancer [239]. Interestingly, treatment with ascorbate is able to activate TET2 in a number of these scenarios, providing further confirmation for the cofactor role of ascorbate. A number of these observations have come from situations that involve low ascorbate status but not deficiency [53,229], suggesting that effects on TET activity are possible within the parameters of the range of

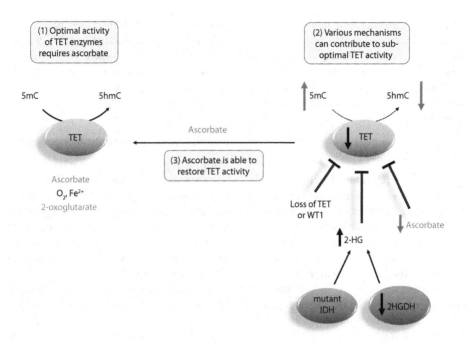

Figure 5.4. The role of ascorbate in ten-eleven translocation (TET) activity. The TET enzymes oxidize methylcytosine (5mC) to hydroxymethylcytosine (5hmC) and require ascorbate for this activity [223–225]. Suboptimal ascorbate concentrations lead to decreased TET activity as evidenced by increased 5mC and decreased 5hmC. In leukemia, low levels of ascorbate are essentially synonymous with heterozygous loss of TET2, and either situation can collaborate with further mutations to generate leukemia [53]. Furthermore, the addition of ascorbate was able to rescue hydroxymethylation by increasing residual TET2 activity [216]. In addition to decreased ascorbate loss of TET, a number of other mechanisms can contribute to decreased TET activity. In leukemia, mutant isocitrate dehydrogenase (mutant IDH) generates the oncometabolite 2-hydroxyglutarate (2-HG) [311]. In clear cell renal carcinoma, decreased levels of 2-HG dehydrogenase (2HGDH) [239] also lead to increased levels of 2-HG. In both instances, *in vitro* models demonstrated decreased TET2 activity, which was restored by treatment with ascorbate.

plasma ascorbate levels in the general population (10–100 μM). Cumulatively, these data suggest that ascorbate is able to mitigate the effect of decreased TET2 activity in models of cancer at least in part by upregulating TET2 activity.

Ascorbate and Other 2-OGDDs in Cancer

The JHDM and AlkB subfamilies of 2-OGDDs are also potentially regulated by changes in the availability of substrates and cofactors. These enzymes regulate DNA repair and gene expression and the greater than 20 JMJC enzymes demethylate mono-, di-, and trimethylated histone lysine or arginine residues via oxidation of the methyl group and the spontaneous removal of formaldehyde (Table 5.1) [73,240]. Together these groups of 2-OGDDs affect genetic processes in all cells, and their potential regulation by ascorbate could provide a broad range of potential targets in cancer cells, many of which are currently uncharacterized.

ASCORBATE AS A COFACTOR FOR COPPER-CONTAINING MONOOXYGENASES

Ascorbate is required for the reactions catalyzed by Cu-containing monooxygenases that facilitate the production of dopamine and norepinephrine, and the synthesis of amidated neuropeptide hormones [241–248]. These compounds have critical functions in our biology, and it is notable that there are very high ascorbate concentrations in the adrenal medulla and pituitary glands where the monooxygenases are active [5,36,114]. Unlike the situation with the 2-OGDD reaction cycle that can proceed to some extent without consuming ascorbate [93], the Cu-monooxygenases utilize the reducing capacity of ascorbate, and this results in the generation of semidehydroascorbate or dehydroascorbate [5]. This results in high turnover of ascorbate in the adrenals and pituitary and may help explain the exacerbated loss of ascorbate under conditions of high stress [249].

There are two Cu-containing monooxygenases in mammals that together are responsible for the synthesis of norepinephrine and amidated neuropeptide hormones with wide-ranging biological impact. The requirement for ascorbate as a cofactor for these enzymes has been extensively documented [241–248].

Dopamine β-Hydroxylase

Dopamine β-hydroxylase is found in the chromaffin granules of the adrenal medulla and sympathetic neurons, where it catalyzes the hydroxylation of the catecholamine neurotransmitter dopamine to generate norepinephrine (Table 5.1 and Figure 5.5) [250]. The reaction proceeds in two distinct steps; ascorbate-mediated reduction of Cu^{2+} in the enzyme active site followed by Cu^+-mediated activation of oxygen to hydroxylate dopamine to the norepinephrine product (Figure 5.5). Ascorbate is consumed stoichiometrically in this reaction cycle with equimolar production of semidehydroascorbate and dopamine [251–254]. The K_m for ascorbate is ~0.6 mM [243,251], and the adrenals and neurons supporting this

activity contain 10–20 mM ascorbate, the highest concentrations in the body [5,36,37,49,51]. These tissues accumulate ascorbate very efficiently and retain their content when plasma supply is limited [114], suggesting adaptation to ensure ongoing optimal dopamine β-hydroxylase activity.

Peptidylglycine α-Amidating Monooxygenase

Many neuropeptide hormones and neurotransmitters contain a C-terminal amide that is necessary for biological activity [241,255]. The final step in the amidation of glycine-extended precursors of these neuropeptides is catalyzed by peptidylglycine α-amidating monooxygenase (PAM), in a reaction that requires copper, ascorbate, and molecular oxygen (Figure 5.5 and Table 5.1) [241,248]. PAM is a multifunctional enzyme with two distinct catalytic domains that generate the amidated products in two sequential steps (Figure 5.5). In the first step, a hydroxylated intermediate is generated by peptidylglycine α-hydroxylating monooxygenase (PHM). This reaction is Cu-dependent, consumes ascorbate and O_2 (Figure 5.5), and is followed by conversion of the intermediate to the α-amidated

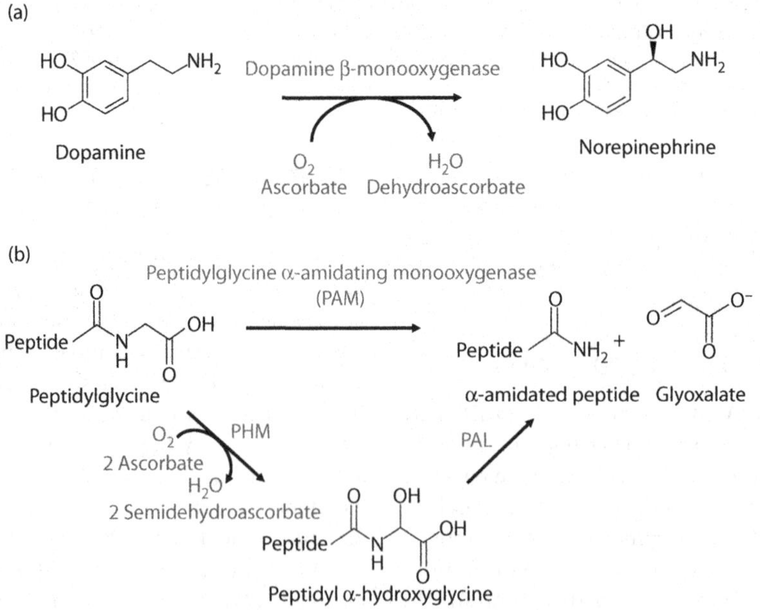

Figure 5.5. The cofactor activity of ascorbate for the Cu-containing monooxygenases. (a) Dopamine β-hydroxylase generates norepinephrine from dopamine, and (b) peptidylglycine α-amidating monooxygenase (PAM) is a bifunctional enzyme that generates amidated neuropeptide hormones in a two-phase reaction. Ascorbate is oxidized in the process of these reactions and would require regeneration by semidehydroascorbate reductase or glutathione-mediated processes to avoid irreversible breakdown.

peptide product and glyoxylate by peptidyl-α-hydroxyglycine α-amidating lyase (PAL) (Figure 5.5) [241,248,256].

PAM is the only enzyme known to produce amidated peptide hormones, is highly conserved, and is found in many tissues including the pituitary, neurons, ovaries, testes, eyes, adrenals, placenta, thymus, and pancreas [248,255,257]. As observed with dopamine β-hydroxylase, the tissue distribution of this enzyme coincides with high local ascorbate concentrations, and intracellular levels of 10–20 mM have been measured [36,37,49,51]. PAM is responsible for the production of oxytocin, vasopressin, thyrotropin-releasing hormone, and substance P, among many others [248,255,257,258], and its activity is therefore essential for many life processes.

EFFECT OF ASCORBATE AVAILABILITY ON COPPER-CONTAINING MONOOXYGENASE-MEDIATED PROCESSES

The early realization that ascorbate was an essential cofactor for dopamine β-hydroxylase and PAM led to numerous studies that together provide a compelling case for the maintenance of optimal ascorbate status in the regulation of these enzyme activities. Much of this information has come from animal studies that have determined the effects of the peptides and catecholamines produced by these enzymes. Recent human data are also particularly worth consideration for their clinical relevance. Some examples are given in the following sections.

Oxytocin

Oxytocin is synthesized in the hypothalamus and stored in the pituitary from where it is released as a fast-acting hormone that stimulates positive emotions [241,255]. It is also synthesized in the corpus luteum and in Leydig cells in the testis [259–261], in line with its known functions in fertility and reproduction, lactation, and childbirth [259–261]. These tissues contain very high concentrations of ascorbate; interestingly, ascorbate concentrations in the ovaries of cattle and sheep varied substantially during the fertility cycle, with ascorbate levels peaking at the time of maximum expression of PAM and oxytocin production [257,260–262]. Ascorbate

administration was also shown to increase plasma oxytocin in rats [263].

Vasopressor Functions—Vasopressin and Norepinephrine

Norepinephrine and vasopressin are able to regulate blood flow, heart rate, and energy metabolism. Vasopressin is released from the pituitary in response to increased osmolality or decreased blood pressure and acts on the kidney and smooth muscle cells to regulate the reabsorption of water and constrict arterioles [264]. Effects of ascorbate availability on vasopressin were first demonstrated in animals in which ascorbate increased plasma vasopressin in association with changes in urinary volume and sodium excretion [263,265,266].

Vasopressor requirements are a major consideration in the clinical care of the critically ill, with management of hypotension presenting a major challenge for the doctors of patients with acute sepsis [267,268]. In response to strong evidence for severely depleted plasma vitamin C levels in these patients [269–274], a number of studies have been undertaken to determine the effect of ascorbate supplementation on vasopressor use and patient outcome in patients in intensive care. Early indications suggest that the recommended ascorbate levels in parenteral nutrition are inadequate and do not compensate for the accelerated turnover in these patients [272]. Substantial clinical benefit of additional vitamin C supplementation, manifest as decreased dependency on vasopressors [275], less multiple organ failure and a dramatic reduction in mortality [276] are suggested. These data have initiated significant interest and discussion, and more targeted clinical trials are anticipated [267,275,277–281].

Tetrahydrobiopterin Recycling

Ascorbate provides reducing equivalents for the recycling of that tetrahydrobiopterin, the cofactor for tyrosine hydroxylase that synthesizes dihydroxyphenylalanine (L-dopa), the precursor of dopamine, and tryptophan hydroxylase, which is involved in the synthesis of serotonin [282,283]. In addition, tetrahydrobiopterin is a cofactor for the nitric oxide synthases, and ascorbate-mediated recycling has been suggested to be a mechanism for observed cardiovascular health benefits of supplementation [284].

Substance P

Substance P is an amidated peptide hormone synthesized by PAM [258] that is virtually ubiquitous in its distribution and action. Its receptors are found in many cell types, and it is known to be involved in stress responses, inflammation, pain, mood disorders, and vasodilation [285–287]. Antagonists of substance P are proposed as possible therapies for many inflammatory diseases, but the effects of vitamin insufficiency have not been investigated.

SUMMARY

The past two decades have seen the discovery of numerous additional enzymes for which ascorbate is a cofactor. The enzymes identified to date belong to two groups, the Fe-containing 2-OGDDs and the Cu-containing monooxygenases. Ascorbate is essential for the regeneration of the active site Fe in the 2-OGDDs and is a required cofactor for the Cu-containing enzymes. Alternative biological reducing compounds such as glutathione do not appear to be able to substitute. The K_m values reported for ascorbate for these enzymes are between 0.2 and 0.6 mM, and the co-localization of these activities with high tissue ascorbate levels is a strong indication that enzyme cofactor activity is the primary biological role of this essential vitamin. The ascorbate-dependent enzymes represent an impressively diverse number of biological functions. These include the generation of catecholamine neurotransmitters and many neuropeptide hormones; the synthesis of carnitine; the posttranslational modification of collagens and the HIF transcription factors; the demethylation of DNA, RNAs, and histones; and DNA repair. Together with improved understanding of ascorbate uptake and turnover kinetics, the recognition that the vitamin has a role in the maintenance of so many biological functions will undoubtedly improve our future management of intake according to clinical need.

REFERENCES

1. Willimott, S. G. and Wokes, F. 1926. The vitamin C of lemon rind. *Biochem J.* 20, 1013–1015.

2. Bracewell, M. F. and Zilva, S. S. 1931. Vitamin C in the orange and the grape fruit. *Biochem J.* 25, 1081–1089.

3. Svirbely, J. L. and Szent-Gyorgyi, A. 1932. The chemical nature of vitamin C. *Biochem. J.* 26, 865–870.

4. Svirbely, J. L. and Szent-Gyorgyi, A. 1933. The chemical nature of vitamin C. *Biochem. J.* 27, 279–285.

5. Du, J., Cullen, J. J. and Buettner, G. R. 2012. Ascorbic acid: Chemistry, biology and the treatment of cancer. *Biochim Biophys Acta.* 1826, 443–457.

6. Smirnoff, N. 2000. Ascorbic acid: Metabolism and functions of a multi-facetted molecule. *Curr. Opin. Plant. Biol.* 3, 229–235.

7. Smirnoff, N. 2018. Ascorbic acid metabolism and functions: A comparison of plants and mammals. *Free Radic. Biol. Med.* 122, 116–129.

8. Yew, M. L. 1973. "Recommended daily allowances" for vitamin C. *Proc. Natl. Acad. Sci. USA* 70, 969-972.

9. Burnett-Hartman, A. N., Fitzpatrick, A. L., Gao, K., Jackson, S. A. and Schreiner, P. J. 2009. Supplement use contributes to meeting recommended dietary intakes for calcium, magnesium, and vitamin C in four ethnicities of middle-aged and older Americans: The multi-ethnic study of atherosclerosis. *J. Am. Diet Assoc.* 109, 422–429.

10. Banhegyi, G., Braun, L., Csala, M., Puskas, F. and Mandl, J. 1997. Ascorbate metabolism and its regulation in animals. *Free Radic. Biol. Med.* 23, 793–803.

11. Levine, M., Cantilena, C. C. and Dhariwal, K. R. 1993. In situ kinetics and ascorbic acid requirements. *World Rev. Nutr. Diet.* 72, 114–127.

12. Levine, M., Dhariwal, K. R., Welch, R. W., Wang, Y. and Park, J. B. 1995. Determination of optimal vitamin C requirements in humans. *Am. J. Clin. Nutr.* 62, 1347S–1356S.

13. Gershoff, S. N. 1993. Vitamin C (ascorbic acid): New roles, new requirements? *Nutr. Rev.* 51, 313–326.

14. Frei, B., Birlouez-Aragon, I. and Lykkesfeldt, J. 2012. Authors' perspective: What is the optimum intake of vitamin C in humans? *Crit. Rev. Food Sci. Nutr.* 52, 815–829.

15. Riemersma, R. A., Carruthers, K. F., Elton, R. A. and Fox, K. A. 2000. Vitamin C and the risk of acute myocardial infarction. *Am. J. Clin. Nutr.* 71, 1181–1186.

16. Nyyssonen, K., Parviainen, M. T., Salonen, R., Tuomilehto, J. and Salonen, J. T. 1997. Vitamin C deficiency and risk of myocardial infarction: Prospective population study of men from eastern Finland. Brit. Med. J. 314, 634–638.

17. Myint, P. K., Luben, R. N., Welch, A. A., Bingham, S. A., Wareham, N. J. and Khaw, K. T. 2008. Plasma vitamin C concentrations predict risk of incident stroke over 10 y in 20 649 participants of the European Prospective Investigation into cancer Norfolk prospective population study. Am. J. Clin. Nutr. 87, 64–69.

18. Myint, P. K., Sinha, S., Luben, R. N., Bingham, S. A., Wareham, N. J. and Khaw, K. T. 2008. Risk factors for first-ever stroke in the EPIC-Norfolk prospective population-based study. Eur. J. Cardiovasc. Prev. Rehab. 15, 663–669.

19. Khaw, K. T., Wareham, N., Luben, R., Bingham, S., Oakes, S., Welch, A. and Day, N. 2001. Glycated haemoglobin, diabetes, and mortality in men in Norfolk cohort of European prospective investigation of cancer and nutrition (EPIC-Norfolk). Brit. Med. J. 322, 15–18.

20. Carter, P., Gray, L. J., Morris, D. H., Davies, M. J. and Khunti, K. 2013. South Asian individuals at high risk of type 2 diabetes have lower plasma vitamin C levels than white Europeans. J. Nutr. Sci. 2, e21.

21. Wilson, R., Willis, J., Gearry, R., Skidmore, P., Fleming, E., Frampton, C. and Carr, A. 2017. Inadequate vitamin C status in Prediabetes and type 2 diabetes mellitus: Associations with glycaemic control, obesity, and smoking. Nutrients 9.

22. Fleischauer, A. T., Simonsen, N. and Arab, L. 2003. Antioxidant supplements and risk of breast cancer recurrence and breast cancer-related mortality among postmenopausal women. Nutr. Can. 46, 15–22.

23. Nechuta, S., Lu, W., Chen, Z., Zheng, Y., Gu, K., Cai, H., Zheng, W. and Shu, X. O. 2011. Vitamin supplement use during breast cancer treatment and survival: A prospective cohort study. Can Epidemiol. Biomarkers Prev. 20, 262–271.

24. Greenlee, H., Hershman, D. L. and Jacobson, J. S. 2009. Use of antioxidant supplements during breast cancer treatment: A comprehensive review. Breast Can. Res. Treatment 115, 437–452.

25. Greenlee, H., Kwan, M. L., Ergas, I. J., Sherman, K. J., Krathwohl, S. E., Bonnell, C., Lee, M. M. and Kushi, L. H. 2009. Complementary and alternative therapy use before and after breast cancer diagnosis: The Pathways Study. Breast Can. Res. Treatment 117, 653–665.

26. Harris, H. R., Bergkvist, L. and Wolk, A. 2013. Vitamin C intake and breast cancer mortality in a cohort of Swedish women. Br. J. Cancer. 109, 257–264.

27. McEligot, A. J., Largent, J., Ziogas, A., Peel, D. and Anton-Culver, H. 2006. Dietary fat, fiber, vegetable, and micronutrients are associated with overall survival in postmenopausal women diagnosed with breast cancer. Nutr. Cancer. 55, 132–140.

28. Poole, E. M., Shu, X., Caan, B. J., Flatt, S. W., Holmes, M. D., Lu, W., Kwan, M. L., Nechuta, S. J., Pierce, J. P. and Chen, W. Y. 2013. Postdiagnosis supplement use and breast cancer prognosis in the after breast cancer pooling project. Breast Can. Res. Treatment 139, 529–537.

29. Rohan, T. E., Hiller, J. E. and McMichael, A. J. 1993. Dietary factors and survival from breast cancer. Nutr. Cancer. 20, 167–177.

30. Harris, H. R., Orsini, N. and Wolk, A. 2014. Vitamin C and survival among women with breast cancer: A meta-analysis. Eur. J. Cancer. 50, 1223–1231.

31. Mamede, A. C., Tavares, S. D., Abrantes, A. M., Trindade, J., Maia, J. M. and Botelho, M. F. 2011. The role of vitamins in cancer: A review. Nutr. Cancer. 63, 479–494.

32. Khaw, K. T., Bingham, S., Welch, A., Luben, R., Wareham, N., Oakes, S. and Day, N. 2001. Relation between plasma ascorbic acid and mortality in men and women in EPIC-Norfolk prospective study: A prospective population study. European Prospective Investigation into cancer and nutrition. Lancet 357, 657–663.

33. Enstrom, J. E., Kanim, L. E. and Klein, M. A. 1992. Vitamin C intake and mortality among a sample of the United States population. Epidemiology 3, 194–202.

34. Loria, C. M., Klag, M. J., Caulfield, L. E. and Whelton, P. K. 2000. Vitamin C status and mortality in US adults. Am. J. Clin. Nutr. 72, 139–145.

35. Kubota, Y., Iso, H., Date, C., Kikuchi, S., Watanabe, Y., Wada, Y., Inaba, Y. and Tamakoshi, A. 2011. Dietary intakes of antioxidant vitamins and mortality from cardiovascular disease: The Japan Collaborative Cohort Study (JACC) study. Stroke 42, 1665–1672.

36. Hornig, D. 1975. Distribution of ascorbic acid, metabolites and analogues in man and animals. Ann. N. Y. Acad. Sci. 258, 103–118.

37. Keith, M. O. and Pelletier, O. 1974. Ascorbic acid concentrations in leukocytes and selected organs of guinea pigs in response to increasing ascorbic acid intake. *Am. J. Clin. Nutr.* 27, 368–372.

38. Levine, M., Conry-Cantilena, C., Wang, Y., Welch, R. W., Washko, P. W., Dhariwal, K. R., Park, J. B. et al. 1996. Vitamin C pharmacokinetics in healthy volunteers: Evidence for a recommended dietary allowance. *Proc. Natl. Acad. Sci. USA* 93, 3704–3709.

39. Liang, W. J., Johnson, D. and Jarvis, S. M. 2001. Vitamin C transport systems of mammalian cells. *Molec. Membr. Biol.* 18, 87–95.

40. May, J. M. 2012. Vitamin C transport and its role in the central nervous system. *Sub-cell Biochem.* 56, 85–103.

41. Savini, I., Rossi, A., Pierro, C., Avigliano, L. and Catani, M. V. 2008. SVCT1 and SVCT2: Key proteins for vitamin C uptake. *Amino Acids* 34, 347–355.

42. Sotiriou, S., Gispert, S., Cheng, J., Wang, Y., Chen, A., Hoogstraten-Miller, S., Miller, G. F. et al. 2002. Ascorbic-acid transporter Slc23a1 is essential for vitamin C transport into the brain and for perinatal survival. *Nat Med.* 8, 514–517.

43. May, J. M. and Qu, Z. C. 2009. Ascorbic acid efflux and re-uptake in endothelial cells: Maintenance of intracellular ascorbate. *Mol. Cell. Biochem.* 325, 79–88.

44. Evans, R. M., Currie, L. and Campbell, A. 1982. The distribution of ascorbic acid between various cellular components of blood, in normal individuals, and its relation to the plasma concentration. *Br. J Nutr.* 47, 473–482.

45. Parker, W. H., Qu, Z. C. and May, J. M. 2015. Ascorbic acid transport in brain microvascular pericytes. *Biochem. Biophys. Res. Commun.* 458, 262–267.

46. Cullen, E. I., May, V. and Eipper, B. A. 1986. Transport and stability of ascorbic acid in pituitary cultures. *Molec. Cell Endocrinol.* 48, 239–250.

47. Siushansian, R. and Wilson, J. X. 1995. Ascorbate transport and intracellular concentration in cerebral astrocytes. *J. Neurochem.* 65, 41–49.

48. Taylor, A., Jacques, P. F., Nowell, T., Perrone, G., Blumberg, J., Handelman, G., Jozwiak, B. and Nadler, D. 1997. Vitamin C in human and guinea pig aqueous, lens and plasma in relation to intake. *Curr. Eye Res.* 16, 857–864.

49. Hornig, D. 1981. Metabolism and requirements of ascorbic acid in man. *South African Med. J.* 60, 818–823.

50. Meur, S. K., Sanwal, P. C. and Yadav, M. C. 1999. Ascorbic acid in buffalo ovary in relation to oestrous cycle. *Ind. J. Biochem. Biophys.* 36, 134–135.

51. Harrison, F. E. and May, J. M. 2009. Vitamin C function in the brain: Vital role of the ascorbate transporter SVCT2. *Free Radic. Biol. Med.* 46, 719–730.

52. Lloyd, J. V., Davis, P. S., Emery, H. and Lander, H. 1972. Platelet ascorbic acid levels in normal subjects and in disease. *J. Clin. Pathol.* 25, 478–483.

53. Agathocleous, M., Meacham, C. E., Burgess, R. J., Piskounova, E., Zhao, Z., Crane, G. M., Cowin, B. L. et al. 2017. Ascorbate regulates haematopoietic stem cell function and leukaemogenesis. *Nature.* 549, 476–481.

54. Buettner, G. R. 1988. In the absence of catalytic metals ascorbate does not autoxidize at pH 7: Ascorbate as a test for catalytic metals. *J. Biochem. Biophys. Meth.* 16, 27–40.

55. Buettner, G. R. and Chamulitrat, W. 1990. The catalytic activity of iron in synovial fluid as monitored by the ascorbate free radical. *Free Radic. Biol. Med.* 8, 55–56.

56. Halliwell, B. and Foyer, C. 1976. Ascorbic acid, metal ions and the superoxide radical. *Biochem. J.* 155, 697–700.

57. Islam, M. S., Leissing, T. M., Chowdhury, R., Hopkinson, R. J. and Schofield, C. J. 2018. 2-oxoglutarate-dependent oxygenases. *Ann. Rev. Biochem.* 87, 585–620.

58. Loenarz, C. and Schofield, C. J. 2011. Physiological and biochemical aspects of hydroxylations and demethylations catalyzed by human 2-oxoglutarate oxygenases. *Trends Biochem. Sci.* 36, 7–18.

59. Hausinger, R. P. 2004. FeII/alpha-ketoglutarate-dependent hydroxylases and related enzymes. *Crit. Rev. Biochem. Mol. Biol.* 39, 21–68.

60. Stetten, M. R. 1949. Some aspects of the metabolism of hydroxyproline, studied with the aid of isotopic nitrogen. *J. Biol. Chem.* 181, 31.

61. Stetten, M. R. and Schoenheimerz, R. 1944. The metabolism of l(-)-Proline studied with the aid of deuterium and isotopic nitrogen. *J. Biol. Chem.* 153, 113–132.

62. Gould, B. S. and Woessner, J. F. 1957. Biosynthesis of collagen; the influence of ascorbic acid on the proline, hydroxyproline, glycine, and collagen content of regenerating guinea pig skin. *J. Biol. Chem.* 226, 289–300.

63. Gustavson, K. H. 1955. The function of hydroxyproline in collanges. *Nature* 175, 70–74.

64. Hall, D. A. and Reed, R. 1957. Hydroxyproline and thermal stability of collagen. *Nature* 180, 243.

65. Peterkofsky, B. and Udenfriend, S. 1965. Enzymatic hydroxylation of proline in microsomal polypeptide leading to formation of collagen. *Proc. Natl. Acad. Sci. USA* 53, 335–342.

66. Hutton, J. J., Jr., Trappel, A. L. and Udenfriend, S. 1966. Requirements for alpha-ketoglutarate, ferrous ion and ascorbate by collagen proline hydroxylase. *Biochem. Biophys. Res. Commun.* 24, 179–184.

67. Pihlajaniemi, T., Myllyla, R. and Kivirikko, K. I. 1991. Prolyl 4-hydroxylase and its role in collagen synthesis. *J. Hepatol.* 13 (Suppl 3), S2–S7.

68. Tiainen, P., Pasanen, A., Sormunen, R. and Myllyharju, J. 2008. Characterization of recombinant human prolyl 3-hydroxylase isoenzyme 2, an enzyme modifying the basement membrane collagen IV. *J. Biol. Chem.* 283, 19432–19439.

69. Masson, N. and Ratcliffe, P. J. 2003. HIF prolyl and asparaginyl hydroxylases in the biological response to intracellular O(2) levels. *J. Cell Sci.* 116, 3041–3049.

70. Stolze, I. P., Mole, D. R. and Ratcliffe, P. J. 2006. Regulation of HIF: Prolyl hydroxylases. *Novartis. Found Symp.* 272, 15–25; discussion 25–36.

71. Pugh, C. W. and Ratcliffe, P. J. 2017. New horizons in hypoxia signaling pathways. *Exp. Cell Res.* 356, 116–121.

72. Yang, M., Chowdhury, R., Ge, W., Hamed, R. B., McDonough, M. A., Claridge, T. D., Kessler, B. M., Cockman, M. E., Ratcliffe, P. J. and Schofield, C. J. 2011. Factor-inhibiting hypoxia-inducible factor (FIH) catalyses the post-translational hydroxylation of histidinyl residues within ankyrin repeat domains. *FEBS J.* 278, 1086–1097.

73. Monfort, A. and Wutz, A. 2013. Breathing-in epigenetic change with vitamin C. *EMBO Rep.* 14, 337–346.

74. Horton, J. R., Upadhyay, A. K., Qi, H. H., Zhang, X., Shi, Y. and Cheng, X. 2010. Enzymatic and structural insights for substrate specificity of a family of Jumonji histone lysine demethylases. *Nat. Struct. Mol. Biol.* 17, 38–43.

75. Klose, R. J., Kallin, E. M. and Zhang, Y. 2006. JmjC-domain-containing proteins and histone demethylation. *Nat. Rev. Genetics.* 7, 715–727.

76. Klose, R. J., Yamane, K., Bae, Y., Zhang, D., Erdjument-Bromage, H., Tempst, P., Wong, J. and Zhang, Y. 2006. The transcriptional repressor JHDM3A demethylates trimethyl histone H3 lysine 9 and lysine 36. *Nature* 442, 312–316.

77. Ng, S. S., Kavanagh, K. L., McDonough, M. A., Butler, D., Pilka, E. S., Lienard, B. M., Bray, J. E. et al. 2007. Crystal structures of histone demethylase JMJD2A reveal basis for substrate specificity. *Nature* 448, 87–91.

78. Kondo, Y. 2014. Targeting histone methyltransferase EZH2 as cancer treatment. *J. Biochem.* 156, 249–257.

79. Thinnes, C. C., England, K. S., Kawamura, A., Chowdhury, R., Schofield, C. J. and Hopkinson, R. J. 2014. Targeting histone lysine demethylases—progress, challenges, and the future. *Biochim. Biophys. Acta* 1839, 1416–1432.

80. Islam, M. S., McDonough, M. A., Chowdhury, R., Gault, J., Khan, A., Pires, E. and Schofield, C. J. 2019. Biochemical and structural investigations clarify the substrate selectivity of the 2-oxoglutarate oxygenase JMJD6. *J. Biol. Chem.* 294, 11637–11652.

81. Poulard, C., Corbo, L. and Le Romancer, M. 2016. Protein arginine methylation/demethylation and cancer. *Oncotarget* 7, 67532–67550.

82. Poulard, C., Rambaud, J., Lavergne, E., Jacquemetton, J., Renoir, J. M., Tredan, O., Chabaud, S., Treilleux, I., Corbo, L. and Le Romancer, M. 2015. Role of JMJD6 in breast tumorigenesis. *PLOS ONE.* 10, e0126181.

83. Chang, B., Chen, Y., Zhao, Y. and Bruick, R. K. 2007. JMJD6 is a histone arginine demethylase. *Science* 318, 444–447.

84. Tahiliani, M., Koh, K. P., Shen, Y., Pastor, W. A., Bandukwala, H., Brudno, Y., Agarwal, S. et al. 2009. Conversion of 5-methylcytosine to 5-hydroxymethylcytosine in mammalian DNA by MLL partner TET1. *Science.* 324, 930–935.

85. Ficz, G., Branco, M. R., Seisenberger, S., Santos, F., Krueger, F., Hore, T. A., Marques, C. J., Andrews, S. and Reik, W. 2011. Dynamic regulation of 5-hydroxymethylcytosine in mouse ES cells and during differentiation. *Nature* 473, 398–402.

86. Duncan, T., Trewick, S. C., Koivisto, P., Bates, P. A., Lindahl, T. and Sedgwick, B. 2002. Reversal of DNA alkylation damage by two human dioxygenases. *Proc. Natl. Acad. Sci. USA* 99, 16660–16665.

87. Aas, P. A., Otterlei, M., Falnes, P. O., Vagbo, C. B., Skorpen, F., Akbari, M., Sundheim, O. et al. 2003. Human and bacterial oxidative demethylases repair alkylation damage in both RNA and DNA. *Nature* 421, 859–863.

88. Stefansson, O. A., Hermanowicz, S., van der Horst, J., Hilmarsdottir, H., Staszczak, Z., Jonasson, J. G., Tryggvadottir, L., Gudjonsson, T.

and Sigurdsson, S. 2017. CpG promoter methylation of the ALKBH3 alkylation repair gene in breast cancer. *BMC Cancer* 17, 469.

89. Ozer, A. and Bruick, R. K. 2007. Non-heme dioxygenases: Cellular sensors and regulators jelly rolled into one? *Nat. Chem. Biol.* 3, 144–153.

90. McDonough, M. A., Loenarz, C., Chowdhury, R., Clifton, I. J. and Schofield, C. J. 2010. Structural studies on human 2-oxoglutarate dependent oxygenases. *Curr. Op. Struct. Biol.* 20, 659–672.

91. Kaufman, S. 1966. Coenzymes and hydroxylases: Ascorbate and dopamine-beta-hydroxylase; tetrahydropteridines and phenylalanine and tyrosine hydroxylases. *Pharmacol Rev.* 18, 61–69.

92. de Jong, L., Albracht, S. P. and Kemp, A. 1982. Prolyl 4-hydroxylase activity in relation to the oxidation state of enzyme-bound iron. The role of ascorbate in peptidyl proline hydroxylation. *Biochim. Biophys. Acta* 704, 326–332.

93. Myllyla, R., Kuutti-Savolainen, E. R. and Kivirikko, K. I. 1978. The role of ascorbate in the prolyl hydroxylase reaction. *Biochem. Biophys. Res. Commun.* 83, 441–448.

94. Dickson, K. M., Gustafson, C. B., Young, J. I., Zuchner, S. and Wang, G. 2013. Ascorbate-induced generation of 5-hydroxymethylcytosine is unaffected by varying levels of iron and 2-oxoglutarate. *Biochem. Biophys. Res. Commun.* 439, 522–527.

95. Flashman, E., Davies, S. L., Yeoh, K. K. and Schofield, C. J. 2010. Investigating the dependence of the hypoxia-inducible factor hydroxylases (factor inhibiting HIF and prolyl hydroxylase domain 2) on ascorbate and other reducing agents. *Biochem. J.* 427, 135–142.

96. Nietfeld, J. J. and Kemp, A. 1981. The function of ascorbate with respect to prolyl 4-hydroxylase activity. *Biochim. Biophys. Acta* 657, 159–167.

97. Puistola, U., Turpeenniemi-Hujanen, T. M., Myllyla, R. and Kivirikko, K. I. 1980. Studies on the lysyl hydroxylase reaction. II. Inhibition kinetics and the reaction mechanism. *Biochim. Biophys. Acta* 611, 51–60.

98. Puistola, U., Turpeenniemi-Hujanen, T. M., Myllyla, R. and Kivirikko, K. I. 1980. Studies on the lysyl hydroxylase reaction. I. Initial velocity kinetics and related aspects. *Biochim. Biophys. Acta* 611, 40–50.

99. Vissers, M. C., Kuiper, C. and Dachs, G. U. 2014. Regulation of the 2-oxoglutarate-dependent dioxygenases and implications for cancer. *Biochem. Soc. Trans* 42, 945–951.

100. Myllyla, R., Majamaa, K., Gunzler, V., Hanauske-Abel, H. M. and Kivirikko, K. I. 1984. Ascorbate is consumed stoichiometrically in the uncoupled reactions catalyzed by prolyl 4-hydroxylase and lysyl hydroxylase. *J. Biol. Chem.* 259, 5403–5405.

101. Counts, D. F., Cardinale, G. J. and Udenfriend, S. 1978. Prolyl hydroxylase half reaction: Peptidyl prolyl-independent decarboxylation of alpha-ketoglutarate. *Proc. Natl. Acad. Sci. USA* 75, 2145–2149.

102. De Jong, L. and Kemp, A. 1984. Stoichiometry and kinetics of the prolyl 4-hydroxylase partial reaction. *Biochim. Biophys. Acta* 787, 105–111.

103. Fukumori, F. and Hausinger, R. P. 1993. Purification and characterization of 2,4-dichlorophenoxyacetate/alpha- ketoglutarate dioxygenase. *J. Biol. Chem.* 268, 24311–24317.

104. Holme, E., Lindstedt, S. and Nordin, I. 1984. Uncoupling and isotope effects in gamma-butyrobetaine hydroxylation. *Biosci. Rep.* 4, 433–440.

105. Hsu, C. A., Saewert, M. D., Polsinelli, L. F., Jr. and Abbott, M. T. 1981. Uracil's uncoupling of the decarboxylation of alpha-ketoglutarate in the thymine 7-hydroxylase reaction of *Neurospora crassa. J. Biol. Chem.* 256, 6098–6101.

106. Trewick, S. C., Henshaw, T. F., Hausinger, R. P., Lindahl, T. and Sedgwick, B. 2002. Oxidative demethylation by *Escherichia coli* AlkB directly reverts DNA base damage. *Nature* 419, 174–178.

107. Welford, R. W., Schlemminger, I., McNeill, L. A., Hewitson, K. S. and Schofield, C. J. 2003. The selectivity and inhibition of AlkB. *J. Biol. Chem.* 278, 10157–10161.

108. Myllyharju, J. and Kivirikko, K. I. 1997. Characterization of the iron- and 2-oxoglutarate-binding sites of human prolyl 4-hydroxylase. *EMBO J.* 16, 1173–1180.

109. Koivunen, P., Hirsila, M., Gunzler, V., Kivirikko, K. I. and Myllyharju, J. 2004. Catalytic properties of the asparaginyl hydroxylase (FIH) in the oxygen sensing pathway are distinct from those of its prolyl 4-hydroxylases. *J. Biol. Chem.* 279, 9899–9904.

110. Koivunen, P., Hirsila, M., Kivirikko, K. I. and Myllyharju, J. 2006. The length of peptide substrates has a marked effect on hydroxylation by the hypoxia-inducible factor prolyl 4-hydroxylases. *J. Biol. Chem.* 281, 28712–28720.

111. Pullar, J. M., Carr, A. C. and Vissers, M. C. M. 2017. The roles of vitamin C in skin health. *Nutrients* 9.

112. Chatterjee, I. B. 1973. Evolution and the biosynthesis of ascorbic acid. *Science* 182, 1271–1272.

113. Chatterjee, I. B., Majumder, A. K., Nandi, B. K. and Subramanian, N. 1975. Synthesis and some major functions of vitamin C in animals. *Ann. N Y. Acad. Sci.* 258, 24–47.

114. Hasselholt, S., Tveden-Nyborg, P. and Lykkesfeldt, J. 2015. Distribution of vitamin C is tissue specific with early saturation of the brain and adrenal glands following differential oral dose regimens in guinea pigs. *Brit. J. Nutr.* 113, 1539–1549.

115. Lindblad, M., Tveden-Nyborg, P. and Lykkesfeldt, J. 2013. Regulation of vitamin C homeostasis during deficiency. *Nutrients* 5, 2860–2879.

116. Vissers, M. C. M., Bozonet, S. M., Pearson, J. F. and Braithwaite, L. J. 2011. Dietary ascorbate affects steady state tissue levels in vitamin C-deficient mice: Tissue deficiency after sub-optimal intake and superior bioavailability from a food source (kiwifruit). *Am. J. Clin. Nutr* 93, 292–301.

117. Kivirikko, K. I. and Myllyharju, J. 1998. Prolyl 4-hydroxylases and their protein disulfide isomerase subunit. *Matrix Biol.* 16, 357–368.

118. Byers, P. H. and Murray, M. L. 2012. Heritable collagen disorders: The paradigm of the Ehlers-Danlos syndrome. *J. Invest. Dermatol.* 132, E6–11.

119. Stephen, R. and Utecht, T. 2001. Scurvy identified in the emergency department: A case report. *J. Emerg. Med.* 21, 235–237.

120. Alexandrescu, D. T., Dasanu, C. A. and Kauffman, C. L. 2009. Acute scurvy during treatment with interleukin-2. *Clin. Exp. Dermatol.* 34, 811–814.

121. Wang, K., Jiang, H., Li, W., Qiang, M., Dong, T. and Li, H. 2018. Role of vitamin C in skin diseases. *Front. Physiol.* 9, 819.

122. Wambier, C. G., Cappel, M. A., Werner, B., Rodrigues, E., Schumacher Welling, M. S., Montemor Netto, M. R. and de Farias Wambier, S. P. 2017. Dermoscopic diagnosis of scurvy. *J. Am. Acad. Dermatol.* 76, S52–s54.

123. Mutgi, K. A., Ghahramani, G., Wanat, K. and Ciliberto, H. 2016. Perifollicular petechiae and easy bruising. *J. Fam. Pract.* 65, 927–930.

124. Peterkofsky, B. 1991. Ascorbate requirement for hydroxylation and secretion of procollagen: Relationship to inhibition of collagen synthesis in scurvy. *Am. J. Clin. Nutr.* 54, 1135S–1140S.

125. Parsons, K. K., Maeda, N., Yamauchi, M., Banes, A. J. and Koller, B. H. 2006. Ascorbic acid-independent synthesis of collagen in mice. *Am. J. Physiol. Endocrinol. Metab.* 290, E1131–1139.

126. Ellinger, S. and Stehle, P. 2009. Efficacy of vitamin supplementation in situations with wound healing disorders: Results from clinical intervention studies. *Curr. Opin. Clin. Nutr. Metab. Care* 12, 588–595.

127. Silverstein, R. J. and Landsman, A. S. 1999. The effects of a moderate and high dose of vitamin C on wound healing in a controlled guinea pig model. *J. Foot Ankle Surg.* 38, 333–338.

128. Lima, C. C., Pereira, A. P., Silva, J. R., Oliveira, L. S., Resck, M. C., Grechi, C. O., Bernardes, M. T. et al. 2009. Ascorbic acid for the healing of skin wounds in rats. *Braz. J. Biol.* 69, 1195–1201.

129. Rasik, A. M. and Shukla, A. 2000. Antioxidant status in delayed healing type of wounds. *Int. J. Exp. Pathol.* 81, 257–263.

130. Taylor, T. V., Rimmer, S., Day, B., Butcher, J. and Dymock, I. W. 1974. Ascorbic acid supplementation in the treatment of pressure-sores. *Lancet* 2, 544–546.

131. ter Riet, G., Kessels, A. G. and Knipschild, P. G. 1995. Randomized clinical trial of ascorbic acid in the treatment of pressure ulcers. *J. Clin. Epidemiol.* 48, 1453–1460.

132. Blass, S. C., Goost, H., Tolba, R. H., Stoffel-Wagner, B., Kabir, K., Burger, C., Stehle, P. and Ellinger, S. 2012. Time to wound closure in trauma patients with disorders in wound healing is shortened by supplements containing antioxidant micronutrients and glutamine: A PRCT. *Clin. Nutr.* 31, 469–475.

133. Kim, M., Otsuka, M., Yu, R., Kurata, T. and Arakawa, N. 1994. The distribution of ascorbic acid and dehydroascorbic acid during tissue regeneration in wounded dorsal skin of guinea pigs. *Int. J. Vitam. Nutr. Res.* 64, 56–59.

134. Hulse, J. D., Ellis, S. R. and Henderson, L. M. 1978. Carnitine biosynthesis. beta-Hydroxylation of trimethyllysine by an alpha-ketoglutarate-dependent mitochondrial dioxygenase. *J. Biol. Chem.* 253, 1654–1659.

135. Dunn, W. A., Rettura, G., Seifter, E. and England, S. 1984. Carnitine biosynthesis from gamma-butyrobetaine and from exogenous protein-bound 6-N-trimethyl-L-lysine by the perfused guinea pig liver. Effect of ascorbate deficiency on the in situ activity of gamma-butyrobetaine hydroxylase. *J. Biol. Chem.* 259, 10764–10770.

136. Lindstedt, S. and Nordin, I. 1984. Multiple forms of gamma-butyrobetaine hydroxylase (EC 1.14.11.1). *Biochem. J.* 223, 119–127.

137. Rebouche, C. J. 1991. Ascorbic acid and carnitine biosynthesis. *Am. J. Clin. Nutr.* 54, 1147S–1152S.

138. Nelson, P. J., Pruitt, R. E., Henderson, L. L., Jenness, R. and Henderson, L. M. 1981. Effect of ascorbic acid deficiency on the in vivo synthesis of carnitine. *Biochim. Biophys. Acta* 672, 123–127.

139. Johnston, C. S., Solomon, R. E. and Corte, C. 1996. Vitamin C depletion is associated with alterations in blood histamine and plasma free carnitine in adults. *J. Am. Coll. Nutr.* 15, 586–591.

140. Johnston, C. S., Corte, C. and Swan, P. D. 2006. Marginal vitamin C status is associated with reduced fat oxidation during submaximal exercise in young adults. *Nutr. Metabol.* 3, 35.

141. Jones, E. and Hughes, R. E. 1982. Influence of oral carnitine on the body weight and survival time of avitaminotic-C guinea pigs. *Nutr. Rep. Int.* 25, 201–203.

142. Noland, R. C., Koves, T. R., Seiler, S. E., Lum, H., Lust, R. M., Ilkayeva, O., Stevens, R. D., Hegardt, F. G. and Muoio, D. M. 2009. Carnitine insufficiency caused by aging and overnutrition compromises mitochondrial performance and metabolic control. *J. Biol. Chem.* 284, 22840–22852.

143. Pearson, J. F., Pullar, J. M., Wilson, R., Spittlehouse, J. K., Vissers, M. C. M., Skidmore, P. M. L., Willis, J., Cameron, V. A. and Carr, A. C. 2017. Vitamin C status correlates with markers of metabolic and cognitive health in 50-year-olds: Findings of the CHALICE Cohort Study. *Nutrients* 9.

144. Ipsen, D. H., Tveden-Nyborg, P. and Lykkesfeldt, J. 2014. Does vitamin C deficiency promote fatty liver disease development? *Nutrients* 6, 5473–5499.

145. Carr, A. C., Bozonet, S. M., Pullar, J. M. and Vissers, M. C. 2013. Mood improvement in young adult males following supplementation with gold kiwifruit, a high-vitamin C food. *J. Nutr. Sci.* 2, e24.

146. Ratcliffe, P. J. 2013. Oxygen sensing and hypoxia signalling pathways in animals: The implications of physiology for cancer. *J. Physiol.* 591, 2027–2042.

147. Semenza, G. L. 2010. HIF-1: Upstream and downstream of cancer metabolism. *Curr. Opin. Genet. Dev.* 20, 51–56.

148. Schödel, J., Oikonomopoulos, S., Ragoussis, J., Pugh, C. W., Ratcliffe, P. J. and Mole, D. R. 2011. High-resolution genome-wide mapping of HIF-binding sites by ChIP-seq. *Blood* 117, e207–217.

149. Colgan, S. P. and Taylor, C. T. 2010. Hypoxia: An alarm signal during intestinal inflammation. *Nat. Rev. Gastro. Hepatol.* 7, 281–287.

150. Masson, N. and Ratcliffe, P. J. 2014. Hypoxia signaling pathways in cancer metabolism: The importance of co-selecting interconnected physiological pathways. *Cancer Metabol.* 2, 3.

151. Semenza, G. L. 2017. Hypoxia-inducible factors: Coupling glucose metabolism and redox regulation with induction of the breast cancer stem cell phenotype. *EMBO J.* 36, 252–259.

152. Majmundar, A. J., Wong, W. J. and Simon, M. C. 2010. Hypoxia-inducible factors and the response to hypoxic stress. *Mol. Cell* 40, 294–309.

153. Chan, M. C., Holt-Martyn, J. P., Schofield, C. J. and Ratcliffe, P. J. 2016. Pharmacological targeting of the HIF hydroxylases—A new field in medicine development. *Molec. Asp. Med.* 47–48, 54–75.

154. Zhang, C., Samanta, D., Lu, H., Bullen, J. W., Zhang, H., Chen, I., He, X. and Semenza, G. L. 2016. Hypoxia induces the breast cancer stem cell phenotype by HIF-dependent and ALKBH5-mediated m(6)A-demethylation of NANOG mRNA. *Proc. Natl. Acad. Sci. USA* 113, E2047–E2056.

155. Chaturvedi, P., Gilkes, D. M., Wong, C. C., Luo, W., Zhang, H., Wei, H., Takano, N., Schito, L., Levchenko, A. and Semenza, G. L. 2013. Hypoxia-inducible factor-dependent breast cancer-mesenchymal stem cell bidirectional signaling promotes metastasis. *J. Clin. Invest.* 123, 189–205.

156. Choudhry, H. and Harris, A. L. 2018. Advances in hypoxia-inducible factor biology. *Cell Metabol.* 27, 281–298.

157. Gonzalez, F. J., Xie, C. and Jiang, C. 2018. The role of hypoxia-inducible factors in metabolic diseases. *Nat. Rev. Endocrinol.* 15, 21–32.

158. Rius, J., Guma, M., Schachtrup, C., Akassoglou, K., Zinkernagel, A. S., Nizet, V., Johnson, R. S., Haddad, G. G. and Karin, M. 2008. NF-kappaB links innate immunity to the hypoxic response through transcriptional regulation of HIF-1alpha. *Nature* 453, 807–811.

159. D'Ignazio, L., Bandarra, D. and Rocha, S. 2016. NF-kappaB and HIF crosstalk in immune responses. *FEBS J.* 283, 413–424.

160. Roscigno, G., Puoti, I., Giordano, I., Donnarumma, E., Russo, V., Affinito, A., Adamo, A. et al. 2017. MiR-24 induces chemotherapy resistance and hypoxic advantage in breast cancer. *Oncotarget* 8, 19507–19521.

161. Fratantonio, D., Cimino, F., Speciale, A. and Virgili, F. 2018. Need (more than) two to Tango: Multiple tools to adapt to changes in oxygen availability. *Biofactors* 44, 207–218.

162. Semenza, G. L. 2016. The hypoxic tumor microenvironment: A driving force for breast cancer progression. *Biochim. Biophys. Acta* 1863, 382–391.

163. Semenza, G. L. 2015. Regulation of the breast cancer stem cell phenotype by hypoxia-inducible factors. *Clin Sci (Lond).* 129, 1037–1045.

164. Deb, S., Johansson, I., Byrne, D., Nilsson, C., Investigators, k., Constable, L., Fjallskog, M. L., Dobrovic, A., Hedenfalk, I. and Fox, S. B. 2014. Nuclear HIF1A expression is strongly prognostic in sporadic but not familial male breast cancer. *Mod. Pathol.* 27, 1223–1230.

165. Li, M., Xiao, D., Zhang, J., Qu, H., Yang, Y., Yan, Y., Liu, X. et al. 2016. Expression of LPA2 is associated with poor prognosis in human breast cancer and regulates HIF-1alpha expression and breast cancer cell growth. *Oncol. Rep.* 36, 3479–3487.

166. Schoning, J. P., Monteiro, M. and Gu, W. 2017. Drug resistance and cancer stem cells: The shared but distinct roles of hypoxia-inducible factors HIF1alpha and HIF2alpha. *Clin. Exp. Pharm. Phys.* 44, 153–161.

167. Vleugel, M. M., Greijer, A. E., Shvarts, A., van der Groep, P., van Berkel, M., Aarbodem, Y., van Tinteren, H., Harris, A. L., van Diest, P. J. and van der Wall, E. 2005. Differential prognostic impact of hypoxia induced and diffuse HIF-1alpha expression in invasive breast cancer. *J Clin Path.* 58, 172–177.

168. Zhang, H., Wong, C. C., Wei, H., Gilkes, D. M., Korangath, P., Chaturvedi, P., Schito, L. et al. 2012. HIF-1-dependent expression of angiopoietin-like 4 and L1CAM mediates vascular metastasis of hypoxic breast cancer cells to the lungs. *Oncogene* 31, 1757–1770.

169. Cao, D., Hou, M., Guan, Y. S., Jiang, M., Yang, Y. and Gou, H. F. 2009. Expression of HIF-1alpha and VEGF in colorectal cancer: Association with clinical outcomes and prognostic implications. *BMC Cancer* 9, 432.

170. Volinia, S., Galasso, M., Sana, M. E., Wise, T. F., Palatini, J., Huebner, K. and Croce, C. M. 2012. Breast cancer signatures for invasiveness and prognosis defined by deep sequencing of microRNA. *Proc. Natl. Acad. Sci. USA* 109, 3024–3029.

171. Wang, T., Gilkes, D. M., Takano, N., Xiang, L., Luo, W., Bishop, C. J., Chaturvedi, P., Green, J. J. and Semenza, G. L. 2014. Hypoxia-inducible factors and RAB22A mediate formation of microvesicles that stimulate breast cancer invasion and metastasis. *Proc. Natl. Acad. Sci. USA* 111, E3234–3242.

172. Wang, W., He, Y. F., Sun, Q. K., Wang, Y., Han, X. H., Peng, D. F., Yao, Y. W., Ji, C. S. and Hu, B. 2014. Hypoxia-inducible factor 1alpha in breast cancer prognosis. *Clin. Chim. Acta* 428, 32–37.

173. Landazuri, M. O., Vara-Vega, A., Viton, M., Cuevas, Y. and del Peso, L. 2006. Analysis of HIF-prolyl hydroxylases binding to substrates. *Biochem. Biophys. Res. Commun.* 351, 313–320.

174. Peet, D. and Linke, S. 2006. Regulation of HIF: Asparaginyl hydroxylation. *Novartis Found. Symp.* 272, 37–49; discussion 49-53, 131–140.

175. Smirnova, N. A., Hushpulian, D. M., Speer, R. E., Gaisina, I. N., Ratan, R. R. and Gazaryan, I. G. 2012. Catalytic mechanism and substrate specificity of HIF prolyl hydroxylases. *Biochemistry* 77, 1108–1119.

176. Chowdhury, R., Candela-Lena, J. I., Chan, M. C., Greenald, D. J., Yeoh, K. K., Tian, Y. M., McDonough, M. A. et al. 2013. Selective small molecule probes for the Hypoxia Inducible Factor (HIF) prolyl hydroxylases. *ACS Chem. Biol.* 8, 1488–1496.

177. Kuiper, C. and Vissers, M. C. 2014. Ascorbate as a co-factor for Fe- and 2-oxoglutarate dependent dioxygenases: Physiological activity in tumor growth and progression. *Front. Oncol.* 4, 359.

178. Chowdhury, R., Hardy, A. and Schofield, C. J. 2008. The human oxygen sensing machinery and its manipulation. *Chem. Soc. Rev.* 37, 1308–1319.

179. Hirsila, M., Koivunen, P., Gunzler, V., Kivirikko, K. I. and Myllyharju, J. 2003. Characterization of the human prolyl 4-hydroxylases that modify the hypoxia-inducible factor. *J. Biol. Chem.* 278, 30772–30780.

180. Koivunen, P. and Myllyharju, J. 2018. Kinetic analysis of HIF prolyl hydroxylases. *Methods Mol. Biol.* 1742, 15–25.

181. Knowles, H. J., Mole, D. R., Ratcliffe, P. J. and Harris, A. L. 2006. Normoxic stabilization of hypoxia-inducible factor-1alpha by modulation of the labile iron pool in differentiating U937 macrophages: Effect of natural resistance-associated macrophage protein 1. *Cancer Res.* 66, 2600–2607.

182. Knowles, H. J., Raval, R. R., Harris, A. L. and Ratcliffe, P. J. 2003. Effect of ascorbate on the

activity of hypoxia-inducible factor in cancer cells. *Cancer Res.* 63, 1764–1768.

183. Yeoh, K. K., Chan, M. C., Thalhammer, A., Demetriades, M., Chowdhury, R., Tian, Y. M., Stolze, I. et al. 2013. Dual-action inhibitors of HIF prolyl hydroxylases that induce binding of a second iron ion. *Org. Biomolec. Chem.* 11, 732–745.

184. Vissers, M. C., Gunningham, S. P., Morrison, M. J., Dachs, G. U. and Currie, M. J. 2007. Modulation of hypoxia-inducible factor-1 alpha in cultured primary cells by intracellular ascorbate. *Free Radic. Biol. Med.* 42, 765–772.

185. Kaczmarek, M., Cachau, R. E., Topol, I. A., Kasprzak, K. S., Ghio, A. and Salnikow, K. 2009. Metal ions-stimulated iron oxidation in hydroxylases facilitates stabilization of HIF-1 alpha protein. *Toxicol. Sci.* 107, 394–403.

186. Kaczmarek, M., Timofeeva, O. A., Karaczyn, A., Malyguine, A., Kasprzak, K. S. and Salnikow, K. 2007. The role of ascorbate in the modulation of HIF-1alpha protein and HIF-dependent transcription by chromium(VI) and nickel(II). *Free Radic. Biol. Med.* 42, 1246–1257.

187. Lu, H., Dalgard, C. L., Mohyeldin, A., McFate, T., Tait, A. S. and Verma, A. 2005. Reversible inactivation of HIF-1 prolyl hydroxylases allows cell metabolism to control basal HIF-1. *J. Biol. Chem.* 280, 41928–41939.

188. Kuiper, C., Dachs, G. U., Currie, M. J. and Vissers, M. C. 2014. Intracellular ascorbate enhances hypoxia-inducible factor (HIF)-hydroxylase activity and preferentially suppresses the HIF-1 transcriptional response. *Free Radic. Biol. Med.* 69, 308–317.

189. Qiao, H., Li, L., Qu, Z. C. and May, J. M. 2009. Cobalt-induced oxidant stress in cultured endothelial cells: Prevention by ascorbate in relation to HIF-1alpha. *Biofactors* 35, 306–313.

190. Campbell, E. J., Vissers, M. C. and Dachs, G. U. 2016. Ascorbate availability affects tumor implantation-take rate and increases tumor rejection in Gulo-/- mice. *Hypoxia (Auckland, N.Z.).* 4, 41–52.

191. Kuiper, C., Molenaar, I. G., Dachs, G. U., Currie, M. J., Sykes, P. H. and Vissers, M. C. 2010. Low ascorbate levels are associated with increased hypoxia-inducible factor-1 activity and an aggressive tumor phenotype in endometrial cancer. *Cancer Res.* 70, 5749–5758.

192. Kuiper, C., Dachs, G. U., Munn, D., Currie, M. J., Robinson, B. A., Pearson, J. F. and Vissers, M. C. 2014. Increased tumor ascorbate is associated with extended disease-free survival and decreased hypoxia-inducible factor-1 activation in human colorectal cancer. *Front Oncol.* 4, 10.

193. Wohlrab, C., Vissers, M. C. M., Phillips, E., Morrin, H., Robinson, B. A. and Dachs, G. U. 2018. The association between ascorbate and the hypoxia-inducible factors in human renal cell carcinoma requires a functional Von Hippel-Lindau protein. *Front Oncol.* 8, 574.

194. Campbell, E. J., Dachs, G. U., Morrin, H. R., Davey, V. C., Robinson, B. A. and Vissers, M. C. M. 2019. Activation of the hypoxia pathway in breast cancer tissue and patient survival are inversely associated with tumor ascorbate levels. *BMC Cancer* 19, 307.

195. Jozwiak, P., Krzeslak, A., Wieczorek, M. and Lipinska, A. 2015. Effect of glucose on GLUT1-dependent intracellular ascorbate accumulation and viability of thyroid cancer cells. *Nutr Cancer.* 67, 1333–1341.

196. Campbell, E. J., Vissers, M. C., Bozonet, S., Dyer, A., Robinson, B. A. and Dachs, G. U. 2015. Restoring physiological levels of ascorbate slows tumor growth and moderates HIF-1 pathway activity in Gulo(-/-) mice. *Cancer Med.* 4, 303–314.

197. Campbell, E. J., Vissers, M. C. M., Wohlrab, C., Hicks, K. O., Strother, R. M., Bozonet, S. M., Robinson, B. A. and Dachs, G. U. 2016. Pharmacokinetic and anti-cancer properties of high dose ascorbate in solid tumors of ascorbate-dependent mice. *Free Radic. Biol. Med.* 99, 451–462.

198. Gao, P., Zhang, H., Dinavahi, R., Li, F., Xiang, Y., Raman, V., Bhujwalla, Z. M., Felsher, D. W., Cheng, L., Pevsner, J., Lee, L. A., Semenza, G. L. and Dang, C. V. 2007. HIF-dependent antitumorigenic effect of antioxidants in vivo. *Cancer Cell* 12, 230–238.

199. Felsenfeld, G. 2014. A brief history of epigenetics. *Cold Spring Harbor Persp. Biol.* 6.

200. Morris, B. J., Willcox, B. J. and Donlon, T. A. 2019. Genetic and epigenetic regulation of human aging and longevity. *Biochim. Biophys. Acta Mol. Bas. Dis.* 1865, 1718–1744.

201. Banik, A., Kandilya, D., Ramya, S., Stunkel, W., Chong, Y. S. and Dheen, S. T. 2017. Maternal factors that induce epigenetic changes contribute to neurological disorders in offspring. *Genes* 8.

202. Geraghty, A. A., Lindsay, K. L., Alberdi, G., McAuliffe, F. M. and Gibney, E. R. 2015. Nutrition during pregnancy impacts offspring's epigenetic status-evidence from human and animal studies. *Nutr. Metabol Insights* 8, 41–47.

203. Ehrlich, M. and Lacey, M. 2013. DNA hypomethylation and hemimethylation in cancer. *Adv. Exp. Med. Biol.* 754, 31–56.

204. Berman, B. P., Weisenberger, D. J., Aman, J. F., Hinoue, T., Ramjan, Z., Liu, Y., Noushmehr, H. et al. 2011. Regions of focal DNA hypermethylation and long-range hypomethylation in colorectal cancer coincide with nuclear lamina-associated domains. *Nat. Genet.* 44, 40–46.

205. Baylin, S. B. and Jones, P. A. 2011. A decade of exploring the cancer epigenome—Biological and translational implications. *Nat Rev Cancer* 11, 726–734.

206. Baylin, S. B. and Jones, P. A. 2016. Epigenetic determinants of cancer. *Cold Spring Harbor Persp. Biol.* 8.

207. Kohli, R. M. and Zhang, Y. 2013. TET enzymes, TDG and the dynamics of DNA demethylation. *Nature* 502, 472–479.

208. Li, D., Guo, B., Wu, H., Tan, L. and Lu, Q. 2015. TET Family of dioxygenases: Crucial roles and underlying mechanisms. *Cytogen Gen. Res.* 146, 171–180.

209. Kriaucionis, S. and Heintz, N. 2009. The nuclear DNA base 5-hydroxymethylcytosine is present in Purkinje neurons and the brain. *Science* 324, 929–930.

210. Li, Y. and O'Neill, C. 2013. 5'-Methylcytosine and 5'-hydroxymethylcytosine each provide epigenetic information to the mouse zygote. *PLOS ONE.* 8, e63689.

211. Hahn, M. A., Qiu, R., Wu, X., Li, A. X., Zhang, H., Wang, J., Jui, J. et al. 2013. Dynamics of 5-hydroxymethylcytosine and chromatin marks in Mammalian neurogenesis. *Cell Rep.* 3, 291–300.

212. Nakamura, T., Liu, Y. J., Nakashima, H., Umehara, H., Inoue, K., Matoba, S., Tachibana, M., Ogura, A., Shinkai, Y. and Nakano, T. 2012. PGC7 binds histone H3K9me2 to protect against conversion of 5mC to 5hmC in early embryos. *Nature* 486, 415–419.

213. Mellen, M., Ayata, P., Dewell, S., Kriaucionis, S. and Heintz, N. 2012. MeCP2 binds to 5hmC enriched within active genes and accessible chromatin in the nervous system. *Cell* 151, 1417–1430.

214. Huang, Y., Chavez, L., Chang, X., Wang, X., Pastor, W. A., Kang, J., Zepeda-Martinez, J. A. et al. 2014. Distinct roles of the methylcytosine oxidases Tet1 and Tet2 in mouse embryonic stem cells. *Proc. Natl. Acad. Sci. USA* 111, 1361–1366.

215. Li, X., Yue, X., Pastor, W. A., Lin, L., Georges, R., Chavez, L., Evans, S. M. and Rao, A. 2016. Tet proteins influence the balance between neuroectodermal and mesodermal fate choice by inhibiting Wnt signaling. *Proc. Natl. Acad. Sci. USA* 113, E8267–E8276.

216. Cimmino, L., Dolgalev, I., Wang, Y., Yoshimi, A., Martin, G. H., Wang, J., Ng, V. et al. 2017. Restoration of TET2 function blocks aberrant self-renewal and leukemia progression. *Cell* 170, 1079–1095.e1020.

217. Ko, M., An, J., Pastor, W. A., Koralov, S. B., Rajewsky, K. and Rao, A. 2015. TET proteins and 5-methylcytosine oxidation in hematological cancers. *Immunol. Rev.* 263, 6–21.

218. Tan, L. and Shi, Y. G. 2012. Tet family proteins and 5-hydroxymethylcytosine in development and disease. *Development (Cambr, Engl).* 139, 1895–1902.

219. Zhang, P., Huang, B., Xu, X. and Sessa, W. C. 2013. Ten-eleven translocation (Tet) and thymine DNA glycosylase (TDG), components of the demethylation pathway, are direct targets of miRNA-29a. *Biochem. Biophys. Res. Commun.* 437, 368–373.

220. Liu, Y., Peng, W., Qu, K., Lin, X., Zeng, Z., Chen, J., Wei, D. and Wang, Z. 2018. TET2: A novel epigenetic regulator and potential intervention target for atherosclerosis. *DNA Cell Biol.* 37, 517–523.

221. Dawlaty, M. M., Breiling, A., Le, T., Barrasa, M. I., Raddatz, G., Gao, Q., Powell, B. E. et al. 2014. Loss of Tet enzymes compromises proper differentiation of embryonic stem cells. *Develop Cell.* 29, 102–111.

222. Dawlaty, M. M., Breiling, A., Le, T., Raddatz, G., Barrasa, M. I., Cheng, A. W., Gao, Q. et al. 2013. Combined deficiency of Tet1 and Tet2 causes epigenetic abnormalities but is compatible with postnatal development. *Develop Cell* 24, 310–323.

223. Blaschke, K., Ebata, K. T., Karimi, M. M., Zepeda-Martinez, J. A., Goyal, P., Mahapatra, S., Tam, A. et al. 2013. Vitamin C induces Tet-dependent DNA demethylation and a blastocyst-like state in ES cells. *Nature* 500, 222–226.

224. Minor, E. A., Court, B. L., Young, J. I. and Wang, G. 2013. Ascorbate induces ten-eleven translocation (Tet) methylcytosine dioxygenase-mediated generation of 5-hydroxymethylcytosine. *J. Biol. Chem.* 288, 13669–13674.

225. Yin, R., Mao, S. Q., Zhao, B., Chong, Z., Yang, Y., Zhao, C., Zhang, D. et al. 2013. Ascorbic

acid enhances Tet-mediated 5-methylcytosine oxidation and promotes DNA demethylation in mammals. *J. Am. Chem. Soc.* 135, 10396–10403.

226. Hore, T. A., von Meyenn, F., Ravichandran, M., Bachman, M., Ficz, G., Oxley, D., Santos, F., Balasubramanian, S., Jurkowski, T. P. and Reik, W. 2016. Retinol and ascorbate drive erasure of epigenetic memory and enhance reprogramming to naive pluripotency by complementary mechanisms. *Proc. Natl. Acad. Sci. USA* 113, 12202–12207.

227. Madruga de Oliveira, A., Rondo, P. H. and Barros, S. B. 2004. Concentrations of ascorbic acid in the plasma of pregnant smokers and nonsmokers and their newborns. *Int. J. Vitam. Nutr. Res.* 74, 193–198.

228. Juhl, B., Lauszus, F. F. and Lykkesfeldt, J. 2017. Poor vitamin C status late in pregnancy is associated with increased risk of complications in type 1 diabetic women: A cross-sectional study. *Nutrients* 9.

229. Mitsuya, K., Parker, A. N., Liu, L., Ruan, J., Vissers, M. C. M. and Myatt, L. 2017. Alterations in the placental methylome with maternal obesity and evidence for metabolic regulation. *PLOS ONE.* 12, e0186115.

230. Habibzadeh, N., Schorah, C. J. and Smithells, R. W. 1986. The effects of maternal folic acid and vitamin C nutrition in early pregnancy on reproductive performance in the guinea-pig. *Br. J. Nutr.* 55, 23–35.

231. Tveden-Nyborg, P., Hasselholt, S., Miyashita, N., Moos, T., Poulsen, H. E. and Lykkesfeldt, J. 2012. Chronic vitamin C deficiency does not accelerate oxidative stress in ageing brains of guinea pigs. *Basic Clin. Pharmacol. Toxicol.* 110, 524–529.

232. Schjoldager, J. G., Paidi, M. D., Lindblad, M. M., Birck, M. M., Kjaergaard, A. B., Dantzer, V., Lykkesfeldt, J. and Tveden-Nyborg, P. 2015. Maternal vitamin C deficiency during pregnancy results in transient fetal and placental growth retardation in guinea pigs. *Eur. J. Nutr.* 54, 667–676.

233. Tveden-Nyborg, P., Vogt, L., Schjoldager, J. G., Jeannet, N., Hasselholt, S., Paidi, M. D., Christen, S. and Lykkesfeldt, J. 2012. Maternal vitamin C deficiency during pregnancy persistently impairs hippocampal neurogenesis in offspring of guinea pigs. *PLOS ONE.* 7, e48488.

234. Vissers, M. C. M. and Das, A. B. 2018. Potential mechanisms of action for vitamin C in cancer: Reviewing the evidence. *Front. Physiol.* 9, 809.

235. Ley, T. J., Miller, C., Ding, L., Raphael, B. J., Mungall, A. J., Robertson, A., Hoadley, K. et al. 2013. Genomic and epigenomic landscapes of adult de novo acute myeloid leukemia. *N. Engl. J. Med.* 368, 2059–2074.

236. Papaemmanuil, E., Gerstung, M., Bullinger, L., Gaidzik, V. I., Paschka, P., Roberts, N. D., Potter, N. E. et al. 2016. Genomic classification and prognosis in acute myeloid leukemia. *N. Engl. J. Med.* 374, 2209–2221.

237. Figueroa, M. E., Abdel-Wahab, O., Lu, C., Ward, P. S., Patel, J., Shih, A., Li, Y. et al. 2010. Leukemic IDH1 and IDH2 mutations result in a hypermethylation phenotype, disrupt TET2 function, and impair hematopoietic differentiation. *Cancer Cell* 18, 553–567.

238. Rampal, R., Alkalin, A., Madzo, J., Vasanthakumar, A., Pronier, E., Patel, J., Li, Y. et al. 2014. DNA hydroxymethylation profiling reveals that WT1 mutations result in loss of TET2 function in acute myeloid leukemia. *Cell Rep.* 9, 1841–1855.

239. Shenoy, N., Bhagat, T. D., Cheville, J., Lohse, C., Bhattacharyya, S., Tischer, A., Machha, V. et al. 2019. Ascorbic acid-induced TET activation mitigates adverse hydroxymethylcytosine loss in renal cell carcinoma. *J. Clin. Invest.* 130, 1612–1625.

240. Tsukada, Y., Fang, J., Erdjument-Bromage, H., Warren, M. E., Borchers, C. H., Tempst, P. and Zhang, Y. 2006. Histone demethylation by a family of JmjC domain-containing proteins. *Nature* 439, 811–816.

241. Eipper, B. A. and Mains, R. E. 1991. The role of ascorbate in the biosynthesis of neuroendocrine peptides. *Am. J. Clin. Nutr.* 54, 1153S–1156S.

242. Glembotski, C. C. 1986. The characterization of the ascorbic acid-mediated alpha-amidation of alpha-melanotropin in cultured intermediate pituitary lobe cells. *Endocrinology* 118, 1461–1468.

243. Herman, H. H., Wimalasena, K., Fowler, L. C., Beard, C. A. and May, S. W. 1988. Demonstration of the ascorbate dependence of membrane-bound dopamine beta-monooxygenase in adrenal chromaffin granule ghosts. *J. Biol. Chem.* 263, 666–672.

244. Levine, M., Morita, K. and Pollard, H. 1985. Enhancement of norepinephrine biosynthesis by ascorbic acid in cultured bovine chromaffin cells. *J. Biol. Chem.* 260, 12942–12947.

245. Linster, C. L. and Van Schaftingen, E. 2007. Vitamin C. Biosynthesis, recycling and degradation in mammals. *FEBS J.* 274, 1–22.

246. Tateishi, K., Arakawa, F., Misumi, Y., Treston, A. M., Vos, M. and Matsuoka, Y. 1994. Isolation and functional expression of human pancreatic peptidylglycine alpha-amidating monooxygenase. *Biochem. Biophys. Res. Commun.* 205, 282–290.

247. Wimalasena, K. and Wimalasena, D. S. 1995. The reduction of membrane-bound dopamine beta-monooxygenase in resealed chromaffin granule ghosts. Is intragranular ascorbic acid a mediator for extragranular reducing equivalents? *J. Biol. Chem.* 270, 27516–27524.

248. Prigge, S. T., Mains, R. E., Eipper, B. A. and Amzel, L. M. 2000. New insights into copper monooxygenases and peptide amidation: Structure, mechanism and function. *Cell Mol. Life Sci.* 57, 1236–1259.

249. Padayatty, S. J., Doppman, J. L., Chang, R., Wang, Y., Gill, J., Papanicolaou, D. A. and Levine, M. 2007. Human adrenal glands secrete vitamin C in response to adrenocorticotrophic hormone. *Am. J. Clin. Nutr.* 86, 145–149.

250. Rush, R. A. and Geffen, L. B. 1980. Dopamine beta-hydroxylase in health and disease. *Crit. Rev. Clin. Lab. Sci.* 12, 241–277.

251. Levin, E. Y., Levenberg, B. and Kaufman, S. 1960. The enzymatic conversion of 3,4-dihydroxyphenylethylamine to norepinephrine. *J. Biol. Chem.* 235, 2080–2086.

252. Diliberto, E. J., Jr. and Allen, P. L. 1980. Semidehydroascorbate as a product of the enzymic conversion of dopamine to norepinephrine. Coupling of semidehydroascorbate reductase to dopamine-beta-hydroxylase. *Mol. Pharmacol.* 17, 421–426.

253. Diliberto, E. J., Jr. and Allen, P. L. 1981. Mechanism of dopamine-beta-hydroxylation. Semidehydroascorbate as the enzyme oxidation product of ascorbate. *J. Biol. Chem.* 256, 3385–3393.

254. Levine, M., Morita, K., Heldman, E. and Pollard, H. B. 1985. Ascorbic acid regulation of norepinephrine biosynthesis in isolated chromaffin granules from bovine adrenal medulla. *J. Biol. Chem.* 260, 15598–15603.

255. Eipper, B. A., Stoffers, D. A. and Mains, R. E. 1992. The biosynthesis of neuropeptides: Peptide alpha-amidation. *Ann. Rev. Neurosci.* 15, 57–85.

256. Simpson, P. D., Eipper, B. A., Katz, M. J., Gandara, L., Wappner, P., Fischer, R., Hodson, E. J., Ratcliffe, P. J. and Masson, N. 2015. Striking oxygen sensitivity of the Peptidylglycine alpha-Amidating Monooxygenase (PAM) in neuroendocrine cells. *J. Biol. Chem.* 290, 24891–24901.

257. Sheldrick, E. L. and Flint, A. P. 1989. Post-translational processing of oxytocin-neurophysin prohormone in the ovine corpus luteum: Activity of peptidyl glycine alpha-amidating mono-oxygenase and concentrations of its cofactor, ascorbic acid. *J. Endocrinol.* 122, 313–322.

258. Jeng, A. Y., Fujimoto, R. A., Chou, M., Tan, J. and Erion, M. D. 1997. Suppression of substance P biosynthesis in sensory neurons of dorsal root ganglion by prodrug esters of potent peptidylglycine alpha-amidating monooxygenase inhibitors. *J. Biol. Chem.* 272, 14666–14671.

259. Kukucka, M. A. and Misra, H. P. 1992. HPLC determination of an oxytocin-like peptide produced by isolated guinea pig Leydig cells: Stimulation by ascorbate. *Arch. Androl.* 29, 185–190.

260. Luck, M. R. and Jungclas, B. 1988. The time-course of oxytocin secretion from cultured bovine granulosa cells, stimulated by ascorbate and catecholamines. *J. Endocrinol.* 116, 247–258.

261. Miszkiel, G., Skarzynski, D., Bogacki, M. and Kotwica, J. 1999. Concentrations of catecholamines, ascorbic acid, progesterone and oxytocin in the corpora lutea of cyclic and pregnant cattle. *Reprod. Nutr. Dev.* 39, 509–516.

262. Wathes, D. C. and Denning-Kendall, P. A. 1992. Control of synthesis and secretion of ovarian oxytocin in ruminants. *J. Reprod. Fert. Suppl.* 45, 39–52.

263. Giusti-Paiva, A. and Domingues, V. G. 2010. Centrally administered ascorbic acid induces antidiuresis, natriuresis and neurohypophyseal hormone release in rats. *Neuro Endocrinol. Lett.* 31, 87–91.

264. Treschan, T. A. and Peters, J. 2006. The vasopressin system: Physiology and clinical strategies. *Anesthesiology* 105, 599–612; quiz 639–540.

265. Deana, R., Bharaj, B. S., Verjee, Z. H. and Galzigna, L. 1975. Changes relevant to catecholamine metabolism in liver and brain of ascorbic acid deficient guinea-pigs. *Int. J. Vitam. Nutr. Res.* 45, 175–182.

266. Bornstein, S. R., Yoshida-Hiroi, M., Sotiriou, S., Levine, M., Hartwig, H. G., Nussbaum, R. L. and Eisenhofer, G. 2003. Impaired adrenal catecholamine system function in mice with

deficiency of the ascorbic acid transporter (SVCT2). *FASEB J.* 17, 1928–1930.

267. Vasu, T. S., Cavallazzi, R., Hirani, A., Kaplan, G., Leiby, B. and Marik, P. E. 2012. Norepinephrine or dopamine for septic shock: Systematic review of randomized clinical trials. *J. Intens. Care Med.* 27, 172–178.

268. Carr, A. C., Shaw, G. M., Fowler, A. A. and Natarajan, R. 2015. Ascorbate-dependent vasopressor synthesis: A rationale for vitamin C administration in severe sepsis and septic shock? *Crit Care (London, England).* 19, 418.

269. Schorah, C. J., Downing, C., Piripitsi, A., Gallivan, L., Al-Hazaa, A. H., Sanderson, M. J. and Bodenham, A. 1996. Total vitamin C, ascorbic acid, and dehydroascorbic acid concentrations in plasma of critically ill patients. *Am. J. Clin. Nutr.* 63, 760–765.

270. Bonham, M. J., Abu-Zidan, F. M., Simovic, M. O., Sluis, K. B., Wilkinson, A., Winterbourn, C. C. and Windsor, J. A. 1999. Early ascorbic acid depletion is related to the severity of acute pancreatitis. *Brit. J. Surg.* 86, 1296–1301.

271. Evans-Olders, R., Eintracht, S. and Hoffer, L. J. 2010. Metabolic origin of hypovitaminosis C in acutely hospitalized patients. *Nutrition* 26, 1070–1074.

272. Carr, A. C., Rosengrave, P. C., Bayer, S., Chambers, S., Mehrtens, J. and Shaw, G. M. 2017. Hypovitaminosis C and vitamin C deficiency in critically ill patients despite recommended enteral and parenteral intakes. *Crit. Care (London, England).* 21, 300.

273. de Grooth, H. J., Manubulu-Choo, W. P., Zandvliet, A. S., Spoelstra-de Man, A. M. E., Girbes, A. R., Swart, E. L. and Oudemans-van Straaten, H. M. 2018. Vitamin C pharmacokinetics in critically Ill patients: A randomized trial of four IV regimens. *Chest* 153, 1368–1377.

274. Marik, P. E. and Hooper, M. H. 2018. Doctor-your septic patients have scurvy! *Crit. Care (London, England).* 22, 23.

275. Zabet, M. H., Mohammadi, M., Ramezani, M. and Khalili, H. 2016. Effect of high-dose ascorbic acid on vasopressor's requirement in septic shock. *J. Res. Pharm. Pract.* 5, 94–100.

276. Marik, P. E., Khangoora, V., Rivera, R., Hooper, M. H. and Catravas, J. 2017. Hydrocortisone, vitamin C, and thiamine for the treatment of severe sepsis and septic shock: A retrospective before-after study. *Chest* 151, 1229–1238.

277. Taeb, A. M., Hooper, M. H. and Marik, P. E. 2017. Sepsis: Current definition, pathophysiology, diagnosis, and management. *Nutr. Clin. Pract.* 32, 296–308.

278. Moskowitz, A., Andersen, L. W., Huang, D. T., Berg, K. M., Grossestreuer, A. V., Marik, P. E., Sherwin, R. L. et al. 2018. Ascorbic acid, corticosteroids, and thiamine in sepsis: A review of the biologic rationale and the present state of clinical evaluation. *Crit Care (London, England).* 22, 283.

279. Marik, P. E. 2018. Vitamin C for the treatment of sepsis: The scientific rationale. *Pharmacol. Ther.* 189, 63–70.

280. Hooper, M. H., Carr, A. and Marik, P. E. 2019. The adrenal-vitamin C axis: From fish to guinea pigs and primates. *Crit Care (London, England).* 23, 29.

281. Wang, Y., Lin, H., Lin, B. W. and Lin, J. D. 2019. Effects of different ascorbic acid doses on the mortality of critically ill patients: A meta-analysis. *Ann. Intens. Care.* 9, 58.

282. Schou-Pedersen, A. M. V., Hansen, S. N., Tveden-Nyborg, P. and Lykkesfeldt, J. 2016. Simultaneous quantification of monoamine neurotransmitters and their biogenic metabolites intracellularly and extracellularly in primary neuronal cell cultures and in sub-regions of guinea pig brain. *J. Chromatog.* 1028, 222–230.

283. Mortensen, A., Hasselholt, S., Tveden-Nyborg, P. and Lykkesfeldt, J. 2013. Guinea pig ascorbate status predicts tetrahydrobiopterin plasma concentration and oxidation ratio in vivo. *Nutr. Res.* 33, 859–867.

284. May, J. M. and Harrison, F. E. 2013. Role of vitamin C in the function of the vascular endothelium. *Antioxid. Redox. Signal* 19, 2068–2083.

285. Argyropoulos, S. V. and Nutt, D. J. 2000. Substance P antagonists: Novel agents in the treatment of depression. *Expert Opin. Investig. Drugs* 9, 1871–1875.

286. O'Connor, T. M., O'Connell, J., O'Brien, D. I., Goode, T., Bredin, C. P. and Shanahan, F. 2004. The role of substance P in inflammatory disease. *J. Cell Physiol.* 201, 167–180.

287. Schwarz, M. J. and Ackenheil, M. 2002. The role of substance P in depression: Therapeutic implications. *Dialogues Clin. Neurosci.* 4, 21–29.

288. Majamaa, K., Gunzler, V., Hanauske-Abel, H. M., Myllyla, R. and Kivirikko, K. I. 1986. Partial identity of the 2-oxoglutarate and ascorbate binding sites of prolyl 4-hydroxylase. *J. Biol. Chem.* 261, 7819–7823.

289. Schofield, C. J. and Ratcliffe, P. J. 2004. Oxygen sensing by HIF hydroxylases. *Nat. Rev. Mol. Cell Biol.* 5, 343–354.

290. Jaakkola, P. M. and Rantanen, K. 2013. The regulation, localization, and functions of oxygen-sensing prolyl hydroxylase PHD3. *Biol. Chem.* 394, 449–457.

291. Lando, D., Peet, D. J., Gorman, J. J., Whelan, D. A., Whitelaw, M. L. and Bruick, R. K. 2002. FIH-1 is an asparaginyl hydroxylase enzyme that regulates the transcriptional activity of hypoxia-inducible factor. *Genes Dev.* 16, 1466–1471.

292. Hewitson, K. S., McNeill, L. A., Riordan, M. V., Tian, Y. M., Bullock, A. N., Welford, R. W., Elkins, J. M. et al. 2002. Hypoxia-inducible factor (HIF) asparagine hydroxylase is identical to factor inhibiting HIF (FIH) and is related to the cupin structural family. *J. Biol. Chem.* 277, 26351–26355.

293. Singleton, R. S., Trudgian, D. C., Fischer, R., Kessler, B. M., Ratcliffe, P. J. and Cockman, M. E. 2011. Quantitative mass spectrometry reveals dynamics of factor-inhibiting hypoxia-inducible factor-catalyzed hydroxylation. *J. Biol. Chem.* 286, 33784–33794.

294. Cockman, M. E., Webb, J. D., Kramer, H. B., Kessler, B. M. and Ratcliffe, P. J. 2009. Proteomics-based identification of novel factor inhibiting hypoxia-inducible factor (FIH) substrates indicates widespread asparaginyl hydroxylation of ankyrin repeat domain-containing proteins. *Mol. Cell Proteomics* 8, 535–546.

295. Simmons, J. M., Muller, T. A. and Hausinger, R. P. 2008. Fe(II)/alpha-ketoglutarate hydroxylases involved in nucleobase, nucleoside, nucleotide, and chromatin metabolism. *Dalton Trans (Cambridge, Engl).* 5132–5142.

296. Feng, T., Yamamoto, A., Wilkins, S. E., Sokolova, E., Yates, L. A., Munzel, M., Singh, P. et al. 2014. Optimal translational termination requires C4 lysyl hydroxylation of eRF1. *Mol. Cell.* 53, 645–654.

297. Kato, M., Araiso, Y., Noma, A., Nagao, A., Suzuki, T., Ishitani, R. and Nureki, O. 2011. Crystal structure of a novel JmjC-domain-containing protein, TYW5, involved in tRNA modification. *Nucl. Acids Res.* 39, 1576–1585.

298. Perche-Letuvee, P., Molle, T., Forouhar, F., Mulliez, E. and Atta, M. 2014. Wybutosine biosynthesis: Structural and mechanistic overview. *RNA Biol.* 11, 1508–1518.

299. Songe-Moller, L., van den Born, E., Leihne, V., Vagbo, C. B., Kristoffersen, T., Krokan, H. E., Kirpekar, F., Falnes, P. O. and Klungland, A. 2010. Mammalian ALKBH8 possesses tRNA methyltransferase activity required for the biogenesis of multiple wobble uridine modifications implicated in translational decoding. *Mol. Cell Biol.* 30, 1814–1827.

300. Zhang, X., Wei, L. H., Wang, Y., Xiao, Y., Liu, J., Zhang, W., Yan, N. et al. 2019. Structural insights into FTO's catalytic mechanism for the demethylation of multiple RNA substrates. *Proc. Natl. Acad. Sci. USA* 116, 2919–2924.

301. Zou, S., Toh, J. D., Wong, K. H., Gao, Y. G., Hong, W. and Woon, E. C. 2016. N(6)-Methyladenosine: A conformational marker that regulates the substrate specificity of human demethylases FTO and ALKBH5. *Sci. Rep.* 6, 25677.

302. Landfors, M., Nakken, S., Fusser, M., Dahl, J. A., Klungland, A. and Fedorcsak, P. 2016. Sequencing of FTO and ALKBH5 in men undergoing infertility work-up identifies an infertility-associated variant and two missense mutations. *Fertil. Steril.* 105, 1170–1179.e1175.

303. Shen, L., Song, C. X., He, C. and Zhang, Y. 2014. Mechanism and function of oxidative reversal of DNA and RNA methylation. *Ann. Rev. Biochem.* 83, 585–614.

304. Peng, D., Ge, G., Gong, Y., Zhan, Y., He, S., Guan, B., Li, Y. et al. 2018. Vitamin C increases 5-hydroxymethylcytosine level and inhibits the growth of bladder cancer. *Clin. Epigen.* 10, 94.

305. Lesniak, R. K., Markolovic, S., Tars, K. and Schofield, C. J. 2016. Human carnitine biosynthesis proceeds via (2S,3S)-3-hydroxy-N(epsilon)-trimethyllysine. *Chem. Commun (Camb).* 53, 440–442.

306. Lindstedt, G. and Lindstedt, S. 1970. Cofactor requirements of gamma-butyrobetaine hydroxylase from rat liver. *J. Biol. Chem.* 245, 4178–4186.

307. Reddy, Y. V., Al Temimi, A. H., White, P. B. and Mecinovic, J. 2017. Evidence that trimethyllysine hydroxylase catalyzes the formation of (2S,3S)-3-hydroxy-N(epsilon)-trimethyllysine. *Org. Lett.* 19, 400–403.

308. Patel, N., Khan, A. O., Mansour, A., Mohamed, J. Y., Al-Assiri, A., Haddad, R., Jia, X., Xiong, Y., Megarbane, A., Traboulsi, E. I. and Alkuraya, F. S. 2014. Mutations in ASPH cause facial dysmorphism, lens dislocation, anterior-segment abnormalities, and spontaneous filtering blebs, or Traboulsi syndrome. *Am. J. Hum. Gen.* 94, 755–759.

309. Diliberto, E. J., Jr., Daniels, A. J. and Viveros, O. H. 1991. Multicompartmental secretion of

ascorbate and its dual role in dopamine beta-hydroxylation. *Am. J. Clin. Nutr.* 54, 1163S–1172S.

310. Glombotski, C. C., Manaker, S., Winokur, A. and Gibson, T. R. 1986. Ascorbic acid increases the thyrotropin-releasing hormone content of hypothalamic cell cultures. *J. Neurosci.* 6, 1796–1802.

311. Mingay, M., Chaturvedi, A., Bilenky, M., Cao, Q., Jackson, L., Hui, T., Moksa, M. et al. 2018. Vitamin C-induced epigenomic remodelling in IDH1 mutant acute myeloid leukaemia. *Leukemia* 32, 11–20.

CHAPTER SIX

Role of Ascorbate and Dehydroascorbic Acid in Metabolic Integration of the Cell

Gábor Bánhegyi, András Szarka, and József Mandl

DOI: 10.1201/9780429442025-6

CONTENTS

INTRODUCTION

Cellular metabolism is determined by contributions of different intracellular organelles, which are also separated metabolic compartments. This metabolism has different functions, trends, and directions in its adaptation to various bioenergetic and signaling stimuli demands. All cellular processes including cell death and autophagy are controlled by intimate cross talk among the various organelles. Intracellular redox homeostasis is an essential part of the cellular metabolic integration. Major determinants and actors of intracellular redox homeostasis are also formed in different intracellular compartments.

There are major differences in redox conditions among the various organelles, which also depend on their cellular functions. Metabolic integration also involves integration of redox conditions in the cytosol, in mitochondria, in the luminal compartment of the endoplasmic reticulum (ER), or in lysosomes. Transports of various metabolites and redox active components are fundamental in the regulation of intermediary metabolism and in the maintenance of the organelle redox homeostasis, and they are essential elements of the interorganelle cross talk.

Ascorbate has several functions in basic cellular functions and regulations [7,9,10,13,52]. It is also noteworthy that different organelles participate in *de novo* ascorbate synthesis [7]. Nevertheless, it is also well-known that several species, including humans, are unable to form ascorbate in spite of its essential role in redox homeostasis, as the last enzyme of the hexuronic acid pathway for ascorbate synthesis has been lost in humans and in some other species. However, it is very instructive to keep these facts in mind. We want to emphasize the role and function of glycogen particles and endoplasmic reticulum in ascorbate formation [7,10,52]. These facts suggest a special integratory role of ascorbate, and the significance of the intracellular transport processes.

VITAMIN C IN METABOLISM: REDOX CONTEXT

Ascorbate is a characteristic component in redox homeostasis. This special role is connected to its participation in redox reactions transferring one or two electrons, as well and its ability to regulate oxidoreductions around different iron (Fe) (and copper [Cu]) ions. These redox reactions are connected directly or indirectly to oxygen supply, underlying its essential biological role.

Almost all vitamin C–dependent reactions are based on the easy electron transfer properties of ascorbate and its oxidized products, monodehydroascorbate (MDHA) and dehydroascorbic acid (DHA). These three compounds are interconnected by a variety of redox reactions, which serve also the recycling of ascorbate (see Figure 6.1). The metabolic network ensures that although ascorbate is an excellent electron donor, its oxidation seldom results in the loss of the vitamin [7], which is especially important in species unable to synthesize ascorbate. Besides the well-known antioxidant role, ascorbate is a cofactor for several enzymes; that role is also based on its redox properties.

ASCORBATE-REQUIRING ENZYMES AND THEIR SUBCELLULAR COMPARTMENTALIZATION

Ascorbate-dependent enzymes have been reported from various locations within the eukaryotic cells, suggesting that ascorbate should be present in all organelles.

Fe[II]/2-Oxoglutarate–Dependent Dioxygenases

In the lumen of the ER, three enzymes are present that are required for the posttranslational modification of proline and lysine during the synthesis of collagen (prolyl-3-hydroxylase, prolyl-4-hydroxylase, lysyl hydroxylase). Hydroxylation is required for the triple helical structure of the collagen molecule; thus, ascorbate is essential to the development and maintenance of the extracellular matrix in blood vessel wall, scar, and connective tissues [50,58].

Other prolyl hydroxylases: hypoxia-inducible factor (HIF) prolyl hydroxylases (three isoforms)

[a] $\quad A + R\bullet \rightarrow MDHA\bullet + R$

[b] $\quad A + Fe^{3+} (\text{or } Cu^{2+}) \rightarrow MDHA\bullet + Fe^{2+} (\text{or } Cu^+)$

[c] $\quad 2\,MDHA\bullet \rightarrow A + DHA$

[d] $\quad MDHA\bullet + e^- \rightarrow A$

[e] $\quad DHA + 2\,RSH \rightarrow A + RSSR$

[f] $\quad A + RH + O_2 \rightarrow DHA + ROH + H_2O$

[g] $\quad RH + O_2 + \text{2-oxoglutarate} \rightarrow ROH + CO_2 + \text{succinate}$

[h] $\quad R\text{-}Me + O_2 + \text{2-oxoglutarate} \rightarrow RH + CH_2O + CO_2 + \text{succinate}$

Figure 6.1. Electron transfer with the involvement of ascorbate and its metabolites. Ascorbate (A) can react with free radicals and other reactive species in spontaneous nonenzymatic reactions [a]. A can reduce transition metal ions either in free or protein-bound form [b]. The above reactions produce the resonance-stabilized monodehydroascorbate (MDHA) radical that can be disproportionate to A and dehydroascorbic acid (DHA) [c], or enzymatically reduced back to A with an electron supplied by NADH [d]. DHA can be reduced back to A at the expense of protein and nonprotein thiols both in enzymatic and nonenzymatic reactions [e]. A is an electron donor for copper-containing monooxygenases, such as dopamine β-hydroxylase and peptidylglycine α-hydroxylating monooxygenase. Ascorbate reduces the copper sites to the catalytically active Cu^+ form [f]. A is also required for the regeneration of Fe^{2+} in hydroxylation [g] or demethylation [h] reactions of different substrates, catalyzed by Fe(II)/2OG-dependent dioxygenases.

are localized in the cytosol. They hydroxylate HIF1α, a key transcriptional factor in the regulation of hypoxic response together with FIH (factor inhibiting HIF1), an aspartate hydroxylase. The outcome of the hydroxylation is the proteasomal degradation of HIF1α and thus the cutoff of the transcription of HIF1α-regulated genes [39].

ε-N-trimethyl-L-lysine hydroxylase and β-butyrobetaine hydroxylase, enzymes necessary for synthesis of carnitine, seem to be localized to the mitochondria and cytosol. Carnitine is essential for the transport of fatty acids into mitochondria for β-oxidation and consequent ATP generation [66]. However, the subcellular localization of 4-hydroxyphenylpyruvate dioxygenase, the enzyme that participates in the catabolism of tyrosine, is less known. In the nucleoplasm Fe(II)/2-oxoglutarate–dependent dioxygenases are involved in the epigenetic regulation of gene expression by demethylating histones and DNA [55]. Enzymes that demethylate histones mainly belong to the Jumonji protein family conserved from yeast to humans with a common jmjC functional domain [82]. DNA demethylation occurs at the methyl group of 5-methylcytosine via subsequent oxidative steps catalyzed by the dioxygenases of the ten-eleven translocation (TET) family [35,45].

Copper Type II Ascorbate-Dependent Monooxygenases

Dopamine β-monooxygenase and peptidylglycine α-hydroxylating monooxygenase are localized in the chromaffin granules, synaptic and secretory vesicles. They are involved in catecholamine and peptide hormone biosynthesis, respectively [21,51,63].

SUBCELLULAR TRANSPORT OF VITAMIN C

Since ascorbate-dependent reactions are present in many subcellular compartments and ascorbate and its oxidized products are charged and water-soluble molecules, transporters must be present in intracellular membranes to keep the optimal organellar concentrations. AA/DHA transporters have been described and characterized initially in the plasma membrane [19]. Ascorbate can be taken up and intracellularly accumulates by secondary active transport mediated by sodium-dependent vitamin C transporters SVCT1 (*SLC23A1*) and SVCT2 (*SLC23A2*). While the expression of SVCT2 is relatively

widespread, protecting metabolically active tissues from oxidative stress or satisfying special metabolic demands, SVCT1 is expressed in epithelial tissues in intestine and kidney and is involved in dietary absorption and renal reabsorption of ascorbate, contributing to the maintenance of whole-body vitamin C levels [15,86].

DHA is taken up by facilitated diffusion mediated by the members of the mammalian facilitative hexose transporter GLUT family [84]. Although under physiologic conditions DHA transport can be less significant since DHA is present at low concentrations in the extracellular space, GLUT-dependent uptake becomes more intense in oxidative stress and can be stimulated by extracellular oxidants or oxidase activities. The DHA-transporting capabilities of GLUT transporters have not been systematically investigated, but it is clear that many—if not all—members of the GLUT family (e.g., GLUT1-4) are able to mediate DHA transport in the plasma membrane [11].

While the presence of these transporters in the plasma membrane is well documented and their participation in vitamin C transport is more or less characterized, their distribution and functioning in the endomembranes remained an enigma for long time. Recent observations shed some light on their role in the intracellular transport of vitamin C.

SVCTs in the Intracellular Membranes

SVCTs require a sodium gradient for their transport activity; this condition has been seldom reported in case of the intracellular membrane. The sole notable exception is the mitochondrial inner membrane, where a proton/sodium exchanger (NHE) promotes sodium efflux and generates an inward gradient [38]. In line with this observation, sodium-dependent ascorbate transport mediated by SVCT2 has been reported in mitochondria [6,57]. These experimental observations are also supported by computational prediction tools [75]. Mitochondrial SVCT2 is active at low millimolar sodium concentration, which allows transport under intracellular conditions [28] and ensures ascorbate accumulation in the mitochondrial matrix—a compartment that requires antioxidant protection [16].

Recently, a new mitochondrial ascorbate transport activity has been described and partially characterized [70]. The protein with a molecular mass of 28–35 kDa catalyzes saturable, temperature- and proton-gradient-dependent, unidirectional ascorbate transport. This proton/ascorbate symporter shows biochemical features different from those of GLUTs or SVCTs.

GLUT Family Members in the Intracellular Membranes

GLUT transporters are primarily glucose transporters in the plasma membrane [56]. They reach their final destination via the secretory pathway; thus, their presence at different compartments can be simply explained by traveling along the pathway [80]. The appearance of GLUTs in the mitochondrial membranes observed by some studies is more surprising.

The protein family of facilitative glucose transporters (SLC2A, GLUT) comprises 14 isoforms in the human proteome. Based on their sequence homology, three classes can be distinguished. The characterization of the more recently described class II and III isoforms revealed that despite their structural similarities, they show a distinct tissue-specific expression pattern, specific subcellular localization, and affinities for a variety of ligands other than glucose.

At least six members of the GLUT family (GLUT1-4, -8, -10) have been recognized as DHA transporters [18,46,53,61,68,69]. Half of them have been reported in the internal membranes: GLUT1 in the Golgi apparatus [60] and the mitochondrial inner membrane [42], GLUT8 in the lysosomes [3,26], and GLUT10 in the nuclear envelope [20], in the mitochondrial inner membrane [46], in the Golgi apparatus [46], and in the ER [30,71]. Thus, vitamin C can be transported into many intracellular compartments in the form of DHA, with the mediation of GLUT transporters [8].

Mitochondrial Vitamin C Transport

The early observation of Ingebretsen and Normann [34] that ascorbate concentration in mammalian mitochondria could be increased by dietary vitamin C supplementation suggested the existence of vitamin C transport through the mitochondrial inner membrane. Later this observation was reinforced by other in vivo studies [48,65]. Mitochondria and the mitochondrial electron transfer chain (ETC) play a more than important role in both the biosynthesis and the recycling of vitamin C in plant cells [12,78]; thus, it is not surprising that mitochondrial vitamin C transport was first characterized in plant cells [77].

This previous finding that vitamin C can enter plant mitochondria as DHA [77] was confirmed later by Golde and coworkers [42] in animal cells. The detailed characterization of the transport shed light to many features. The stereoselective mitochondrial D-glucose uptake that competed with the uptake of DHA suggested the involvement of glucose transporters, especially the members of the GLUT family. Indeed, the in silico analysis of the N-terminal sequences of human GLUT isoforms gave high probability for the mitochondrial localization of GLUT1.

The localization predicted by computational analysis was verified by the expression of GLUT1 green fluorescent protein, immunoblot analysis, and cellular immunolocalization [42]. At this point, all the observations suggest that vitamin C enters mitochondria in its oxidized form by the mediation of GLUT1 in a similar fashion to its cellular uptake and to the plant mitochondrial analog. Five years later the mitochondrial localization of another member of the GLUT family, GLUT10, was also reported in mouse aortic smooth muscle cells and insulin-stimulated adipocytes [46]. GLUT10 could elevate the mitochondrial uptake of labeled DHA in H_2O_2-treated smooth muscle cells that resulted in lower reactive oxygen species (ROS) levels. This protecting effect of vitamin C could be suspended by glucose pretreatment or by RNA interference of GLUT10 mRNA expression in smooth muscle cells [46].

The reduced form, ascorbate, was ruled out as the transported form of vitamin C through the inner mitochondrial membrane since the mitochondrial uptake and accumulation of this form could not be observed in mitochondria from both human kidney cells or from rat liver tissue [42,48]. Until 2013 DHA was considered to be the transported form of vitamin C, and GLUT transporters were considered solely to be the transporters responsible for vitamin C transport through the inner mitochondrial membrane. However, 6 years ago, Cantoni and coworkers showed the mitochondrial expression of the sodium-dependent ascorbic acid transporter, SVCT2, in U937 human myeloid leukemia cells [6,32]. This observation raised the possibility of the transport of the reduced form, ascorbate, through the inner mitochondrial membrane. A year later this possibility was strongly supported by further experimental results. The relevance of the transport of ascorbate and the existence of SVCT2 in the mitochondria was confirmed by confocal colocalization experiments and immunoblotting of proteins originated from highly purified mitochondrial fractions of HEK-293 cells [57]. All the investigated 16 different human cell lines (including normal, neoplastic, and primary cells) showed similar localization patterns proving the generality of the mitochondrial localization of SVCT2. The mitochondrial protein abundance of SVCT2 decreased by at least 75% due to the silencing of SVCT2. According to the observed lower mitochondrial abundance of SVCT2, the capacity of the mitochondrial ascorbate transport was only one-fourth of the control cells [57]. Hence, the functionality of mitochondrial SVCT2 transporters was also demonstrated. Interestingly, no GLUT10 expression could be found in HEK-293 cells, and the mitochondrial expression of GLUT1 could not be reinforced in the same experiments [57]. Thus, the role of GLUTs in the mitochondrial vitamin C transport was queried.

According to these controversial results and observations, our picture of mitochondrial vitamin C transport had become quite obscure by the spring of 2014. The quasi-solid role of GLUTs and the transport of DHA seemed to be queried, and the role of the initially excluded SVCTs and the transport of ascorbate in mitochondrial vitamin C transport seemed to be strengthened. At this point, our group enlisted the aid of in silico prediction tools to clarify the in vitro and in vivo results. The possibility of mitochondrial localization for all the members of the GLUT and SVCT families has been investigated by seven different in silico prediction tools. These computational predictions rely only on the protein sequence. They can score the likelihood that a protein belongs to a given compartment. The chosen prediction tools used different algorithms such as neural networks, a linear combination of a number of sequence characteristics such as amino-acid abundance, maximum hydrophobicity, and maximum hydrophobic moment, the k-nearest-neighbor method, a support vector machine-based approach, an n-gram-based Bayesian classifier, or gene ontology to give a prediction score that a protein belongs to a given compartment.

Six out of the seven prediction tools gave a good chance for the mitochondrial localization of GLUT1. The high probability of its mitochondrial localization became more evident if its mitochondrial probability scores were compared to the scores of the other GLUTs and to those of dicarboxylate carrier, a protein with well-known mitochondrial localization [75]. In this way, our in silico results reinforced the GLUT1-EGFP and immunoblot-based findings of KC et al. on the mitochondrial localization of GLUT1 [42]. However, the scores we got for the mitochondrial localization of GLUT10 were by far the lowest out of all the investigated transport proteins that really query the role of GLUT10 as a mitochondrial vitamin C transporter [75]. On the contrary, the mitochondrial localization of SVCT2 reached high scores, in concordance with the confocal colocalization, immunoblotting, and SVCT2 silencing experiments of the earlier mentioned study of Muñoz-Montesino et al. [57]. It is noteworthy that GLUT9 and GLUT11 also scored highly for mitochondrial localization. Unfortunately, the DHA transport ability of these transporters has not yet been investigated. Thus, their role in mitochondrial vitamin C transport—based on these in silico results—cannot be confirmed or ruled out.

To clarify these questions around mitochondrial vitamin C transport, an enriched hydroxyapatite fraction from highly purified rat liver and potato mitochondria was reconstituted in proteoliposomes, and a functional ascorbic acid transporter was found in them [70]. Two main protein bands in the range of 28–35 kDa were found both in rat and potato hydroxyapatite eluates. The presence of SVCT2 or GLUTs was excluded, since they can be characterized by higher molecular masses ranges between 55 and 60 kDa. The protein showed unidirectional transport and could be stimulated by the generation of an inwardly directed proton gradient. In contrast to GLUT1 and SVCT2, which has a pH optimum at pH 7.5–8 [28,83], it has a pH optimum between 6.5 and 7 [70]. The rat and potato proteins showed similar inhibitor sensitivity, and both were highly sensitive to only the lysine blocking reagent pyridoxal phosphate, while neither of them was affected by typical SVCT2 or GLUT inhibitors. Contrary to the ascorbate uptake mediated by SVCTs, ascorbate uptake into proteoliposomes, reconstituted with both potato and rat liver mitochondrial extract, was not influenced by the presence of sodium and can be characterized by much higher K_M (8.4–60 μM versus 1.5 mM) [11,28].

All of these results, the sodium independency, the different molecular weight, pH dependence, inhibitor profile, and K_M, suggest that the observed ascorbate transport in proteoliposomes is not mediated by GLUT1 or SVCT2 [70]. The resulting proposal is that beyond GLUT1 and SVCT2, there is a third potential mitochondrial transport protein. The typical ascorbate concentration in mammalian cells is 0.1–0.8 mM [73,74]. The ascorbate concentration of the human plasma does not reach the concentration of 0.1 mM [47]. The recently described mitochondrial ascorbate transporter can be characterized by 1.5 millimolar K_M values, the SVCT2 by 8.4–60 μM, and the GLUT transporters, which transport DHA, by millimolar K_M values. None of the transporters can be ruled out from having a role in mitochondrial vitamin C transport on the basis of the proposed intracellular concentrations and the K_M values of the transports, per se. Therefore, the picture on mitochondrial vitamin C transport (and transporters) becomes even more obscure.

VITAMIN C IN MITOCHONDRIA

Mitochondrial Reactive Oxygen Species Generation and Elimination

There is a strong relationship between ROS generation, mitochondrial membrane potential ($\Delta\psi m$), and the activity of the complexes of the electron transfer chain [85]. Any block in the electron transfer chain can lead to an increase in mitochondrial ROS (mROS) generation, while the drop in $\Delta\psi m$ stimulated by uncoupling can lead to a reduced rate of free radical production. Reasonably, the hyperpolarization of mitochondria leads to enhanced generation of ROS [1]. On one hand, the generation of mROS could play an essential role in signaling pathways, since ROS can behave as secondary messengers [67]. On the other hand, the level of ROS must be kept below a certain level, otherwise they can cause severe cytotoxicity [67,81]. This nontoxic level of ROS for their signaling role can be achieved by the fine-tuning of ROS production and scavenging pathways [67].

There are two different ways to regulate the amount of mROS in the cells: first by the regulation of their production, and second by scavenging them via different antioxidant mechanisms [22,23,67]. The first way to influence mROS production is to cause or prevent over-reduction of the electron transfer chain. The over-reduction of the electron transfer chain can be avoided, for instance, by the activation of uncoupling

proteins (UCPs) [59]. In case of the malfunction of the first way, electrons can escape from the electron transfer chain and generate free radicals, predominantly superoxide radical [1,67]. At this point, the antioxidant system has an important role as the second means of ROS regulation. This defense mechanism has both enzymatic and nonenzymatic components. Thanks to the rapid spontaneous or enzymatic dismutation of superoxide anion to hydrogen peroxide (H_2O_2) by manganese superoxide dismutase (SOD) in the mitochondrial matrix or by copper-zinc-SOD in the intermembrane space, its lifetime is very short, being not longer than 1 ns [1]. The result of the dismutation of superoxide anion is that the membrane-permeable H_2O_2 can participate in signaling cascades or be degraded by the enzymes catalase, glutathione peroxidase, and peroxiredoxin 3 [29]. As mentioned, above a limit H_2O_2 is dangerous for cells, since it can readily react with transition metals in the Fenton reaction, to form the highly reactive, cytotoxic hydroxyl radical [81]. Beyond the mentioned enzymatic elements of the antioxidant system, the small molecular weight nonenzymatic elements also play an important role in antioxidant defense. The two major water-soluble antioxidants of the cells are the glutathione and the ascorbate.

Vitamin C in Mitochondrial Antioxidant Defense

It is well documented that elevated levels of ROS can induce the collapse of $\Delta\psi m$ that leads to cell death [14]. The decrease of $\Delta\psi m$ due to H_2O_2 treatment could be partially mitigated, and the mitochondrial release of cytochrome c could also be avoided by DHA pretreatment in HL-60 cells [31], in monocytes after FAS-induced apoptosis [64], or in cells undergoing hypoxia-reperfusion [25]. The exposure of U937 leukemia cells to arsenite caused the formation of superoxide with mitochondrial origin and the consequent inhibition of the superoxide sensitive enzyme, aconitase. The cells died by a mixed autophagic and apoptotic mechanism. All the toxic effects mediated by superoxide were entirely dependent on its conversion to H_2O_2 and could be suspended by low micromolar ascorbate treatment [32,33]. The known persistent organic pollutant and liver damage toxicant dichloro-diphenoxy-trichloroethane (DDT) induced the death of HL-7702 cells. DDT treatment caused mitochondrial hyperpolarization, the consequent enhancement of ROS generation, and resulted in the release of cytochrome c into the cytosol. Other apoptotic markers such as the subsequent elevations of Bax and p53, along with suppression of Bcl-2, the activations of caspase-3 and -8 were triggered due to the DDT treatment. The treatment of the cells with vitamin C and vitamin E remarkably suppressed the mitochondrial damage and ROS generation.

Vitamin C treatment could also mitigate the caspase activation and the mitochondrial release of cytochrome c, demonstrating the antiapoptotic and mitochondrial protective role of vitamin C in DDT poison [37]. Similarly, vitamin C treatment of H4 human neuroglioma cells overexpressing human amyloid precursor protein and rat neuroblastoma cells could attenuate the anesthetic isoflurane-induced caspase-3 activation. Furthermore, vitamin C treatment could mitigate the isoflurane-induced elevation of ROS levels, opening of the mitochondrial permeability transition pore, reduction of $\Delta\psi$m, and reduced ATP levels in the cells. The in vitro results could be supported also at an in vivo level since vitamin C ameliorated the isoflurane-induced cognitive impairment in mice via the inhibition of oxidative stress, mitochondrial dysfunction, and reduction in ATP levels [17]. All these observations suggest that mitochondrial ascorbate plays an important role in the preservation of $\Delta\psi$m in different stresses. Vitamin C exerts its mitochondrial protective, antiapoptotic effect through ROS scavenging [17,25,31,37,48,64].

The mitochondrial DNA in the lack of classic protective histones and repair mechanism is more susceptible to different stresses such as ROS. It was undoubtedly shown that ascorbic acid protected mtDNA against the ROS-induced elevation of 8-oxo-dGuanidine and apurinic/apyrimidinic sites. The ascorbic acid preload of the cells significantly mitigated the H_2O_2-induced shearing of mtDNA [48]. These observations were further strengthened by the findings of Jarrett et al. that vitamin C could significantly diminish the H_2O_2-induced lesions in mtDNA in retinal pigment epithelium cells [36].

In a further study that focused on the suppressive effect of ascorbic acid on changes in mitochondrial function and mutagenesis by the radiation-induced bystander effect, the authors found that increases in ROS generation led to the induction of gene mutations. It was observed that the mitochondrial modulations and the mutation inductions by the radiation-induced bystander effect were completely suppressed by treatment with vitamin C in cells treated with conditioned medium, suggesting that mutagenesis induced by mROS can be relieved by ascorbic acid [40,41]. All these findings suggest that vitamin C is predominantly utilized as an antioxidant in the mitochondria.

Vitamin C and Mitochondrial Electron Transfer

Vitamin C is linked to the mitochondrial electron transfer by the means of its synthesis and recycling in plant cells and by the means of its recycling in animal and human cells. The ultimate enzyme of the main ascorbate biosynthetic pathway in plant cells is the L-galactono-1,4-lactone dehydrogenase (GLDH). GLDH delivers electrons to oxidized cytochrome c by oxidizing L-galactono-1,4-lactone to ascorbate. The enzyme shows an absolute requirement for oxidized cytochrome c; therefore, L-galactono-1,4-lactone can be defined as an alternative respiratory substrate in plants cells. Furthermore, GLDH binds to the 420, 470, and 850 kDa complex I assembly intermediates and has an important role in the assembly of complex I [76]. The oxidized form of vitamin C, DHA can reach the mitochondrial matrix both in plant and animal cells by the mediation of one of the transport mechanisms described in section Subcellular transport of vitamin C. Since DHA is very unstable and only ascorbate possesses antioxidant and free radical scavenger properties, DHA taken up or generated in the matrix must be recycled (reduced to ascorbate), otherwise it is lost within minutes under physiologic circumstances [87].

The α-lipoic acid- and thioredoxin reductase-dependent means of mitochondrial DHA reduction are definitely not the major components of the mitochondrial DHA reduction machinery in animal cells [48,88]. However, the loading of mitochondria causes significant decrease in the mitochondrial GSH content, and the depletion of mitochondrial GSH causes significant impairment in DHA reduction; thus, it is likely that GSH-dependent reduction of DHA is one of the major ascorbate recycling pathways in mammalian mitochondria [48].

However, there is another vitamin C recycling pathway that couples the recycling directly to the electron transfer chain. Experiments taken on liver mitochondrial fraction showed that in the absence of respiratory substrates, the organelle loses its ability to maintain or recycle the level of ascorbate [48,49,54], suggesting that the electron transfer chain might contribute to the reduction of DHA. At the same time, respiratory substrates such as succinate, malate, and glycerol-3-phosphate could enhance ascorbate recycling from DHA. Using specific inhibitors of the mitochondrial electron transport chain, the site of ascorbate sparing was localized to complex III [48]. DHA reduction could be significantly enhanced by succinate addition both in mammalian [54] and plant mitochondria [78] that underline the universal role of the electron transfer chain in DHA reduction.

The close relationship of the electron transfer chain and DHA reduction can contribute to the stabilization of mitochondrial redox balance in at least two ways: First, the possible over-reduction of the elements of the electron transfer chain can be alleviated by the donation of electrons to DHA reduction. This way, DHA reduction functions as an electron safety valve, and the electron leakage of the electron transfer chain and subsequent ROS formation can also be mitigated. Second, the recycling of oxidized vitamin C provides ascorbate, the reduced form that may scavenge ROS directly at the site of their generation.

A special case of the over-reduction of the electron transfer chain occurs when one or more complex(es) does/do not work properly due to mutation(s) in its/their gene(s). Because of the earlier arguments, the consequences of the impairment of electron transfer chain elements might be mitigated by ascorbate therapy. The enhanced generation of superoxide—observed in fibroblasts from patients with electron transport chain deficiencies—could be decreased by ascorbate treatment [72]. The activities of I–III and II–III complexes could simultaneously be stimulated by vitamin C supplementation of the cell culture media [72]. There were attempts to use the ascorbate-DHA redox couple and vitamin K as a therapeutic agent in the therapy of mitochondrial diseases. A young woman with mitochondrial myopathy and severe exercise intolerance was treated by vitamin C and menadione to bypass the block in complex III [43,44], because the redox potentials of these electron carriers fit the gap created by the cytochrome c dysfunction [27]. Unfortunately, after an initial improvement documented by ^{31}P nuclear magnetic resonance spectroscopy of muscle [2], this state was not sustained. (Menadione had to be discontinued because of its withdrawal by the U.S. Food and Drug Administration [43].) Except for this case, effective vitamin C therapy in other patients suffering from mitochondrial diseases has not been reported.

ASCORBATE COMPARTMENTALIZATION AND HUMAN DISEASES

Scurvy and Subcellular Scurvy

Scurvy, the generalized shortage of ascorbate, has been known for centuries, although the pathologic mechanism remained enigmatic until the discovery of ascorbate [79]. However, recent observations show that not only global but also local ascorbate concentrations are important with respect to human health. The dramatic differences in "vitamin C proteome," that is, enzymes and transporters utilizing the different redox forms of vitamin C, between organelles suggest that subcellular ascorbate compartmentalization might also be a basis of human diseases. Subcellular scurvy could be generated in a murine model. The authors investigated the effect of the combined loss-of-function mutations of genes encoding the main electron transfer proteins participating in oxidative protein folding in the ER (ERO1α, ERO1β, and peroxiredoxin 4). Surprisingly, the maneuver resulted in a minor impairment of disulfide bond formation. The observation suggested that an alternative electron transfer pathway exists; the involvement of ascorbate was hypothesized since low tissue concentrations were detected. It was found that in the absence of the normal constituents of the electron transfer chain, cysteinyl groups were oxidized to sulfenic acids. In the ER, lumen sulfenic acid and prolyl/lysyl hydroxylases competed for ascorbate, and thus, normal collagen hydroxylation was inhibited. Although mice are able to synthesize vitamin C, the increased local consumption overcame the capacity of synthesis and transport [89,90].

Werner Syndrome

The shortage of ascorbate generated by ER luminal processes seems to also be present in the background of a human disease. Werner syndrome is a premature aging disorder caused by mutations in a RecQ-like DNA helicase (Wrn).

Ascorbate supplementation proved to be beneficial in different models of the disease: in WS fibroblast [41], in Wrn mutant mice [4], and in wrn-1 *Caenorhabditis elegans* [24]. Mutant mice lacking part of the helicase domain of the Wrn ortholog exhibit several phenotypic features of Werner syndrome. Moreover, mislocalization of the Wrn mutant protein to the ER fraction was observed with increased oxidative stress and activation of the ER stress response [4]. When ascorbate synthesis of the mutant mice was abolished by knocking out the gulonolactone oxidase gene, double-mutant mice exhibited small size, sterility, osteopenia, and a severe reduction in life span. High doses of ascorbate improved the phenotype [5]. The results suggest that the ER-mislocalized Wrn protein increases the consumption or inhibits the uptake of ascorbate and DHA, activating the organellar stress. Further studies are needed to explore these possibilities.

Arterial Tortuosity Syndrome

A further possible human example of ascorbate compartmentalization disease is arterial tortuosity syndrome (ATS). This rare congenital connective tissue disorder manifested in the tortuosity of main arteries and thoracic aneurysm formation. The genetic background is the mutation of the GLUT10 gene [20]. The GLUT10 protein has been reported to be present in the endomembranes (ER and Golgi) and mitochondrial membranes [20,30,46,71]. GLUT10 is able to transport DHA, as it was demonstrated in mitochondria [46], in plasma membrane permeabilized fibroblasts, and in reconstituted GLUT10-containing proteoliposomes [61]. The absence of GLUT10 in the ER might explain the alterations of the extracellular matrix due to impaired hydroxylation of proteins in ATS [71]. Its missing function in the nuclear envelope can hamper the nuclear import of vitamin C. A recent report showed that the ascorbate concentration was lower in the nucleus of ATS fibroblasts. Consequently, the global methylation/hydroxymethylation pattern of DNA was also changed, and gene-specific alterations were also found [62]. The findings raise the possibility that epigenetic effects contribute to the pathogenesis as well. In summary, the different hypotheses on ATS pathogenesis are uniform, postulating altered DHA transport into subcellular compartments.

Five years ago, we introduced the term *ascorbate compartmentalization disease* [11], referring to conditions where subcellular distribution of ascorbate is perturbed. Although some results of the last years support the existence of such diseases, the systematic investigation of subcellular vitamin C concentration and redox state is still lacking. Furthermore, analyzing the vitamin C–dependent proteome—under both normal and pathologic conditions—is an additional aim to understand the role of vitamin C compartmentalization in health and disease.

ASCORBATE IN THE ORCHESTRATION OF METABOLISM

Ascorbate-dependent enzymes show themselves in various locations and in different branches of the metabolism. The question arises whether ascorbate plays a role of *maestro* in these processes. Although ascorbate-dependent transcriptional and epigenetic gene regulation has not been widely explored, the metabolic patterns in ascorbate-rich and ascorbate-poor conditions show some consistency. In the presence of normal or high ascorbate supply, oxidative metabolism (e.g., that includes the β-oxidation of fatty acids, and the hydroxylation of methyl groups in histone and DNA) is favored. It should be noted that ascorbate is also a product of oxidative metabolism in vitamin C–synthesizing species. In contrast, in conditions of ascorbate shortage, anaerobic metabolism dominates with increased glucose uptake and glycolysis. The stability of the extracellular matrix (ECM) is also challenged by decreased posttranslational modification of ECM proteins and by stimulated angiogenesis. Thus, ascorbate-poor conditions overlap with requirements for cancer invasiveness.

In summary, there are several characteristics of vitamin C that make it appropriate to play a special role in the integration of intermediary metabolism in the cells at the organelle level.

1. The unique redox moieties (e.g., pro- and antioxidant) of ascorbate/dehydroascorbate redox couple among the various redox active compounds.

2. The coenzyme role of ascorbate in various enzyme complexes, which have either essential regulatory (e.g., HIF) or also structural (e.g., collagen) functions.

3. The different regulations of concentrations of the reduced and oxidized form of the redox pair, ascorbate versus dehydroascorbate. It is noted that the latter is kept low, compared to ascorbate, which might suggest a possible regulatory role of dehydroascorbate. Recent data on their indirect impact on gene expression confirm these speculations.

4. The multiorganelle aspects of processes, where ascorbate is involved (e.g., protein folding).

5. The characteristic role of ascorbate in various organelle functions, and the functions of ascorbate transporters in them.

REFERENCES

1. Angelova, P.R. and Abramov, A.Y. 2018. Role of mitochondrial ROS in the brain: From physiology to neurodegeneration. *FEBS Lett.* 592, 692–702.

2. Argov, Z., Bank, W.J., Maris, J., Eleff, S., Kennaway, N.G., Olson, R.E. and Chance, B. 1986. Treatment of mitochondrial myopathy due to complex III deficiency with vitamins K3 and C: A 31P-NMR follow-up study. *Ann. Neurol.* 19, 598–602.

3. Augustin, R., Riley, J. and Moley, K.H. 2005. GLUT8 contains a [DE]XXXL[LI] sorting motif and localizes to a late endosomal/lysosomal compartment. *Traffic* 6, 1196–1212.

4. Aumailley, L., Dubois, M.J., Garand, C., Marette, A. and Lebel, M. 2015. Impact of vitamin C on the cardiometabolic and inflammatory profiles of mice lacking a functional Werner syndrome protein helicase. *Exp. Gerontol.* 72, 192–203.

5. Aumailley, L., Dubois, M.J., Brennan, T.A., Garand, C., Paquet, E.R., Pignolo, R.J., Marette, A. and Lebel, M. 2018. Serum vitamin C levels modulate the lifespan and endoplasmic reticulum stress response pathways in mice synthesizing a nonfunctional mutant WRN protein. *FASEB J.* 32, 3623–3640.

6. Azzolini, C., Fiorani, M., Cerioni, L., Guidarelli, A. and Cantoni, O. 2013. Sodium-dependent transport of ascorbic acid in U937 cell mitochondria. *IUBMB Life.* 65, 149–153.

7. Bánhegyi, G., Braun, L., Csala, M., Puskás, F. and Mandl, J. 1997. Ascorbate metabolism and its regulation in animals. *Free Radic. Biol. Med.* 23, 793–803.

8. Bánhegyi, G., Marcolongo, P., Puskás, F., Fulceri, R., Mandl, J. and Benedetti A. 1998. Dehydroascorbate and ascorbate transport in rat liver microsomal vesicles. *J. Biol. Chem.* 273, 2758–2762.

9. Bánhegyi, G. and Mandl, J. 2001. The hepatic glycogenoreticular system (minireview). *Pathol. Oncol. Res.* 7, 107–110.

10. Bánhegyi G., Csala, M., Szarka, A., Varsányi, M., Benedetti, A. and Mandl, J. 2003. Role of ascorbate in oxidative protein folding. *Biofactors* 17, 37–46.

11. Bánhegyi, G., Benedetti, A., Margittai, E., Marcolongo, P., Fulceri, R., Németh, C. E. and Szarka A. 2014. Subcellular compartmentation of ascorbate and its variation in disease states. *Biochim. Biophys. Acta* 1843, 1909–1916.

12. Bartoli, C.G., Pastori, G.M. and Foyer, C.H. 2000. Ascorbate biosynthesis in mitochondria is linked to the electron transport chain between complexes III and IV. *Plant. Physiol.* 123, 335–344.

13. Braun, L., Garzó, T., Mandl, J. and Bánhegyi, G. 1994. Ascorbic acid synthesis is stimulated by enhanced glycogenolysis in murine liver. *FEBS Lett.* 352, 4–6.

14. Burke P.J. 2017. Mitochondria, bioenergetics and apoptosis in cancer. *Trends Cancer.* 3, 857–870.

15. Bürzle, M., Suzuki, Y., Ackermann D., Miyazaki, H., Maeda, N., Clémençon, B., Burrier, R. and Hediger, M.A. 2013. The sodium-dependent ascorbic acid transporter family SLC23. *Mol. Aspects Med.* 34, 436–454.

16. Cantoni, O., Guidarelli, A. and Fiorani, M. 2018. Mitochondrial uptake and accumulation of Vitamin C: What can we learn from cell culture studies? *Antioxid. Redox Signal.* 29, 1502–1515.

17. Cheng, B., Zhang, Y., Wang, A., Dong, Y. and Xie, Z. 2015. Vitamin C attenuates isoflurane-induced Caspase-3 activation and cognitive impairment. *Mol Neurobiol.* 52, 1580–1589.

18. Corpe, C.P., Eck, P., Wang J., Al-Hasani, H. and Levine, M. 2013. Intestinal dehydroascorbic acid (DHA) transport mediated by the facilitative sugar transporters, GLUT2 and GLUT8. *J. Biol. Chem.* 288, 9092–9101.

19. Corti, A., Casini, A.F. and Pompella, A. 2010. Cellular pathways for transport and efflux of ascorbate and dehydroascorbate. *Arch. Biochem. Biophys.* 500, 107–115.

20. Coucke, P.J., Willaert, A., Wessels, M.W., Callewaert, B., Zoppi, N., De Backer, J., Fox, J.E. et al. 2006. Mutations in the facilitative glucose

transporter GLUT10 alter angiogenesis and cause arterial tortuosity syndrome. *Nat. Genet.* 38, 452–457.

21. Crivellato, E., Nico, B. and Ribatti, D. 2008. The chromaffin vesicle: Advances in understanding the composition of a versatile, multifunctional secretory organelle. *Anat. Rec. (Hoboken)* 291, 1587–1602.

22. Czobor, Á., Hajdinák, P. and Szarka, A. 2017. Rapid ascorbate response to bacterial elicitor treatment in Arabidopsis thaliana cells. *Acta Physiol. Plant* 39, 62.

23. Czobor, Á., Hajdinák, P., Németh, B., Piros, B., Németh, Á. and Szarka, A. 2019. Comparison of the response of alternative oxidase and uncoupling proteins to bacterial elicitor induced oxidative burst. *PLOS ONE* 14(1), e0210592.

24. Dallaire, A., Proulx, S., Simard, M.J. and Lebel, M. 2014. Expression profile of *Caenorhabditis elegans* mutant for the Werner syndrome gene ortholog reveals the impact of vitamin C on development to increase life span. BMC *Genomics.* 15, 940.

25. Dhar-Mascareño, M., Cárcamo, J.M. and Golde, D.W. 2005. Hypoxia-reoxygenation-induced mitochondrial damage and apoptosis in human endothelial cells are inhibited by vitamin C. *Free Radic. Biol. Med.* 38, 1311–1322.

26. Diril, M.K., Schmidt, S., Krauss M., Gawlik, V., Joost, H.G., Schürmann, A., Haucke, V. and Augustin R. 2009. Lysosomal localization of GLUT8 in the testis—The EXXXLL motif of GLUT8 is sufficient for its intracellular sorting via AP1- and AP2-mediated interaction. *FEBS J.* 276, 3729–3743.

27. Eleff, S., Kennaway, N.G., Buist, N.R., Darley-Usmar, V.M., Capaldi, R.A., Bank, W.J. and Chance, B. 1984. 31P NMR study of improvement in oxidative phosphorylation by vitamins K3 and C in a patient with a defect in electron transport at complex III in skeletal muscle. *Proc. Natl. Acad. Sci. USA* 81, 3529–3533.

28. Fiorani, M., Azzolini, C., Cerioni, L., Scotti, M., Guidarelli, A., Ciacci, C. and Cantoni O. 2015. The mitochondrial transporter of ascorbic acid functions with high affinity in the presence of low millimolar concentrations of sodium and in the absence of calcium and magnesium. *Biochim. Biophys. Acta* 1848, 1393–1401.

29. Forman, H.J., Maiorino, M. and Ursini, F. 2010. Signaling functions of reactive oxygen species. *Biochemistry* 49, 835–842.

30. Gamberucci, A., Marcolongo, P., Németh, C.E., Zoppi, N., Szarka, A., Chiarelli, N., Hegedűs, T. et al. 2017. GLUT10—Lacking in arterial tortuosity syndrome—Is localized to the endoplasmic reticulum of human fibroblasts. *Int. J. Mol. Sci.* 18(8). pii: E1820.

31. Gruss-Fischer, T. and Fabian, I. 2002. Protection by ascorbic acid from denaturation and release of cytochrome c, alteration of mitochondrial membrane potential and activation of multiple caspases induced by H_2O_2, in human leukemia cells. *Biochem. Pharmacol.* 63, 1325–1335.

32. Guidarelli, A., Fiorani, M., Azzolini, C., Cerioni, L., Scotti, M. and Cantoni, O. 2015. U937 cell apoptosis induced by arsenite is prevented by low concentrations of mitochondrial ascorbic acid with hardly any effect mediated by the cytosolic fraction of the vitamin. *Biofactors* 41, 101–110.

33. Guidarelli, A., Carloni, S., Balduini, W., Fiorani, M. and Cantoni, O. 2016. Mitochondrial ascorbic acid prevents mitochondrial O_2^--formation, an event critical for U937 cell apoptosis induced by arsenite through both autophagic-dependent and independent mechanisms. *Biofactors.* 42, 190–200.

34. Ingebretsen, O.C. and Normann, P.T. 1982. Transport of ascorbate into guinea pig liver mitochondria. *Biochim. Biophys. Acta* 684, 21–26.

35. Ito, S., Shen, L., Dai, Q., Wu S.C., Collins, L.B., Swenberg, J.A., He, C. and Zhang, Y. 2011. Tet proteins can convert 5-methylcytosine to 5-formylcytosine and 5-carboxylcytosine. *Science* 333, 1300–1303.

36. Jarrett, S.G., Cuenco, J. and Boulton, M. 2006. Dietary antioxidants provide differential subcellular protection in epithelial cells. *Redox Rep.* 11, 144–152.

37. Jin, X., Song, L., Liu, X., Chen, M., Li, Z., Cheng, L. and Ren, H. 2014. Protective efficacy of vitamins C and E on p,p′-DDT-induced cytotoxicity via the ROS-mediated mitochondrial pathway and NF-κB/FasL pathway. *PLOS ONE* 9(12), e113257.

38. Jung, D.W., Apel, L.M. and Brierley, G.P. 1992. Transmembrane gradients of free Na^+ in isolated heart mitochondria estimated using a fluorescent probe. *Am. J. Physiol.* 262, C1047–C1055.

39. Kaelin, W.G. Jr. and Ratcliffe, P.J. 2008. Oxygen sensing by metazoans: The central role of the HIF hydroxylase pathway. *Mol. Cell* 30, 393–402.

40. Kashino, G., Tamari, Y., Kumagai, J., Tano, K. and Watanabe, M. 2013. Suppressive effect of ascorbic acid on the mutagenesis induced by the bystander effect through mitochondrial function. *Free Radic. Res.* 47, 474–479.

41. Kashino, G., Kodama, S., Nakayama Y., Suzuki, K., Fukase, K., Goto, M. and Watanabe, M. 2003. Relief of oxidative stress by ascorbic acid delays cellular senescence of normal human and Werner syndrome fibroblast cells. *Free Rad. Biol. Med.* 35, 438–443.

42. KC, S., Cárcamo, J.M. and Golde, D.W. 2005. Vitamin C enters mitochondria via facilitative glucose transporter 1 (Glut1) and confers mitochondrial protection against oxidative injury. *FASEB J.* 19, 1657–1667.

43. Keightley, J.A., Anitori, R., Burton, M.D., Quan, F., Buist, N.R. and Kennaway, N.G. 2000. Mitochondrial encephalomyopathy and complex III deficiency associated with a stop-codon mutation in the cytochrome b gene. *Am. J.Hum. Genet.* 67, 1400–1410.

44. Kennaway, N.G., Buist, N.R., Darley-Usmar, V.M., Papadimitriou, A., Dimauro, S., Kelley, R.I., Capaldi, R.A., Blank, N.K. and D'Agostino, A. 1984. Lactic acidosis and mitochondrial myopathy associated with deficiency of several components of complex III of the respiratory chain. *Pediatr. Res.* 18, 991–999.

45. Kohli, R.M. and Zhang, Y. 2013. TET enzymes, TDG and the dynamics of DNA demethylation. *Nature* 502, 472–479.

46. Lee, Y.C., Huang, H.Y., Chang, C.J., Cheng, C.H. and Chen, Y.T. 2010. Mitochondrial GLUT10 facilitates dehydroascorbic acid import and protects cells against oxidative stress: Mechanistic insight into arterial tortuosity syndrome. *Hum. Mol. Genet.* 19, 3721–3733.

47. Levine, M., Padayatty, S.J. and Espey, M.G. 2011. Vitamin C: A concentration-function approach yields pharmacology and therapeutic discoveries. *Adv. Nutr.* 2, 78–88.

48. Li, X., Cobb, C.E., Hill, K.E., Burk, R.F. and May, J.M. 2001. Mitochondrial uptake and recycling of ascorbic acid. *Arch. Biochem. Biophys.* 387, 143–153.

49. Li, X., Cobb, C.E. and May, J.M. 2002. Mitochondrial recycling of ascorbic acid from dehydroascorbic acid: Dependence on the electron transport chain. *Arch. Biochem. Biophys.* 403, 103–110.

50. Loenarz, C. and Schofield C.J. 2008. Expanding chemical biology of 2-oxoglutarate oxygenases. *Nat. Chem. Biol.* 4, 152–156.

51. MacPherson, I.S. and Murphy M.E. 2007. Type-2 copper-containing enzymes. *Cell. Mol. Life Sci.* 64, 2887–2899.

52. Mandl, J., Szarka, A. and Bánhegyi, G. 2009. Vitamin C: Update on physiology and pharmacology. *Br. J. Pharmacol.* 157, 1097–1110.

53. Mardones, L., Ormazabal, V., Romo, X., Jaña, C., Binder, P., Peña, E., Vergara, M. and Zúñiga, F.A. 2011. The glucose transporter-2 (GLUT2) is a low affinity dehydroascorbic acid transporter. *Biochem. Biophys. Res. Commun.* 410, 7–12.

54. May, J.M., Li, L., Qu, Z.C. and Cobb C.E. 2007. Mitochondrial recycling of ascorbic acid as a mechanism for regenerating cellular ascorbate. *Biofactors* 30, 35–48.

55. Monfort, A. and Wutz, A. 2013. Breathing-in epigenetic change with vitamin C. *EMBO Rep.* 14, 337–346.

56. Mueckler, M. and Thorens, B. 2013. The SLC2 (GLUT) family of membrane transporters. *Mol. Aspects Med.* 34, 121–138.

57. Muñoz-Montesino, C., Roa, F.J., Peña, E., González, M., Sotomayor, K., Inostroza, E. A., Muñoz, C. et al. 2014. Mitochondrial ascorbic acid transport is mediated by a low-affinity form of the sodium-coupled ascorbic acid transporter-2. *Free Rad Biol Med.* 70, 241–254.

58. Myllyharju, J. 2008. Prolyl 4-hydroxylases, key enzymes in the synthesis of collagens and regulation of the response to hypoxia, and their roles as treatment targets. *Ann. Med.* 40, 402–417.

59. Negre-Salvayre, A., Hirtz, C., Carrera, G., Cazenave, R., Troly, M., Salvayre, R., Penicaud, L. and Casteilla, L. 1997. A role for uncoupling protein-2 as a regulator of mitochondrial hydrogen peroxide generation. *FASEB J.* 11, 809–815.

60. Nemeth, B.A., Tsang, S.W., Geske R.S. and Haney, P.M. 2000. Golgi targeting of the GLUT1 glucose transporter in lactating mouse mammary gland. *Pediatr. Res.* 47, 444–450.

61. Németh, C.E., Marcolongo, P., Gamberucci A., Fulceri, R., Benedetti, A., Zoppi, N., Ritelli, M. et al. 2016. Glucose transporter type 10—lacking in arterial tortuosity syndrome—facilitates dehydroascorbic acid transport. *FEBS Lett.* 590, 1630–1640.

62. Németh, E.C., Nemoda, Z., Lőw P., Szabó, P., Horváth, E.Z., Willaert, A., Boel, A. et al. 2019.

Decreased nuclear ascorbate accumulation accompanied with altered genomic methylation pattern in fibroblasts from arterial tortuosity syndrome patients. *Ox. Med. Cell. Longev.* Article ID 8156592.

63. Patak, P., Willenberg, H.S. and Bornstein, S.R. 2004. Vitamin C is an important cofactor for both adrenal cortex and adrenal medulla. *Endocr. Res.* 30, 871–875.

64. Perez-Cruz, I., Carcamo, J.M. and Golde, D.W. 2003. Vitamin C inhibits FAS-induced apoptosis in monocytes and U937 cells. *Blood* 102, 336–343.

65. Ramanathan, K., Shila, S., Kumaran, S. and Panneerselvam, C. 2003. Ascorbic acid and α-tocopherol as potent modulators on arsenic induced toxicity in mitochondria. *J Nutr Biochem* 14, 416–420.

66. Rebouche, C.J. 1991. Ascorbic acid and carnitine biosynthesis. *Am. J. Clin. Nutr.* 54, 1147S–1152S.

67. Rigoulet, M., Yoboue, E.D. and Devin, A. 2011. Mitochondrial ROS generation and its regulation: Mechanisms involved in H(2)O(2) signaling. *Antioxid. Redox Signal.* 14, 459–468.

68. Rumsey, S.C., Kwon, O., Xu, G.W., Burant, C.F., Simpson, I. and Levine, M. 1997. Glucose transporter isoforms GLUT1 and GLUT3 transport dehydroascorbic acid. *J. Biol. Chem.* 272, 18982–18989.

69. Rumsey, S.C., Daruwala, R., Al-Hasani, H., Zarnowski, M.J., Simpson, I.A. and Levine M. 2000. Dehydroascorbic acid transport by GLUT4 in Xenopus oocytes and isolated rat adipocytes. *J. Biol. Chem.* 275, 28246–28253.

70. Scalera, V., Giangregorio, N., De Leonardis, S., Console, L., Carulli, E.S. and Tonazzi, A. 2018. Characterization of a novel mitochondrial ascorbate transporter from rat liver and potato mitochondria. *Front. Mol. Biosci.* 5, 58. eCollection 2018.

71. Segade, F. 2010. Glucose transporter 10 and arterial tortuosity syndrome: The vitamin C connection. *FEBS Lett.* 584, 2990–2994.

72. Sharma, P. and Mongan, P.D. 2001. Ascorbate reduces superoxide production and improves mitochondrial respiratory chain function in human fibroblasts with electron transport chain deficiencies. *Mitochondrion* 1, 191–198.

73. Siushansian, R. and Wilson, J.X. 1995. Ascorbate transport and intracellular concentration in cerebral astrocytes. *J Neurochem.* 65, 41–49.

74. Sturgeon, B.E., Sipe, H.J. Jr., Barr, D.P., Corbett, J.T., Martinez, J.G. and Mason, R.P. 1998. The fate of the oxidizing tyrosyl radical in the presence of glutathione and ascorbate. Implications for the radical sink hypothesis. *J. Biol. Chem.* 273, 30116–30121.

75. Szarka, A. and Balogh, T. 2015. In silico aided thoughts on mitochondrial vitamin C transport. *J. Theor. Biol.* 365, 181–189.

76. Szarka, A., Bánhegyi, G. and Asard, H. 2013. The inter-relationship of ascorbate transport, metabolism and mitochondrial, plastidic respiration. *Antioxid. Redox Signal.* 19, 1036–1044.

77. Szarka, A., Horemans, N., Bánhegyi, G. and Asard, H. 2004. Facilitated glucose and dehydroascorbate transport in plant mitochondria. *Arch. Biochem. Biophys.* 428, 73–80.

78. Szarka, A., Horemans, N., Kovács, Z., Gróf, P., Mayer, M. and Bánhegyi, G. 2007. Dehydroascorbate reduction in plant mitochondria is coupled to the respiratory electron transfer chain. *Physiol. Plant* 129, 225–232.

79. Szent-Györgyi, A. and Haworth, W.N. 1933. "Hexuronic Acid" (ascorbic acid) as the antiscorbutic factor. *Nature* 131, 24.

80. Takanaga, H. and Frommer, W.B. 2010. Facilitative plasma membrane transporters function during ER transit. *FASEB J.* 24, 2849–2858.

81. Tóth, Sz., Lőrincz, T. and Szarka, A. 2018. Concentration does matter: The beneficial and potentially harmful effects of ascorbate in humans and plants. *Antioxid. Redox Signal.* 29, 1516–1533.

82. Tsukada, Y., Fang, J., Erdjument-Bromage, H., Warren, M.E., Borchers, C.H., Tempst, P. and Zhang, Y. 2006. Histone demethylation by a family of JmjC domain-containing proteins. *Nature* 439, 811–816.

83. Tsukaguchi, H., Tokui, T., Mackenzie, B., Berger, U. V., Chen, X.Z., Wang, Y., Brubaker, R.F., and Hediger, M.A. 1999. A family of mammalian Na$^+$-dependent L-ascorbic acid transporters. *Nature* 399, 70–75.

84. Vera, J.C., Rivas, C.I., Fischbarg, J. and Golde, D.W. 1993. Mammalian facilitative hexose transporters mediate the transport of dehydroascorbic acid. *Nature* 364, 79–82.

85. Votyakova, T.V. and Reynolds, I.J. 2001. DeltaPsi(m)-Dependent and -independent production of reactive oxygen species by rat brain mitochondria. *J. Neurochem.* 79, 266–277.

86. Wang, Y., Mackenzie, B., Tsukaguchi, H., Weremowicz, S., Morton, C.C. and Hediger, M.A.

2000. Human vitamin C (L-ascorbic acid) transporter SVCT1. *Biochem. Biophys. Res. Commun.* 267, 488–494.

87. Winkler, B.S. 1987. In vitro oxidation of ascorbic acid and its prevention by GSH. *Biochim. Biophys. Acta* 925, 258–264.

88. Xu, D.P. and Wells, W.W. 1996. α-Lipoic acid dependent regeneration of ascorbic acid from dehydroascorbic acid in rat liver mitochondria. *J. Bioenerg. Biomembr.* 28, 77–85.

89. Zito, E. 2013. PRDX4, an endoplasmic reticulum-localized peroxiredoxin at the crossroads between enzymatic oxidative protein folding and nonenzymatic protein oxidation. *Antioxid. Redox Signal.* 18, 1666–1674.

90. Zito, E., Hansen, H.G., Yeo G.S., Fujii, J. and Ron, D. 2012. Endoplasmic reticulum thiol oxidase deficiency leads to ascorbic acid depletion and noncanonical scurvy in mice. *Mol. Cell.* 48, 39–51.

PART THREE

Vitamin C and Immune Function

CHAPTER SEVEN

Vitamin C in Pneumonia and Sepsis

Anitra C. Carr

DOI: 10.1201/9780429442025-7

CONTENTS

INTRODUCTION

In the early literature, one of the most striking symptoms reported for the vitamin C deficiency disease scurvy was a marked susceptibility to infections, particularly pneumonia (reviewed in [1]). Autopsy findings from the 1920s indicated that pneumonia was one of the most frequent complications of scurvy and was the prevailing cause of death. Infantile scurvy was also observed to predispose children to infections, particularly of the respiratory tract. According to Hemilä [1], these findings supported observations of the disappearance of a pneumonia epidemic in Sudan when antiscorbutic treatment was given to the numerous cases of scurvy that occurred at the same time. Conversely, there have also been reports of scurvy following infectious epidemics, suggesting that infections can severely deplete vitamin C levels in the body [1]. Case reports indicate that children have developed scurvy symptoms after, or concurrently with, respiratory infections [2,3]. The authors stated that "possibly the increased metabolic needs associated with this infection unmasked a subclinical vitamin C deficiency"

[2] and that "scurvy occurred as a result of their increased requirement of vitamin C due to stress of illness combined with poor dietary intake. It is therefore recommended that during illness one should be careful about the intake of vitamin C, keeping in mind that acute illness rapidly depletes stores of ascorbic acid. Those already malnourished are more prone to this development" [3]. Similarly, others have reported scurvy symptoms following confirmed or suspected respiratory infection, stating that "sepsis of either digestive or pulmonary origin, leading to sustained metabolic demand, might have acted as a precipitating factor" [4,5].

The anecdotal and epidemiological observations of a connection between vitamin C status and infections have been supported by animal studies using vitamin C–requiring guinea pigs and mice deficient in L-gulono-g-lactone oxidase (Gulo$^{-/-}$), the rate-limiting enzyme in vitamin C synthesis. Identification of bacteria in the tissues of scorbutic guinea pigs in the early literature led to the erroneous hypothesis by some researchers that scurvy may in fact be an infectious disease [1]. Research has indicated an increased severity of infections in scorbutic guinea pigs, with higher mortality observed in vitamin C–deficient animals infected with *Pseudomonas aeruginosa* compared with vitamin C–replete animals [6]. Vitamin C–deficient Gulo knockout mice were three times as likely as vitamin C–replete mice to die following infection with *Klebsiella pneumoniae* [7]. The lungs appear to be particularly susceptible to deficiency with vitamin C–deficient Gulo knockout mice exhibiting greater lung pathology following infection with influenza [8]. Conversely, infection of animals with *P. aeruginosa* and influenza A virus resulted in decreased vitamin C levels in tissues and fluids [6,9,10], possibly due to the inflammatory response and enhanced oxidative stress. Interestingly, infection by itself was found to decrease antioxidant capacity more than a vitamin C–deficient diet in guinea pigs, suggesting a high consumption of antioxidants during infection [6]. Although enhanced markers of oxidative stress have been observed in infected mice, there were no significant differences in the levels of the oxidation products in the vitamin C–deficient and –replete mice indicating that vitamin C may be acting via mechanisms other than oxidant scavenging [7]. Overall, the animal studies support a two-way relationship between vitamin C and infection.

PNEUMONIA AND VITAMIN C

Pneumonia is an acute infection of the lungs that can be caused by a range of microorganisms, including those of bacterial, fungal, or viral origin [11]. These microorganisms reach the lower respiratory tract and, dependent on microbial virulence factors, the host's immune defenses, and integrity of barriers, cause inflammation in the alveoli and consequently result in pneumonia. Diagnosis is usually determined through radiographic imaging, indicating shadowing of a lobe or segment of the lung, and the clinician's clinical assessment, and empiric treatment is through prompt antimicrobial intervention. Symptoms include cough, fever, aches, sweating, and shivering, and some patients may present with pleuritic chest pain and confusion [12]. Lower respiratory infections, such as pneumonia, are a leading cause of morbidity and mortality worldwide. In 2016, lower respiratory infections caused nearly 2.4 million deaths worldwide, making lower respiratory infections the sixth

leading cause of mortality for all ages and the leading cause of death among children younger than 5 years [13]. This equated to more than 335 million episodes of lower respiratory infections and more than 65 million hospital admissions in 2016. Lower respiratory infection mortality is high in the elderly, and rates are increasing due to an increasing aging population, with the number of adults older than 70 years increasing by 50% between 2000 and 2016 [14].

Streptococcus pneumoniae is the leading cause of lower respiratory infection morbidity and mortality globally, contributing to more deaths than all other assessed etiologies combined [14]. Research has also indicated that increased pneumonia incidence is associated with higher deprivation and is particularly prevalent in developing countries where poverty is more prevalent [14,15].

Pneumonia has been reported as one of the most common complications and causes of mortality in individuals with scurvy, suggesting an important link between vitamin C status and lower respiratory infection [1]. This premise is supported by meta-analyses of three interventional studies that indicated that prophylactic administration of at least 200 mg/d vitamin C decreased the incidence of pneumonia in the study populations [16–19]. Furthermore, analysis of the vitamin C status of patients with pneumonia and acute respiratory distress syndrome has indicated significantly lower vitamin C concentrations in patients when compared with healthy controls, and levels appeared to inversely correlate with the severity of the condition (Table 7.1) [20,21]. Up to 40% of patients with pneumonia exhibited outright vitamin C deficiency (i.e., plasma vitamin C levels <11 µmol/L), and levels remained low for at least 4 weeks [22,23]. These studies indicate a higher utilization of, and potentially also a higher requirement for, vitamin C during lower respiratory tract infection.

An early report by Klenner indicated that administration of 2–4 g/d intravenous or intramuscular doses of vitamin C to patients with pneumonia decreased the symptoms of nausea, headache, temperature, and cyanosis [24]. Subsequent interventional studies have indicated that administration of oral or intravenous vitamin C decreased the severity of the respiratory symptoms, particularly in the most severely ill, and also decreased hospital length of stay in a dose-dependent manner (Table 7.2) [22,23]. A trend

TABLE 7.1
Vitamin C status of patients with pneumonia

Study Type	Cohort	Vitamin C (μmol/L)	References
Case control	20 Healthy volunteers	66 ± 3	[20]
	11 Pneumonia cases	31 ± 9	
Case control	28 Healthy participants	49 ± 1	[21]
	35 Lobular pneumonia		
	7 Acute—did not survive	17 ± 1	
	15 Acute—survived	24 ± 1	
	13 Convalescent cases	34 ± 1	
Interventional	29 Pneumonia/bronchitis		[22]
(placebo	Week 0	24 ± 5 (40%)[a]	
group)	Week 2	19 ± 3 (37%)	
	Week 4	24 ± 6 (25%)	
Interventional	70 Pneumonia cases		[23]
(control	Day 0	41	
group)	Day 5–10	24–23	
	Day 15–20	32–35	
	Day 30	39	

NOTE: Data represent mean and standard error of the mean (SEM).

[a] Percentage of patients with vitamin C deficiency. Vitamin C status categories: saturating (>70 μmol/L), adequate (>50 μmol/L), hypovitaminosis C (<23 μmol/L), and deficient (<11 μmol/L).

toward decreased mortality was observed in the interventional study by Hunt et al. [22], and a more recent case control study using a higher dose of 6 g/d intravenous vitamin C (in combination with thiamine and hydrocortisone) exhibited a significant (56%) decrease in mortality in patients with severe pneumonia [25]. Interestingly, Cathcart hypothesized that patients with severe respiratory infections and pneumonia had higher requirements for vitamin C based on the observation that they could tolerate more than 10 times the usual bowel tolerance doses of 4–15 g/24 hour [26]. Mochalkin assessed plasma vitamin C levels following intervention and observed that administration of 0.25–0.8 g/d vitamin C was insufficient to maintain the patients' initial vitamin C status; however, administration of 0.5–1.6 g/d vitamin C was able to maintain the patients' plasma vitamin C status for the duration of the study (30 days) [23]. However, these plasma concentrations were still inadequate (i.e., <50 μmol/L), suggesting a requirement of >1.6 g/d for saturating plasma status. Thus, patients with severe infections, such as pneumonia, have higher requirements for

TABLE 7.2
Vitamin C intervention in patients with pneumonia

Patients	Intervention	Outcomes	References
99 Severe pneumonia	IV vitamin C	↓ Hospital mortality	[25]
46 Controls	0 g/d		
53 Treatment	6 g/d		
57 Pneumonia/bronchitis	Oral vitamin C	↓ Respiratory symptom	[22]
29 Placebo	0 g/d	score in most severely ill	
28 Treatment	0.2 g/d		
140 Pneumonia cases	Oral vitamin C	↓ Hospital length of stay	[23]
70 Control	0 g/d	24 days	
39 Low dose	0.25–0.8 g/d	19 days	
31 High dose	0.5–1.6 g/d	15 days	

ABBREVIATIONS: ↓ decrease; IV, intravenous.

vitamin C, and doses of vitamin C that provide adequate to saturating plasma vitamin C status in these patients appear to have beneficial effects on patient outcomes.

SEPSIS AND VITAMIN C

Sepsis is a condition of life-threatening organ dysfunction caused by a dysregulated host response to infection [27]. Sepsis is characterized by profound dysregulation of the circulatory, metabolic, and immune systems, and it is the primary cause of death from infection. Patients who develop septic shock can have hospital mortality rates of up to 50%. Management of sepsis involves fluid resuscitation for hypoperfusion and vasopressor drug administration for those in shock [28]. Global estimates indicate nearly 32 million sepsis cases and more than 5 million deaths annually [29]. Although sepsis mortality rates have been decreasing, particularly in developed countries such as the United States and Australasia, the incidence of sepsis continues to increase, likely due to an aging population [30–34]. Despite huge research efforts toward an attempt to identify effective sepsis therapies, to date most of these have proven futile [35]. Furthermore, patients who survive sepsis can often have long-term physical disabilities, cognitive dysfunction, or psychological issues, such as anxiety, depression, and posttraumatic stress disorder, which significantly affect their quality of life [36].

Case control studies have consistently indicated significantly lower vitamin C status in critically ill patients, particularly those with sepsis (Table 7.3). These critically ill patients have by far the lowest vitamin C status when compared with other common disease states [37,38]. Lower vitamin C status in these patients was associated with increased inflammation (C-reactive protein levels), increased severity of the illness (days in the intensive care unit [ICU]), and multiple organ failure [37,39,40]. Nearly 40% of patients with septic shock were deficient, and almost 90% had hypovitaminosis C, despite receiving recommended enteral and parenteral intakes [39]. Administration of 1 g/d vitamin C to critically ill patients was found to be insufficient to raise the patients' plasma vitamin C concentrations above the hypovitaminosis C cutoff, but 3 g/d resulted in saturating plasma status (i.e., $\sim$70 μmol/L) [41,42]. Recent pharmacokinetic data indicated that administration of 2 g/d vitamin C to critically ill patients, as either bolus or continuous infusions, resulted in plasma concentrations in the normal range, although hypovitaminosis C occurred in some patients following cessation of the intervention, suggesting sustained therapy may be required to prevent this from occurring [43]. Overall, these findings indicate that critically ill patients have vitamin C requirements that are approximately 10-fold higher than healthy individuals, whose plasma vitamin C typically saturates with intakes of 0.2 g/d [44].

TABLE 7.3
Vitamin C status of patients with sepsis

Study Type	Cohort	Vitamin C (μmol/L)	References
Observational	24 Septic shock patients	15 $\pm$ 2 (38%[a], 88%[b])	[39]
Interventional (baseline)	24 Severe sepsis patients	18 $\pm$ 2	[45]
Case control	6 Healthy controls	48 $\pm$ 6	[46]
	19 Severe sepsis	14 $\pm$ 3	
	37 Septic shock	14 $\pm$ 3	
Case control	14 Healthy controls	76 $\pm$ 6	[47]
	11 Septic encephalopathy	19 $\pm$ 11	
Case control	34 Healthy controls	62 (55–72)	[37]
	62 ICU (injury, surgery, sepsis)	11 (8–22)	

NOTE: Data represent mean and standard error of the mean (SEM) or median and interquartile range.

ABBREVIATION: ICU, intensive care unit.

[a] Percentage of patients with vitamin C deficiency.

[b] Hypovitaminosis C. Vitamin C status categories: saturating (>70 μmol/L), adequate (>50 μmol/L), hypovitaminosis C (<23 μmol/L), and deficient (<11 μmol/L).

Critically ill patients treated with ~3 g/d intravenous vitamin C (in combination with various other antioxidant vitamins and minerals) have shown improved outcomes, including decreased organ failure, ICU and hospital length of stay, mortality, inflammation, and infections/sepsis [42,48–51]. However, with combination interventions, it is difficult to know which component(s) are contributing to the various outcomes, particularly since baseline concentrations of the various components are not typically assessed in the patients prior to intervention. In 2014, the first phase I study investigating intravenous vitamin C as monotherapy in patients with sepsis was published [45]. In this study, 24 patients with severe sepsis were treated with 0, 50, or 200 mg/kg body weight intravenous vitamin C per day, which provided a dose-dependent decrease in systemic organ failure and decreased pro-inflammatory (C-reactive protein and procalcitonin) and tissue damage (thrombomodulin) biomarkers (Table 7.4). Although the study was not powered to detect a decrease

TABLE 7.4
Vitamin C intervention in patients with sepsis

Patients	Intervention	Outcomes	References
Vitamin C Administered Alone			
100 Septic shock 50/Group	IV vitamin C: 0 or 6 g/d Duration: until ICU discharge	↓ Vasopressor duration ↓ ICU length of stay X Length of mechanical ventilation X Renal replacement therapy X ICU mortality	[53]
28 Septic shock 14/Group	IV vitamin C: 0 or 100 mg/kg/d Duration: 3 days	↓ Norepinephrine dose and duration X ICU length of stay ↓ 28-day mortality	[52]
24 Severe sepsis 8/Group	IV vitamin C: 0, 50, or 200 mg/kg/d Duration: 4 days	↓ Systemic organ failure ↓ C-reactive protein, procalcitonin, thrombomodulin levels	[45]
Vitamin C plus Thiamine/Hydrocortisone Cocktail			
94 Severe sepsis 47/Group (retrospective)	IV vitamin C: 0 or 6 g/d + thiamine + hydrocortisone Duration: as little as 1 dose or up to 4 days	X ICU or hospital mortality X ICU or hospital length of stay X Renal replacement therapy for AKI X Time to vasopressor independence	[54]
1144 Septic shock 229 Treatment 915 Controls (retrospective)	IV vitamin C: 0 or 6 g/d + thiamine Duration: 1 day only	X 28-day or hospital mortality X ICU or hospital length of stay X Duration of mechanical ventilation X New renal replacement therapy	[55]
24 Septic shock 12/Group	IV vitamin C: 0 or 6 g/d + thiamine + hydrocortisone Duration: 4 days	↓ Vasopressin and noradrenaline requirements ↓ Procalcitonin levels X Systemic organ failure	[56]
94 Severe sepsis 47/Group (retrospective)	IV vitamin C: 0 or 6 g/d + thiamine + hydrocortisone Duration: 4 days or until ICU discharge	↓ Vasopressor duration ↓ Systemic organ failure ↓ Procalcitonin levels ↓ Renal replacement therapy X ICU length of stay ↓ Hospital mortality	[57]

ABBREVIATIONS: ↓ decrease; AKI, acute kidney injury; IV, intravenous; X, no change.

in mortality, fewer participants died in the treatment arms. Subsequently, a randomized controlled trial carried out in 28 patients with septic shock treated with 100 mg/kg body weight intravenous vitamin C per day showed a significant decrease in vasopressor requirements (dose and duration of norepinephrine) and a dramatic (78%) decrease in 28-day mortality [52]. No difference in ICU length of stay was observed. Another recent randomized study administering 6 g/d vitamin C or placebo to 100 septic shock patients also showed decreased requirements for vasopressors and decreased length of ICU stay [53]. However, no differences in ICU mortality, duration of mechanical ventilation, or renal replacement therapy were observed between the two groups.

Several recent studies have investigated the efficacy of administering a cocktail of vitamin C with thiamine (vitamin B1), with or without hydrocortisone (Table 7.4). A before-and-after study was carried out in which 47 patients with severe sepsis were treated with 6 g/d intravenous vitamin C, in combination with 0.4 g/d thiamine (vitamin B1) and hydrocortisone, and were compared with 47 retrospective controls, who also received hydrocortisone at the attending physicians' discretion [57]. This study also showed decreased vasopressor requirements, as well as decreased systemic organ failure and requirement for renal replacement therapy. Furthermore, a dramatic (79%) decrease in hospital mortality was observed in the group who received the intervention. A smaller randomized study administering the same cocktail of vitamin C, thiamine, and hydrocortisone to 24 cardiac surgery patients with septic shock showed decreased vasopressin and norepinephrine requirements and decreased procalcitonin levels in the treatment group, although no difference in sequential organ failure assessment (SOFA) scores was observed between the two groups [56]. A recent retrospective analysis of septic patients administered the same cocktail, however, showed no effect of treatment on any of the assessed outcomes (i.e., hospital and ICU mortality and length of stay, renal replacement therapy for acute kidney injury, or time to vasopressor independence). It should be noted that patients were included in the analysis if they received as little as one dose of the cocktail, and although a subgroup analysis of the 20 patients who did receive the full 4 days (or until discharge) also

showed no significant outcome effects, these numbers would likely be too low to provide appropriate power. Another before-and-after study administering vitamin C and thiamine, but without hydrocortisone, to 229 septic shock patients found no effect on ICU or hospital mortality or length of stay when compared with 915 retrospective controls [55]. However, the treatment was administered for only 1 day, and the treatment group also had significantly higher baseline morbidity than the control group [58]. Despite this, in patients with the most severe organ dysfunction, the treatment did decrease mortality. Thus, most of these small studies have indicated that intravenous doses of ~6–7 g/d vitamin C administered for 3–4 days may improve the outcomes of patients with sepsis and septic shock, including a decreased requirement for vasopressors [59]. Currently, over a dozen registered randomized controlled trials are underway around the world to determine if these encouraging findings are reproducible.

ACUTE LUNG DYSFUNCTION

Acute respiratory infections and sepsis can result in the development of acute lung injury which, in its most severe form, is known as acute respiratory distress syndrome [60]. During acute lung injury, bronchoalveolar barrier function is compromised, resulting in abnormal capillary permeability and pulmonary edema [61]. Sepsis-induced acute lung injury is also associated with diminished expression and function of tight junction proteins in lung epithelium [62]. Furthermore, the iron pumps and channels that normally function to maintain continuous fluid clearance by the lungs can be affected early during sepsis [63]. In a murine model of sepsis-induced acute lung injury, Fisher et al. reported that concurrent administration of 200 mg/kg vitamin C to mice attenuated the resultant lung dysfunction [62]. The authors reported decreased lung water and alveolar epithelial permeability and increased alveolar fluid clearance in the animals that received vitamin C. Furthermore, increased expression of iron pumps and channels and increased expression of tight junction and cytoskeletal connector proteins were observed over and above both sepsis-induced and control levels, suggesting enhanced gene transcription in the presence of vitamin C (Figure 7.1). Other

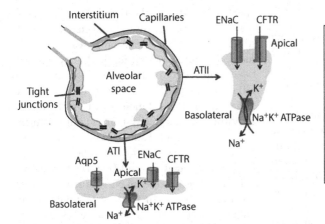

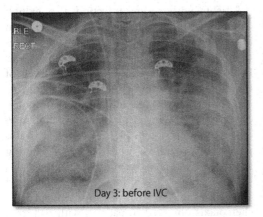
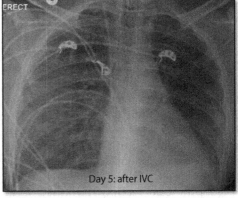

Figure 7.1. Effect of vitamin C administration on sepsis-induced acute lung injury in a mouse model. Vitamin C administration significantly increased expression of the key pumps and channels involved in alveolar fluid transport, including aquaporin 5 (Aqp5), cystic fibrosis transmembrane conductance regulator (CFTR), epithelial sodium channel (ENaC), and sodium-potassium-ATPase (Na⁺-K⁺-ATPase). Vitamin C administration also significantly increased expression of the tight junction proteins claudin-18 and occludin, as well as the cytoskeletal connector protein zona occludens-1. (↓ decreased, ↑ increased.) (The diagram is courtesy of R. Natarajan [personal communication], and the text box is a summary of findings from Fisher, B. J. et al. 2012. *Am. J. Physiol. Lung Cell Mol. Physiol.* 303, L20–L32.)

researchers have reported synergistic interactions between vitamin C and hydrocortisone in attenuating lipopolysaccharide-induced hyperpermeability of human lung microvascular endothelial cells [64]. The combination was found to normalize the expression and activation of proteins associated with actin stress fibers, which play an important role in the formation and maintenance of cell-cell adhesion, including tight junctions. Recently, case reports have been published that demonstrate dramatic lung clearance in septic patients with acute respiratory distress syndrome of bacterial and viral origins following treatment with intravenous vitamin C [65,66]. Vitamin C was administered to the patients at a dose of 200 mg/kg/d, and within 1–2 days there was observable clearance of the lung infiltrate upon chest x-ray (Figure 7.2). Similar lung clearance has been reported for vitamin C combination therapy in cases of noninfectious and aspiration-induced acute respiratory distress syndrome [67,68].

Figure 7.2. Chest x-rays of a septic patient with acute respiratory distress syndrome before and after intravenous vitamin C (IVC) administration. IVC (200 mg/kg/d) was initiated on hospital day 4, and chest x-ray on day 5 revealed significantly improved opacities. (Images from Bharara, A. et al. 2016 *Case Rep. Crit. Care.* Article ID 8560871, 2016, 4p. CC-BY.)

ROLE OF NEUTROPHILS

Infiltration and activation of neutrophils in lung tissue in response to infection represent a primary mechanism for sepsis-induced pulmonary dysfunction and injury [69]. Activated neutrophils release reactive oxygen species, proteolytic enzymes, and other pro-inflammatory mediators that can directly damage tissues and prolong the inflammatory process [70]. Spent neutrophils normally undergo a process of programmed cell death known as *apoptosis*, which facilitates clearance of the neutrophils from sites of inflammation by phagocytosing macrophages [71]. This process prevents excessive tissue damage and supports resolution of inflammation. Attenuated neutrophil apoptosis has been reported in patients with sepsis, and this appears to be related to disease severity [72–74]. Delayed apoptosis may be due to the regulatory effects of pro-inflammatory mediators as well as oxidative inactivation of the redox-sensitive caspase effector enzymes [75,76]. Vitamin C may be able to decrease pro-inflammatory mediator release during sepsis and also protect the oxidant-sensitive caspase enzymes in activated neutrophils, thereby protecting apoptotic cell death pathways [77,78]. In support of this premise, peritoneal neutrophils isolated from vitamin C–deficient Gulo knockout mice exhibited attenuated apoptosis and instead underwent necrotic cell death [79]. These vitamin C–deficient neutrophils were not phagocytosed by macrophages *in vitro* and persisted at inflammatory loci *in vivo*. Furthermore, Fisher et al. reported that administration of 200 mg/kg vitamin C to mice attenuated peritonitis-induced sequestration of neutrophils, and decreased myeloperoxidase mRNA expression, in the lungs of the treated animals [62]. This was also associated with a lower acute lung injury score and decreased mortality in these animals. The clearance of neutrophils from sites of inflammation could conceivably contribute to the enhanced lung clearance observed in patients following vitamin C infusions (Figure 7.2).

Neutrophils that fail to undergo apoptosis instead undergo necrotic cell death. The subsequent release of intracellular components, such as proteases, can cause extensive tissue damage [70]. One recently identified form of neutrophil death, termed *necroptosis*, occurs when caspases are inactivated [80]. Necroptotic signaling pathways can result in the release of "neutrophil extracellular traps" (NETs) composed of neutrophil DNA, histones, and enzymes [81,82]. Although NETs have been proposed as a unique method of microbial killing [83], they have also been implicated in tissue damage and organ failure [84,85]. NET-associated histones can act as damage-associated molecular pattern proteins, activating the immune system and causing further tissue damage [86]. Patients with sepsis, or who go on to develop sepsis, have significantly elevated levels of circulating cell-free DNA, which is believed to be a marker of NET formation [84,87]. Preclinical studies in vitamin C–deficient Gulo knockout mice indicated enhanced NETs in the lungs of septic animals and increased circulating cell-free DNA [88]. The levels of these markers were attenuated in vitamin C–sufficient animals or in vitamin C–deficient animals that were administered vitamin C 30 minutes after induction of sepsis. The same investigators showed that *in vitro* supplementation of human neutrophils with vitamin C attenuated phorbol ester-induced NET formation [88]. Administration of gram doses of vitamin C to septic patients over 4 days, however, did not appear to decrease circulating cell-free DNA levels [89]. The duration of treatment may have been too short, or initiated too late in the inflammatory process, to provide a beneficial effect. It should also be noted that cell-free DNA is not specific for neutrophil-derived DNA, as it may also derive from necrotic tissue; however, the association of neutrophil-specific proteins or enzymes, such as myeloperoxidase, with the DNA can potentially provide an indication of its source [84].

Patients with severe infection also exhibit compromised neutrophil chemotactic activity that can compromise the cells' ability to migrate to sites of infection [90,91]. This neutrophil "paralysis" is believed to be partly due to enhanced levels of immune-suppressive mediators released during the compensatory anti-inflammatory response observed following initial hyperstimulation of the immune system [92]. However, it is also possible that vitamin C depletion observed during severe infection may contribute. Support for this premise comes from studies in the 1980s and 1990s that indicated that the impaired leukocyte chemotaxis observed in patients with recurrent infections could be restored in response to supplementation with gram doses of vitamin C

[93–98]. Furthermore, vitamin C supplementation of neonates with suspected sepsis dramatically improved neutrophil chemotaxis [99]. Patients with recurrent infections can also exhibit impaired neutrophil bacterial killing and phagocytosis, which can be significantly improved following supplementation with gram doses of vitamin C, resulting in long-lasting clinical improvement [93,94,97,98,100].

Neutrophils accumulate vitamin C against a concentration gradient resulting in levels 50- to 100-fold higher than plasma concentrations [101,102]. Active uptake of vitamin C occurs via specialized sodium-dependent vitamin C transporters (SVCTs) and can also occur following activation of the neutrophil oxidative burst and accumulation of the oxidized form of vitamin C, dehydroascorbic acid, via glucose transporters (GLUTs) [103]. It is believed that the accumulation of millimolar vitamin C concentrations indicates an important role for the vitamin within these cells, and depleted levels may compromise vital functions [77]. Vitamin C levels in leukocytes have been reported to decrease by half within 24 hours of subjects contracting an upper respiratory tract infection, and these levels were restored to normal when the infection resolved [104]. The investigators found that administration of 6 g/d of vitamin C during the infection could attenuate the decline in leukocyte vitamin C, although 200 mg/d did not affect the drop in vitamin C levels over the first few days of the infection. However, a randomized controlled trial in patients with acute respiratory infections administered 200 mg/d vitamin C showed repletion of neutrophil and mononuclear cell vitamin C levels within 2 weeks, whereas cells isolated from the placebo group remained low [22]. Thus, repletion of neutrophil vitamin C status via vitamin C administration during infection may enhance vital neutrophil functions [77].

MECHANISMS OF ACTION

Vitamin C is a potent water-soluble antioxidant, able to scavenge a wide range of reactive oxygen species, thus protecting essential cellular structures, metabolic functions, and signaling pathways from oxidative damage [105,106]. Vitamin C also exhibits anti-inflammatory activity with inverse associations observed between vitamin C and pro-inflammatory cytokines and acute phase reactants such as C-reactive protein and

procalcitonin [39,45,56,57,77]. Severe infection and sepsis are characterized by significant oxidative stress and overwhelming inflammatory mediators, sometimes referred to as a "cytokine storm" [107,108]. These stressors can contribute to the pathophysiology of sepsis, such as impaired microcirculatory flow, coagulopathy, capillary plugging, increased endothelial dysfunction and permeability, and multiorgan failure [109,110]. As such, the role of vitamin C in severe infection and sepsis has often focused on its antioxidant and anti-inflammatory functions and effects on signaling pathways [109,110]. However, less attention has been paid to its role as an enzyme cofactor.

Biosynthetic Functions

One of the primary roles of vitamin C in the body is to act as a cofactor for a family of metalloenzymes with various biosynthetic and regulatory roles [111–113]. These enzymes introduce hydroxyl groups into biomolecules and comprise two main categories: iron- and 2-oxoglutarate–dependent dioxygenases and copper-containing monooxygenases. Of the former category, vitamin C has long been known to act as a cofactor for the lysyl and prolyl hydroxylases required for stabilization of the tertiary structure of collagen, an essential component of the vasculature [114]. Vitamin C may also be able to stimulate the expression of collagen mRNA, perhaps through its gene regulatory mechanisms described later [77]. Similarly, vitamin C is a cofactor for the two hydroxylases involved in carnitine biosynthesis, a molecule required for transport of fatty acids into mitochondria for generation of metabolic energy [115]. Mitochondrial dysfunction and depleted ATP levels are observed in sepsis; thus, vitamin C may be able to contribute to metabolic resuscitation via both antioxidant and cofactor mechanisms [116,117].

Vitamin C is also known to facilitate the synthesis of the catecholamines dopamine, norepinephrine, and epinephrine within the sympathetic nervous system and adrenal medulla. These catecholamines are central to the cardiovascular response to severe infection; they increase arterial pressure through binding to α-adrenergic receptors on the smooth muscle cells of the vasculature and can promote increased cardiac contractility and heart rate through binding to β-adrenergic receptors on cardiac muscle [118]. Vitamin C is a cofactor for the copper-containing monooxygenase dopamine β-hydroxylase that introduces a hydroxyl group onto dopamine to form norepinephrine (Figure 7.3) [119,120]. Epinephrine is subsequently synthesized in the adrenal glands via methylation of the amine group of norepinephrine. Recent research indicates that vitamin C may also stimulate the rate-limiting enzyme tyrosine hydroxylase via recycling the enzyme's cofactor, tetrahydrobiopterin, thus facilitating hydroxylation of L-tyrosine to form the dopamine precursor L-dopa (Figure 7.3) [121]. Some evidence also suggests that vitamin C may enhance the synthesis of tyrosine hydroxylase [121]; this is possibly through its gene-regulatory effects, as described below.

It is noteworthy that the tissues where the catecholamines are synthesized (i.e., the brain and adrenal glands) contain the highest levels of vitamin C in the body [122], indicating that the vitamin plays a vital role in these organs. Furthermore, animal models of vitamin C deficiency have shown significant retention of the vitamin in the brain during dietary depletion [123–125], supporting the importance of vitamin C in the central nervous system. Impaired adrenal hormone synthesis has been observed in critically ill patients and is probably a common complication in severe sepsis [126,127]. Interestingly, norepinephrine levels are decreased in vitamin C–deficient animal models, particularly in the adrenal glands [128–130]. Research has shown that vitamin C is also secreted from the adrenal glands as part of the stress response [131], which could conceivably result in adrenal vitamin C depletion under conditions of sustained stress. Thus, appropriate supplementation of vitamin C in sepsis may support endogenous synthesis of vasoactive catecholamines.

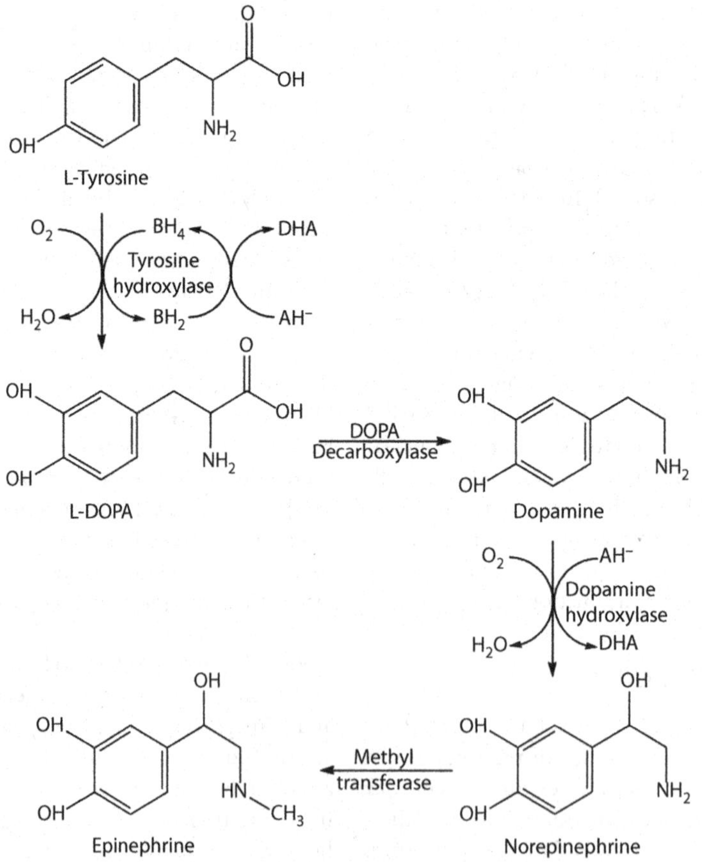

Figure 7.3. Vitamin C–dependent synthesis of the catecholamine vasopressors dopamine, norepinephrine, and epinephrine. Vitamin C acts as a cofactor for the metallo-enzyme dopamine β-hydroxylase and also recycles tetrahydrobiopterin, a cofactor for the rate-limiting enzyme tyrosine hydroxylase. (AH⁻ ascorbate, DHA dehydroascorbic acid, BH_4 tetrahydrobiopterin, BH_2 dihydrobiopterin.)

Vitamin C is also a cofactor for the copper-containing enzyme peptidylglycine α-amidating monooxygenase (PAM) that is required for the synthesis of amidated neuropeptide hormones [132]. The carboxy-terminal amine group of amidated peptides is essential for their biological activities [133]. One of these amidated peptide hormones is vasopressin, also known as arginine vasopressin (AVP) or antidiuretic hormone (ADH), which is synthesized in the hypothalamus, posttranslationally modified by PAM, and then stored in the posterior pituitary [134]. Vasopressin is secreted in response to decreased blood volume or arterial pressure or increased plasma osmolality. It interacts with specific receptors expressed by vascular smooth muscle cells and kidney collecting ducts to cause vasoconstriction and water retention, respectively [135]. The hormone is synthesized as a pre-prohormone that undergoes sequential cleavage steps to produce provasopressin and finally a glycine-extended precursor. The carboxy-terminal glycine residue of the vasopressin precursor subsequently undergoes posttranslational modification by the vitamin C–dependent enzyme PAM to generate the active carboxy-amidated hormone (Figure 7.4). Support for a connection between vitamin C and vasopressin biosynthesis comes from an animal study, whereby centrally administered vitamin C enhanced circulating levels of vasopressin and induced antidiuresis [136].

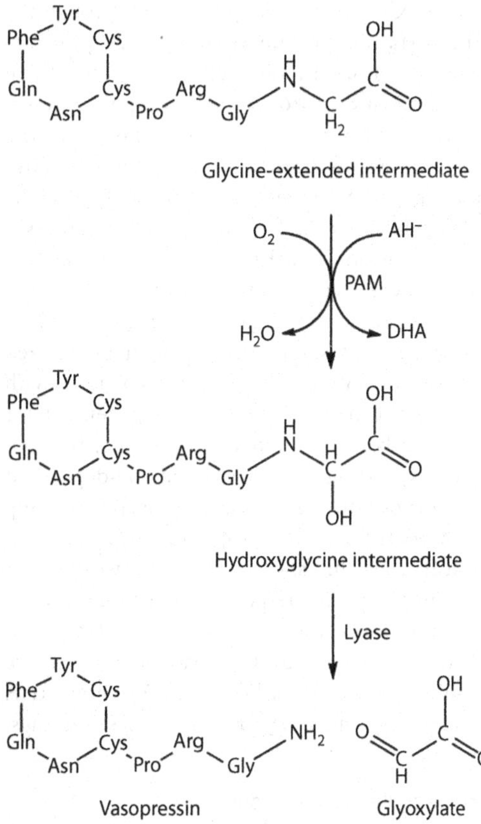

Circulating vasopressin levels increase dramatically during the initial phase of septic shock, but this is followed by a significant decline in the latter phase [137,138]. Patients in late-phase septic shock have significantly lower levels of circulating vasopressin compared with patients in cardiogenic shock, despite similar hypotension [138]. The decline in circulating vasopressin levels after the onset of septic shock is due to depletion of pituitary stores and possibly also impaired vasopressin synthesis [139]. It is of interest to note that the pituitary gland, where the enzyme PAM is abundantly expressed, has the highest levels of vitamin C in the body [122]. Thus, it is conceivable that the depleted vitamin C status of patients with sepsis could contribute to the observed decrease in vasopressin biosynthesis [137–139]. Furthermore, pro-vasopressin, which lacks the carboxyterminal amine of mature vasopressin, is significantly associated with mortality in patients with pneumonia and septic shock, with higher levels

Figure 7.4. Vitamin C–dependent synthesis of mature carboxy-terminal amidated vasopressin. Vitamin C is a cofactor for the metallo-enzyme peptidylglycine α-amidating monooxygenase (PAM). The enzyme comprises two domains: a copper-dependent monooxygenase domain, which converts glycine-extended peptides into hydroxyglycine intermediates; and a lyase domain, which converts the hydroxyglycine intermediates into amidated products. (AH⁻ ascorbate, DHA dehydroascorbic acid.)

observed in nonsurvivors than survivors [140,141]. Although pro-vasopressin has enhanced stability in circulation, the higher ratio of pro-vasopressin to mature vasopressin could also be due to decreased posttranslational activation of the hormone by PAM because of limited cofactor availability. This premise is supported by the observation that other peptide pro-hormones that are substrates of PAM (e.g., pro-adrenomedullin and procalcitonin) are also elevated in severe infectious conditions, particularly in nonsurvivors [141–143]. It is interesting to note that elevated procalcitonin levels decrease following administration of vitamin C to septic patients [45,56,57].

Septic shock is normally managed through the administration of catecholamine vasopressors, primarily norepinephrine, to elevate mean arterial pressure to ≥65 mm Hg [144]. Vasopressin administration is also recommended in the Surviving Sepsis Campaign guidelines to raise mean arterial pressure to target or to decrease the norepinephrine dose [28]. Exogenous vasopressor administration to patients with septic shock, however, can result in adverse side effects such as tissue ischemia and resultant necrosis. Based on the evidence presented earlier, we hypothesized that administration of vitamin C to patients with septic shock may decrease the requirement for exogenously administered vasopressors, through acting as a cofactor for in vivo enzyme-dependent synthesis of norepinephrine and vasopressin [59]. Our hypothesis has been supported by four recent clinical trials that showed deceased vasopressor and norepinephrine requirements (both dose and duration) following administration of 6–7 g/d of intravenous vitamin C to patients with severe sepsis and septic shock [52,53,56,57]. Other trials are currently underway to confirm these findings.

Gene Regulatory Functions

Recent research has uncovered new roles for vitamin C in the regulation of transcription factor activity and epigenetic marks, thus influencing gene transcription and cell signaling pathways [112,113]. For example, the iron- and 2-oxoglutarate–dependent asparagyl and prolyl hydroxylases required for the downregulation of the transcription factor hypoxia-inducible factor-1α (HIF-1α) utilize vitamin C as a cofactor [112]. HIF-1α is a constitutively expressed transcription factor that regulates numerous genes, including those involved in energy metabolism, angiogenic signaling, and vasomotor regulation [145]. Under normoxic conditions, HIF-1α is downregulated via hydroxylase-mediated posttranslational modifications that prevent coactivator binding and target HIF for proteosomal degradation [146]. During the hypoxia and ischemia observed with acute lung infections and sepsis, HIF is upregulated due to the absence of substrates (e.g., oxygen) and cofactors (e.g., vitamin C) required for hydroxylase-dependent downregulation. Although initially beneficial to the host, prolonged upregulation of HIF can result in pulmonary hypertension and edema [147]. HIF also facilitates neutrophil survival at hypoxic loci through delaying apoptosis [148]. In vitamin C–deficient Gulo knockout mice, upregulation of HIF-1α was observed under normoxic conditions, along with attenuated neutrophil apoptosis and clearance by macrophages [79]. HIF-1α has also been proposed as a regulator of NET release by neutrophils [149], thus providing a potential mechanism by which vitamin C could downregulate NET generation by these cells [88].

More recently, an important role for vitamin C has emerged in the regulation of DNA and histone demethylation by acting as a cofactor for iron- and 2-oxoglutarate–dependent enzymes that hydroxylate methylated epigenetic marks [113]. Vitamin C acts as a cofactor for the ten-eleven translocation (TET) dioxygenases that hydroxylate methylated cytosine moieties in DNA [150–152]. The hydroxymethylcytosine mark can be further oxidized and subsequently removed through both active and passive DNA repair mechanisms but may also represent an epigenetic mark in its own right [153]. Vitamin C is also a cofactor for several Jumonji C domain-containing histone demethylases (JHDMs) that catalyze histone demethylation [154]. Methylation of lysine and arginine residues on histones is closely associated with activation or silencing of transcription. JHDMs can hydroxylate mono-, di-, and trimethylated histone lysine and arginine residues, resulting in demethylation, and vitamin C is required for optimal catalytic activity and demethylation by JHDMs [155,156]. The involvement of vitamin C in JHDM-dependent histone demethylation was confirmed in somatic cell reprogramming [157,158]. It is likely that the gene regulatory functions of vitamin C play major roles in its immune-regulating functions. For example, preliminary evidence indicates that vitamin C can regulate T-cell maturation via epigenetic mechanisms involving the TETs and histone demethylation [159–161]. Supplementation of healthy volunteers with vitamin C was found to modulate ex vivo lipopolysaccharide-stimulated gene expression in mononuclear cells, specifically enhancing synthesis of the anti-inflammatory cytokine interleukin-10 [162]. Due to the thousands of genes regulated via both DNA and histone demethylation, epigenomic regulation by vitamin C likely plays a major role in its pleiotropic health-promoting and disease-modifying effects.

CONCLUSIONS AND FUTURE DIRECTIONS

Vitamin C has a myriad of functions that appear to contribute to beneficial outcomes in severe respiratory infections, such as pneumonia, and the potentially life-threatening condition of sepsis. These include acting as a scavenger of reactive oxygen species and a modulator of inflammatory mediators, as well as a cofactor for a variety of biosynthetic and regulatory enzymes, with the potential to modulate the transcription of thousands of genes and numerous cell signaling pathways. Observational studies indicate that patients with severe respiratory infections and sepsis have depleted vitamin C status during their illness, including a high prevalence of deficiency, despite recommended intakes. This suggests that these patients have higher requirements for vitamin C, which has been borne out in small interventional studies indicating that gram doses of vitamin C are required to normalize the vitamin C status of critically ill patients. Although optimal dietary intakes of vitamin C (i.e., 200 mg/d) may decrease the risk of developing a respiratory infection [16,163], it appears that successively higher amounts of vitamin C are required as infectious diseases progress in severity (Figure 7.5). A handful of small randomized controlled trials indicate that administration of gram doses of vitamin C to patients with pneumonia and sepsis improves multiorgan function, particularly the pulmonary, cardiovascular, and renal systems, as well as potentially decreases mortality rates. Currently, larger interventional studies are underway to confirm the effects of vitamin C on mortality. If these trials support the initially promising findings of the smaller studies, there would then be strong justification to introduce vitamin C administration into routine clinical practice for these conditions, as well as to assess the effects of vitamin C administration on related infectious conditions. Unlike many drugs that target only one specific biochemical pathway, vitamin C targets multiple pathways and hence can exhibit body-wide effects.

To date, there have been no studies that have assessed the effects of vitamin C intervention on the long-term quality-of-life outcomes of patients with pneumonia and sepsis (Figure 7.5). The reported vitamin C–dependent decreases in organ dysfunction, including decreased acute renal failure, which normally requires ongoing dialysis or a transplant, would be expected to significantly improve long-term patient quality of life. The human brain has a particularly high requirement for vitamin C, and both observational and interventional studies have shown inverse associations between vitamin C and cognitive dysfunction and psychiatric disorders such as depression and anxiety [165,166]. Furthermore, vitamin C may be able to modulate the stress response with inverse associations observed between vitamin C status and cortisol levels,

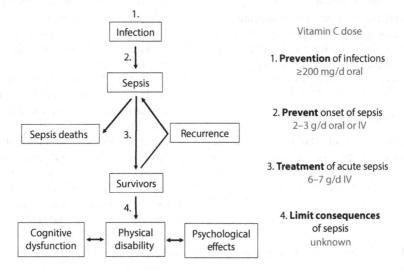

Figure 7.5. Vitamin C requirements during severe infection. The requirements for vitamin C increase during the progression from prevention of infection to treatment of acute sepsis. (Adapted from Carr, A. C. 2018. *Crit. Care* 22, 247.)

and decreased cortisol levels reported following vitamin C intervention [167]. Most of the vitamin C interventional studies carried out in critically ill patients have been for a total duration of 4 days, which may not be sufficient to have a dramatic effect on quality of life at 3 or 6 months later due to the rapid return of vitamin C levels to baseline following withdrawal of supplementation [43,168]. Because vitamin C is water soluble and is not retained by the body, sustained long-term quality-of-life effects will likely require ongoing vitamin C supplementation.

Vitamin C administration has been shown to improve the quality of life and decrease the symptoms of oncology patients, likely through repletion of inadequate vitamin C levels and decrease of the off-target toxicity and adverse side effects of chemotherapeutic drugs [169]. Animal studies have indicated that administration of drugs such as sedatives, analgesics, and muscle relaxants to vitamin C–synthesizing animals stimulates an increased synthesis/excretion of the vitamin, suggesting a higher requirement for vitamin C due to enhanced metabolism when drugs are administered [170,171]. Klenner also reported that penicillin had a retarding effect on the action of vitamin C in patients with pneumonia, and in one case, beneficial effects were not obtained until the penicillin was discontinued [24]. Therefore, it is conceivable that administration of additional vitamin C to patients with sepsis may counteract some of the adverse effects of the many drugs that are administered to these patients during intensive care. The effects of vitamin C on drug-related side effects and long-term patient quality of life should be assessed in further clinical trials of patients with severe respiratory infections and sepsis.

ACKNOWLEDGMENTS

Thank you to Dr. Paul Marik for critically reviewing the manuscript. The author is supported by a Health Research Council of New Zealand Sir Charles Hercus Health Research Fellowship.

REFERENCES

1. Hemilä, H. 2017. Vitamin C and infections. Nutrients 9, E339.
2. Duggan, C. P., Westra, S. J. and Rosenberg, A. E. 2007. Case records of the Massachusetts General Hospital. Case 23-2007. A 9-year-old boy with bone pain, rash, and gingival hypertrophy. N. Engl. J. Med. 357, 392–400.
3. Khalid, M. M. 2009. Scurvy; radiological diagnosis. Prof. Med. J. 16, 466–468.
4. Doll, S. and Ricou, B. 2013. Severe vitamin C deficiency in a critically ill adult: A case report. Eur. J. Clin. Nutr. 67, 881–882.
5. Ramar, S., Sivaramakrishnan, V. and Manoharan, K. 1993. Scurvy—A forgotten disease. Arch. Phys. Med. Rehabil. 74, 92–95.
6. Jensen, P. O., Lykkesfeldt, J., Bjarnsholt, T., Hougen, H. P., Hoiby, N. and Ciofu, O. 2012. Poor antioxidant status exacerbates oxidative stress and inflammatory response to pseudomonas aeruginosa lung infection in guinea pigs. Basic Clin. Pharmacol. Toxicol. 110, 353–358.
7. Gaut, J. P., Belaaouaj, A., Byun, J., Roberts, L. J. 2nd, Maeda, N., Frei, B. and Heinecke, J. W. 2006. Vitamin C fails to protect amino acids and lipids from oxidation during acute inflammation. Free Radic. Biol. Med. 40, 1494–1501.
8. Li, W., Maeda, N. and Beck, M. A. 2006. Vitamin C deficiency increases the lung pathology of influenza virus-infected gulo−/− mice. J. Nutr. 136, 2611–2616.
9. Buffinton, G. D., Christen, S., Peterhans, E. and Stocker, R. 1992. Oxidative stress in lungs of mice infected with influenza A virus. Free Radic. Res. Commun. 16, 99–110.
10. Hennet, T., Peterhans, E. and Stocker, R. 1992. Alterations in antioxidant defences in lung and liver of mice infected with influenza A virus. J. Gen. Virol. 73 (Pt 1), 39–46.
11. Musher, D. M. and Thorner, A. R. 2014. Community-acquired pneumonia. N. Engl. J. Med. 371, 1619–1628.
12. Levy, M. L., Le Jeune, I., Woodhead, M. A., Macfarlaned, J. T. and Lim, W. S. 2010. Primary care summary of the British Thoracic Society Guidelines for the management of community acquired pneumonia in adults: 2009. update. Endorsed by the Royal College of General Practitioners and the Primary Care Respiratory Society UK. Prim. Care Respir. J. 19, 21–27.
13. GBD 2016 Causes of Death Collaborators. 2017. Global, regional, and national age-sex specific mortality for 264 causes of death, 1980-2016: A systematic analysis for the Global Burden of Disease Study 2016. Lancet 390, 1151–1210.
14. GBD 2016 Lower Respiratory Infections Collaborators. 2018. Estimates of the global, regional, and national morbidity, mortality, and

aetiologies of lower respiratory infections in 195 countries, 1990–2016: A systematic analysis for the Global Burden of Disease Study 2016. *Lancet Infect. Dis* 18(11), 1191–1210.

15. Burton, D. C., Flannery, B., Bennett, N. M., Farley, M. M., Gershman, K., Harrison, L. H., Lynfield, R. et al. 2010. Socioeconomic and racial/ethnic disparities in the incidence of bacteremic pneumonia among US adults. *Am. J. Public Health* 100, 1904–1911.

16. Hemilä, H. and Louhiala, P. 2013. Vitamin C for preventing and treating pneumonia. *Cochrane database Syst. Rev.* 8, Cd005532.

17. Glazebrook, A. J. and Thomson, S. 1942. The administration of vitamin C in a large institution and its effect on general health and resistance to infection. *J. Hyg. (Lond.)* 42, 1–19.

18. Kimbarowski, J. A. and Mokrow, N. J. 1967. Colored precipitation reaction of the urine according to Kimbarowski (FARK) as an index of the effect of ascorbic acid during treatment of viral influenza (English translation: http://www.mv.helsinki.fi/home/hemila/T4.pdf). *Dtsch. Gesundheitsw.* 22, 2413–2418.

19. Pitt, H. A. and Costrini, A. M. 1979. Vitamin C prophylaxis in marine recruits. *JAMA* 241, 908–911.

20. Bakaev, V. V. and Duntau, A. P. 2004. Ascorbic acid in blood serum of patients with pulmonary tuberculosis and pneumonia. *Int. J. Tuberc. Lung Dis.* 8, 263–266.

21. Chakrabarti, B. and Banerjee, S. 1955. Dehydroascorbic acid level in blood of patients suffering from various infectious diseases. *Proc. Soc. Exp. Biol. Med.* 88, 581–583.

22. Hunt, C., Chakravorty, N. K., Annan, G., Habibzadeh, N. and Schorah, C. J. 1994. The clinical effects of vitamin C supplementation in elderly hospitalised patients with acute respiratory infections. *Int. J. Vitam. Nutr. Res.* 64, 212–219.

23. Mochalkin, N. I. 1970. Ascorbic acid in the complex therapy of acute pneumonia (English translation: http://www.mv.helsinki.fi/home/hemila/T5.pdf). *Voen. Med. Zh.* 9, 17–21.

24. Klenner, F. R. 1948. Virus pneumonia and its treatment with vitamin C. *South. Med. Surg.* 110, 36–38.

25. Kim, W. Y., Jo, E. J., Eom, J. S., Mok, J., Kim, M. H., Kim, K. U., Park, H. K., Lee, M. K. and Lee, K. 2018. Combined vitamin C, hydrocortisone,

and thiamine therapy for patients with severe pneumonia who were admitted to the intensive care unit: Propensity score-based analysis of a before-after cohort study. *J. Crit. Care.* 47, 211–218.

26. Cathcart, R. F. 1981. Vitamin C, titrating to bowel tolerance, anascorbemia, and acute induced scurvy. *Med. Hypotheses* 7, 1359–1376.

27. Singer, M., Deutschman, C. S., Seymour, C. W., Shankar-Hari, M., Annane, D., Bauer, M., Bellomo, R. et al. 2016. The third international consensus definitions for sepsis and septic shock (Sepsis-3). *JAMA* 315, 801–810.

28. Rhodes, A., Evans, L. E., Alhazzani, W., Levy, M. M., Antonelli, M., Ferrer, R., Kumar, A. et al. 2017. Surviving sepsis campaign: International guidelines for management of sepsis and septic shock: 2016. *Crit. Care Med.* 45, 486–552.

29. Fleischmann, C., Scherag, A., Adhikari, N. K., Hartog, C. S., Tsaganos, T., Schlattmann, P., Angus, D. C. and Reinhart, K. 2016. Assessment of global incidence and mortality of hospital-treated sepsis. Current estimates and limitations. *Am. J. Respir. Crit. Care Med.* 193, 259–272.

30. Vincent, J. L., Marshall, J. C., Namendys-Silva, S. A., Francois, B., Martin-Loeches, I., Lipman, J., Reinhart, K. et al. 2014. Assessment of the worldwide burden of critical illness: The Intensive Care Over Nations (ICON) audit. *Lancet Respir. Med.* 2, 380–386.

31. Gaieski, D. F., Edwards, J. M., Kallan, M. J. and Carr, B. G. 2013. Benchmarking the incidence and mortality of severe sepsis in the United States. *Crit. Care Med.* 41, 1167–1174.

32. Stevenson, E. K., Rubenstein, A. R., Radin, G. T., Wiener, R. S. and Walkey, A. J. 2014. Two decades of mortality trends among patients with severe sepsis: A comparative meta-analysis. *Crit. Care Med.* 42, 625–631.

33. Kaukonen, K. M., Bailey, M., Suzuki, S., Pilcher, D. and Bellomo, R. 2014. Mortality related to severe sepsis and septic shock among critically ill patients in Australia and New Zealand, 2000–2012. *JAMA* 311, 1308–1316.

34. Finfer, S., Bellomo, R., Lipman, J., French, C., Dobb, G. and Myburgh, J. 2004. Adult-population incidence of severe sepsis in Australian and New Zealand intensive care units. *Intensive Care Med.* 30, 589–596.

35. Gotts, J. E. and Matthay, M. A. 2016. Sepsis: Pathophysiology and clinical management. *BMJ* 353, i1585.

36. Prescott, H. C. and Angus, D. C. 2018. Enhancing recovery from sepsis: A review. *JAMA* 319, 62–75.

37. Schorah, C. J., Downing, C., Piripitsi, A., Gallivan, L., Al-Hazaa, A. H., Sanderson, M. J. and Bodenham, A. 1996. Total vitamin C, ascorbic acid, and dehydroascorbic acid concentrations in plasma of critically ill patients. *Am. J. Clin. Nutr.* 63, 760–765.

38. McGregor, G. P. and Biesalski, H. K. 2006. Rationale and impact of vitamin C in clinical nutrition. *Curr. Opin. Clin. Nutr. Metab. Care* 9, 697–703.

39. Carr, A. C., Rosengrave, P. C., Bayer, S., Chambers, S., Mehrtens, J. and Shaw, G. M. 2017. Hypovitaminosis C and vitamin C deficiency in critically ill patients despite recommended enteral and parenteral intakes. *Crit. Care.* 21, 300.

40. Borrelli, E., Roux-Lombard, P., Grau, G. E., Girardin, E., Ricou, B., Dayer, J. and Suter, P. M. 1996. Plasma concentrations of cytokines, their soluble receptors, and antioxidant vitamins can predict the development of multiple organ failure in patients at risk. *Crit. Care Med.* 24, 392–397.

41. Long, C. L., Maull, K. I., Krishnan, R. S., Laws, H. L., Geiger, J. W., Borghesi, L., Franks, W., Lawson, T. C. and Sauberlich, H. E. 2003. Ascorbic acid dynamics in the seriously ill and injured. *J. Surg. Res.* 109, 144–148.

42. Nathens, A. B., Neff, M. J., Jurkovich, G. J., Klotz, P., Farver, K., Ruzinski, J. T., Radella, F., Garcia, I. and Maier, R. V. 2002. Randomized, prospective trial of antioxidant supplementation in critically ill surgical patients. *Ann. Surg.* 236, 814–822.

43. de Grooth, H. J., Manubulu-Choo, W. P., Zandvliet, A. S., Spoelstra-de Man, A. M. E., Girbes, A. R., Swart, E. L. and Oudemans-van Straaten, H. M. 2018. Vitamin-C pharmacokinetics in critically ill patients: A randomized trial of four intravenous regimens. *Chest.* 153(6), 1368–1377.

44. Levine, M., Conry-Cantilena, C., Wang, Y., Welch, R. W., Washko, P. W., Dhariwal, K. R., Park, J. B. et al. 1996. Vitamin C pharmacokinetics in healthy volunteers: Evidence for a recommended dietary allowance. *Proc. Natl. Acad. Sci. USA* 93, 3704–3709.

45. Fowler, A. A., Syed, A. A., Knowlson, S., Sculthorpe, R., Farthing, D., DeWilde, C., Farthing, C. A. et al. 2014. Phase I safety trial of intravenous ascorbic acid in patients with severe sepsis. *J. Transl. Med.* 12, 32.

46. Doise, J. M., Aho, L. S., Quenot, J. P., Guilland, J. C., Zeller, M., Vergely, C., Aube, H., Blettery, B. and Rochette, L. 2008. Plasma antioxidant status in septic critically ill patients: A decrease over time. *Fundam. Clin. Pharmacol.* 22, 203–209.

47. Voigt, K., Kontush, A., Stuerenburg, H. J., Muench-Harrach, D., Hansen, H. C. and Kunze, K. 2002. Decreased plasma and cerebrospinal fluid ascorbate levels in patients with septic encephalopathy. *Free Radic. Res.* 36, 735–739.

48. Collier, B. R., Giladi, A., Dossett, L. A., Dyer, L., Fleming, S. B. and Cotton, B. A. 2008. Impact of high-dose antioxidants on outcomes in acutely injured patients. *JPEN J. Parenter. Enteral Nutr.* 32, 384–388.

49. Berger, M. M., Soguel, L., Shenkin, A., Revelly, J. P., Pinget, C., Baines, M. and Chiolero, R. L. 2008. Influence of early antioxidant supplements on clinical evolution and organ function in critically ill cardiac surgery, major trauma, and subarachnoid hemorrhage patients. *Crit. Care* 12, R101.

50. Giladi, A. M., Dossett, L. A., Fleming, S. B., Abumrad, N. N. and Cotton, B. A. 2011. High-dose antioxidant administration is associated with a reduction in post-injury complications in critically ill trauma patients. *Injury* 42, 78–82.

51. Sandesc, M., Rogobete, A. F., Bedreag, O. H., Dinu, A., Papurica, M., Cradigati, C. A., Sarandan, M. et al. 2018. Analysis of oxidative stress-related markers in critically ill polytrauma patients: An observational prospective single-center study. *Bosn. J. Basic Med. Sci.* 18(2), 191–197.

52. Zabet, M. H., Mohammadi, M., Ramezani, M. and Khalili, H. 2016. Effect of high-dose ascorbic acid on vasopressor's requirement in septic shock. *J. Res. Pharm. Pract.* 5, 94–100.

53. Nabil Habib, T. and Ahmed, I. 2017. Early adjuvant intravenous vitamin C treatment in septic shock may resolve the vasopressor dependence. *Int. J. Microbiol. Adv. Immunol.* 05, 77–81.

54. Litwak, J. J., Cho, N., Nguyen, H. B., Moussavi, K. and Bushell, T. 2019. Vitamin C, hydrocortisone, and thiamine for the treatment of severe sepsis and septic shock: A retrospective analysis of real-world application. *J. Clin. Med.* 8.

55. Shin, T. G., Kim, Y. J., Ryoo, S. M., Hwang, S. Y., Jo, I. J., Chung, S. P., Choi, S. H., Suh, G. J. and Kim, W. Y. 2019. Early vitamin C and thiamine administration to patients with septic shock in

emergency departments: Propensity score-based analysis of a before-and-after cohort study. *J. Clin. Med.* 8.

56. Balakrishnan, M., Gandhi, H., Shah, K., Pandya, H., Patel, R., Keshwani, S. and Yadav, N. 2018. Hydrocortisone, vitamin C and thiamine for the treatment of sepsis and septic shock following cardiac surgery. *Indian J. Anaesth.* 62, 934–939.

57. Marik, P. E., Khangoora, V., Rivera, R., Hooper, M. H. and Catravas, J. 2017. Hydrocortisone, vitamin C, and thiamine for the treatment of severe sepsis and septic shock: A retrospective before-after study. *Chest* 151, 1229–1238.

58. Carr, A. C. 2019. Duration of intravenous vitamin C therapy is a critical consideration. *Crit. Care Resusc.* 21(3), 220–221.

59. Carr, A. C., Shaw, G. M., Fowler, A. A. and Natarajan, R. 2015. Ascorbate-dependent vasopressor synthesis: A rationale for vitamin C administration in severe sepsis and septic shock? *Crit. Care* 19, e418.

60. Fein, A. M. and Calalang-Colucci, M. G. 2000. Acute lung injury and acute respiratory distress syndrome in sepsis and septic shock. *Crit. Care Clin.* 16, 289–317.

61. Zemans, R. L. and Matthay, M. A. 2004. Bench-to-bedside review: The role of the alveolar epithelium in the resolution of pulmonary edema in acute lung injury. *Crit. Care* 8, 469–477.

62. Fisher, B. J., Kraskauskas, D., Martin, E. J., Farkas, D., Wegelin, J. A., Brophy, D., Ward, K. R., Voelkel, N. F., Fowler, A. A.3rd, and Natarajan, R. 2012. Mechanisms of attenuation of abdominal sepsis induced acute lung injury by ascorbic acid. *Am. J. Physiol. Lung Cell Mol. Physiol.* 303, L20–L32.

63. Mutlu, G. M. and Sznajder, J. I. 2005. Mechanisms of pulmonary edema clearance. *Am. J. Physiol. Lung Cell Mol. Physiol.* 289, L685–L695.

64. Barabutis, N., Khangoora, V., Marik, P. E. and Catravas, J. D. 2017. Hydrocortisone and ascorbic acid synergistically prevent and repair lipopolysaccharide-induced pulmonary endothelial barrier dysfunction. *Chest* 152, 954–962.

65. Bharara, A., Grossman, C., Grinnan, D., Syed, A. A., Fisher, B. J., DeWilde, C., Natarajan, R. and Fowler, A. A. 2016. Intravenous vitamin C administered as adjunctive therapy for recurrent acute respiratory distress syndrome. *Case Rep. Crit. Care* Article ID 8560871, 2016, 4p.

66. Fowler, A. A., Kim, C., Lepler, L., Malhotra, R., Debesa, O., Natarajan, R., Fisher, B. J. et al. 2017. Intravenous vitamin C as adjunctive therapy for enterovirus/rhinovirus induced acute respiratory distress syndrome. *World J. Crit. Care Med.* 6, 85–90.

67. Marik, P. E. and Long, A. 2018. ARDS complicating pustular psoriasis: Treatment with low-dose corticosteroids, vitamin C and thiamine. *BMJ Case Rep.* 2018.

68. Gurganus, M. M., Marik, P. E. and Varon, J. 2019. The successful treatment of severe aspiration pneumonitis with the combination of hydrocortisone, ascorbic acid, and thiamine. *Crit. Care Shock* 22, 57–61.

69. Brown, K. A., Brain, S. D., Pearson, J. D., Edgeworth, J. D., Lewis, S. M. and Treacher, D. F. 2006. Neutrophils in development of multiple organ failure in sepsis. *Lancet* 368, 157–169.

70. Pechous, R. D. 2017. With friends like these: The complex role of neutrophils in the progression of severe pneumonia. *Front. Cell. Infect. Microbiol.* 7, 160.

71. Fox, S., Leitch, A. E., Duffin, R., Haslett, C. and Rossi, A. G. 2010. Neutrophil apoptosis: Relevance to the innate immune response and inflammatory disease. *J. Innate. Immun.* 2, 216–227.

72. Taneja, R., Parodo, J., Jia, S. H., Kapus, A., Rotstein, O. D. and Marshall, J. C. 2004. Delayed neutrophil apoptosis in sepsis is associated with maintenance of mitochondrial transmembrane potential and reduced caspase-9 activity. *Crit. Care Med.* 32, 1460–1469.

73. Tamayo, E., Gomez, E., Bustamante, J., Gomez-Herreras, J. I., Fonteriz, R., Bobillo, F., Bermejo-Martin, J. F. et al. 2012. Evolution of neutrophil apoptosis in septic shock survivors and nonsurvivors. *J. Crit. Care* 27, 415.e1–415.e11.

74. Fialkow, L., Fochesatto Filho, L., Bozzetti, M. C., Milani, A. R., Rodrigues Filho, E. M., Ladniuk, R. M., Pierozan, P. et al. 2006. Neutrophil apoptosis: A marker of disease severity in sepsis and sepsis-induced acute respiratory distress syndrome. *Crit. Care* 10, R155.

75. Colotta, F., Re, F., Polentarutti, N., Sozzani, S. and Mantovani, A. 1992. Modulation of granulocyte survival and programmed cell death by cytokines and bacterial products. *Blood* 80, 2012–2020.

76. Wilkie, R. P., Vissers, M. C., Dragunow, M. and Hampton, M. B. 2007. A functional NADPH oxidase prevents caspase involvement in the

clearance of phagocytic neutrophils. *Infect. Immun.* 75, 3256–3263.

77. Carr, A. C. and Maggini, S. 2017. Vitamin C and immune function. *Nutrients* 9.

78. Vissers, M. C. and Hampton, M. B. 2004. The role of oxidants and vitamin C on neutrophil apoptosis and clearance. *Biochem. Soc. Trans.* 32, 499–501.

79. Vissers, M. C. and Wilkie, R. P. 2007. Ascorbate deficiency results in impaired neutrophil apoptosis and clearance and is associated with up-regulation of hypoxia-inducible factor 1alpha. *J. Leukoc. Biol.* 81, 1236–1244.

80. Wang, X., Yousefi, S. and Simon, H. U. 2018. Necroptosis and neutrophil-associated disorders. *Cell Death Dis.* 9, 111.

81. Desai, J., Mulay, S. R., Nakazawa, D. and Anders, H. J. 2016. Matters of life and death. How neutrophils die or survive along NET release and is "NETosis" = necroptosis? *Cell Mol. Life Sci.* 73, 2211–2219.

82. Zawrotniak, M. and Rapala-Kozik, M. 2013. Neutrophil extracellular traps (NETs)—Formation and implications. *Acta Biochim. Pol.* 60, 277–284.

83. Brinkmann, V., Reichard, U., Goosmann, C., Fauler, B., Uhlemann, Y., Weiss, D. S., Weinrauch, Y. and Zychlinsky, A. 2004. Neutrophil extracellular traps kill bacteria. *Science* 303, 1532–1535.

84. Czaikoski, P. G., Mota, J. M., Nascimento, D. C., Sonego, F., Castanheira, F. V., Melo, P. H., Scortegagna, G. T. et al. 2016. Neutrophil extracellular traps induce organ damage during experimental and clinical sepsis. *PLOS ONE* 11, e0148142.

85. Camicia, G., Pozner, R. and de Larranaga, G. 2014. Neutrophil extracellular traps in sepsis. *Shock* 42, 286–294.

86. Silk, E., Zhao, H., Weng, H. and Ma, D. 2017. The role of extracellular histone in organ injury. *Cell Death Dis.* 8, e2812.

87. Margraf, S., Logters, T., Reipen, J., Altrichter, J., Scholz, M. and Windolf, J. 2008. Neutrophil-derived circulating free DNA (cf-DNA/NETs): A potential prognostic marker for posttraumatic development of inflammatory second hit and sepsis. *Shock* 30, 352–358.

88. Mohammed, B. M., Fisher, B. J., Kraskauskas, D., Farkas, D., Brophy, D. F., Fowler, A. A. and Natarajan, R. 2013. Vitamin C: A novel regulator of neutrophil extracellular trap formation. *Nutrients* 5, 3131–3151.

89. Natarajan, R., Fisher, B. J., Syed, A. A. and Fowler, A. A. 2014. Impact of intravenous ascorbic acid infusion on novel biomarkers in patients with severe sepsis. *J. Pulm. Respir. Med.* 4, 8p.

90. Demaret, J., Venet, F., Friggeri, A., Cazalis, M. A., Plassais, J., Jallades, L., Malcus, C. et al. 2015. Marked alterations of neutrophil functions during sepsis-induced immunosuppression. *J. Leukoc. Biol.* 98, 1081–1090.

91. Alves-Filho, J. C., Spiller, F. and Cunha, F. Q. 2010. Neutrophil paralysis in sepsis. *Shock* 34 (Suppl 1), 15–21.

92. Hotchkiss, R. S., Monneret, G. and Payen, D. 2013. Sepsis-induced immunosuppression: From cellular dysfunctions to immunotherapy. *Nat. Rev. Immunol.* 13, 862–874.

93. Rebora, A., Crovato, F., Dallegri, F. and Patrone, F. 1980. Repeated staphylococcal pyoderma in two siblings with defective neutrophil bacterial killing. *Dermatologica.* 160, 106–112.

94. Patrone, F., Dallegri, F., Bonvini, E., Minervini, F. and Sacchetti, C. 1982. Disorders of neutrophil function in children with recurrent pyogenic infections. *Med. Microbiol. Immunol.* 171, 113–122.

95. Boura, P., Tsapas, G., Papadopoulou, A., Magoula, I. and Kountouras, G. 1989. Monocyte locomotion in anergic chronic brucellosis patients: The in vivo effect of ascorbic acid. *Immunopharmacol. Immunotoxicol.* 11, 119–129.

96. Anderson, R. and Theron, A. 1979. Effects of ascorbate on leucocytes: Part III. In vitro and in vivo stimulation of abnormal neutrophil motility by ascorbate. *S. Afr. Med. J.* 56, 429–433.

97. Corberand, J., Nguyen, F., Fraysse, B. and Enjalbert, L. 1982. Malignant external otitis and polymorphonuclear leukocyte migration impairment. Improvement with ascorbic acid. *Arch. Otolaryngol.* 108, 122–124.

98. Levy, R. and Schlaeffer, F. 1993. Successful treatment of a patient with recurrent furunculosis by vitamin C: Improvement of clinical course and of impaired neutrophil functions. *Int. J. Dermatol.* 32, 832–834.

99. Vohra, K., Khan, A. J., Telang, V., Rosenfeld, W. and Evans, H. E. 1990. Improvement of neutrophil migration by systemic vitamin C in neonates. *J. Perinatol.* 10, 134–136.

100. Rebora, A., Dallegri, F. and Patrone, F. 1980. Neutrophil dysfunction and repeated infections: Influence of levamisole and ascorbic acid. *Br. J. Dermatol.* 102, 49–56.

101. Washko, P., Rotrosen, D. and Levine, M. 1989. Ascorbic acid transport and accumulation in human neutrophils. J. Biol. Chem. 264, 18996–19002.

102. Evans, R. M., Currie, L. and Campbell, A. 1982. The distribution of ascorbic acid between various cellular components of blood, in normal individuals, and its relation to the plasma concentration. Br. J. Nutr. 47, 473–482.

103. Corpe, C. P., Lee, J. H., Kwon, O., Eck, P., Narayanan, J., Kirk, K. L. and Levine, M. 2005. 6-Bromo-6-deoxy-L-ascorbic acid: An ascorbate analog specific for Na+-dependent vitamin C transporter but not glucose transporter pathways. J. Biol. Chem. 280, 5211–5220.

104. Hume, R. and Weyers, E. 1973. Changes in leucocyte ascorbic acid during the common cold. Scott. Med. J. 18, 3–7.

105. Carr, A. and Frei, B. 1999. Does vitamin C act as a pro-oxidant under physiological conditions? FASEB J. 13, 1007–1024.

106. Sen, C. K. and Packer, L. 1996. Antioxidant and redox regulation of gene transcription. FASEB J. 10, 709–720.

107. Macdonald, J., Galley, H. F. and Webster, N. R. 2003. Oxidative stress and gene expression in sepsis. Br. J. Anaesth. 90, 221–232.

108. Chousterman, B. G., Swirski, F. K. and Weber, G. F. 2017. Cytokine storm and sepsis disease pathogenesis. Semin. Immunopathol. 39, 517–528.

109. Wilson, J. X. 2013. Evaluation of vitamin C for adjuvant sepsis therapy. Antioxid. Redox Signal. 19, 2129–2140.

110. Oudemans-van Straaten, H. M., Spoelstra-de Man, A. M. and de Waard, M. C. 2014. Vitamin C revisited. Crit. Care 18, 460.

111. England, S. and Seifter, S. 1986. The biochemical functions of ascorbic acid. Annu. Rev. Nutr. 6, 365–406.

112. Kuiper, C. and Vissers, M. C. 2014. Ascorbate as a co-factor for Fe- and 2-oxoglutarate dependent dioxygenases: Physiological activity in tumor growth and progression. Front. Oncol. 4, 359.

113. Young, J. I., Zuchner, S. and Wang, G. 2015. Regulation of the epigenome by vitamin C. Annu. Rev. Nutr. 35, 545–564.

114. May, J. M. and Harrison, F. E. 2013. Role of vitamin C in the function of the vascular endothelium. Antioxid. Redox Signal. 19, 2068–2083.

115. Rebouche, C. J. 1991. Ascorbic acid and carnitine biosynthesis. Am. J. Clin. Nutr. 54, 1147S–1152S.

116. Brealey, D., Brand, M., Hargreaves, I., Heales, S., Land, J., Smolenski, R., Davies, N. A., Cooper, C. E. and Singer, M. 2002. Association between mitochondrial dysfunction and severity and outcome of septic shock. Lancet 360, 219–223.

117. Moskowitz, A., Andersen, L. W., Huang, D. T., Berg, K. M., Grossestreuer, A. V., Marik, P. E., Sherwin, R. L. et al. 2018. Ascorbic acid, corticosteroids, and thiamine in sepsis: A review of the biologic rationale and the present state of clinical evaluation. Crit. Care 22, 283.

118. De Backer, D. and Scolletta, S. 2013. Clinical management of the cardiovascular failure in sepsis. Curr. Vasc. Pharmacol. 11, 222–242.

119. Levine, M. 1986. Ascorbic acid specifically enhances dopamine beta-monooxygenase activity in resting and stimulated chromaffin cells. J. Biol. Chem. 261, 7347–7356.

120. May, J. M., Qu, Z. C., Nazarewicz, R. and Dikalov, S. 2013. Ascorbic acid efficiently enhances neuronal synthesis of norepinephrine from dopamine. Brain Res. Bull. 90, 35–42.

121. May, J. M., Qu, Z. C. and Meredith, M. E. 2012. Mechanisms of ascorbic acid stimulation of norepinephrine synthesis in neuronal cells. Biochem. Biophys. Res. Commun. 426, 148–152.

122. Hornig, D. 1975. Distribution of ascorbic acid, metabolites and analogues in man and animals. Ann. N. Y. Acad. Sci. 258, 103–118.

123. Vissers, M. C., Bozonet, S. M., Pearson, J. F. and Braithwaite, L. J. 2011. Dietary ascorbate intake affects steady state tissue concentrations in vitamin C-deficient mice: Tissue deficiency after suboptimal intake and superior bioavailability from a food source (kiwifruit). Am. J. Clin. Nutr. 93, 292–301.

124. Hughes, R. E., Hurley, R. J. and Jones, P. R. 1971. The retention of ascorbic acid by guinea-pig tissues. Br. J. Nutr. 26, 433–438.

125. Hasselholt, S., Tveden-Nyborg, P. and Lykkesfeldt, J. 2015. Distribution of vitamin C is tissue specific with early saturation of the brain and adrenal glands following differential oral dose regimens in guinea pigs. Br. J. Nutr. 113, 1539–1549.

126. Nieboer, P., van der Werf, T. S., Beentjes, J. A., Tulleken, J. E., Zijlstra, J. G. and Ligtenberg, J. J. 2000. Catecholamine dependency in a poly-trauma patient: Relative adrenal insufficiency? Intensive Care Med. 26, 125–127.

127. Duggan, M., Browne, I. and Flynn, C. 1998. Adrenal failure in the critically ill. Br. J. Anaesth. 81, 468–470.

128. Hoehn, S. K. and Kanfer, J. N. 1980. Effects of chronic ascorbic acid deficiency on guinea pig lysosomal hydrolase activities. J. Nutr. 110, 2085–2094.

129. Deana, R., Bharaj, B. S., Verjee, Z. H. and Galzigna, L. 1975. Changes relevant to catecholamine metabolism in liver and brain of ascorbic acid deficient guinea-pigs. Int. J. Vitam. Nutr. Res. 45, 175–182.

130. Bornstein, S. R., Yoshida-Hiroi, M., Sotiriou, S., Levine, M., Hartwig, H. G., Nussbaum, R. L. and Eisenhofer, G. 2003. Impaired adrenal catecholamine system function in mice with deficiency of the ascorbic acid transporter (SVCT2). FASEB J. 17, 1928–1930.

131. Padayatty, S. J., Doppman, J. L., Chang, R., Wang, Y., Gill, J., Papanicolaou, D. A. and Levine, M. 2007. Human adrenal glands secrete vitamin C in response to adrenocorticotrophic hormone. Am. J. Clin. Nutr. 86, 145–149.

132. Prigge, S. T., Mains, R. E., Eipper, B. A. and Amzel, L. M. 2000. New insights into copper monooxygenases and peptide amidation: Structure, mechanism and function. Cell. Mol. Life Sci. 57, 1236–1259.

133. Merkler, D. J. 1994. C-terminal amidated peptides: Production by the in vitro enzymatic amidation of glycine-extended peptides and the importance of the amide to bioactivity. Enzyme Microb. Technol. 16, 450–456.

134. Treschan, T. A. and Peters, J. 2006. The vasopressin system: Physiology and clinical strategies. Anesthesiology. 105, 599–612; quiz 639-540.

135. Russell, J. A. 2011. Bench-to-bedside review: Vasopressin in the management of septic shock. Crit. Care. 15, 226.

136. Giusti-Paiva, A. and Domingues, V. G. 2010. Centrally administered ascorbic acid induces antidiuresis, natriuresis and neurohypophyseal hormone release in rats. Neuro Endocrinol. Lett. 31, 87–91.

137. Landry, D. W., Levin, H. R., Gallant, E. M., Ashton, R. C., Jr., Seo, S., D'Alessandro, D., Oz, M. C. and Oliver, J. A. 1997. Vasopressin deficiency contributes to the vasodilation of septic shock. Circulation 95, 1122–1125.

138. Sharshar, T., Blanchard, A., Paillard, M., Raphael, J. C., Gajdos, P. and Annane, D. 2003. Circulating vasopressin levels in septic shock. Crit. Care Med. 31, 1752–1758.

139. Sharshar, T., Carlier, R., Blanchard, A., Feydy, A., Gray, F., Paillard, M., Raphael, J. C., Gajdos, P. and Annane, D. 2002. Depletion of neurohypophyseal content of vasopressin in septic shock. Crit. Care Med. 30, 497–500.

140. Kruger, S., Papassotiriou, J., Marre, R., Richter, K., Schumann, C., von Baum, H., Morgenthaler, N. G., Suttorp, N. and Welte, T. 2007. Pro-atrial natriuretic peptide and pro-vasopressin to predict severity and prognosis in community-acquired pneumonia: Results from the German competence network CAPNETZ. Intensive Care Med. 33, 2069–2078.

141. Guignant, C., Voirin, N., Venet, F., Poitevin, F., Malcus, C., Bohe, J., Lepape, A. and Monneret, G. 2009. Assessment of pro-vasopressin and pro-adrenomedullin as predictors of 28-day mortality in septic shock patients. Intensive Care Med. 35, 1859–1867.

142. Becker, K. L., Nylen, E. S., White, J. C., Muller, B. and Snider, R. H., Jr. 2004. Procalcitonin and the calcitonin gene family of peptides in inflammation, infection, and sepsis: A journey from calcitonin back to its precursors. J. Clin. Endocrinol. Metab. 89, 1512–1525.

143. Arora, S., Singh, P., Singh, P. M. and Trikha, A. 2015. Procalcitonin levels in survivors and nonsurvivors of sepsis: Systematic review and meta-analysis. Shock 43, 212–221.

144. Vasu, T. S., Cavallazzi, R., Hirani, A., Kaplan, G., Leiby, B. and Marik, P. E. 2012. Norepinephrine or dopamine for septic shock: Systematic review of randomized clinical trials. J. Intensive Care Med. 27, 172–178.

145. Schofield, C. J. and Ratcliffe, P. J. 2004. Oxygen sensing by HIF hydroxylases. Nat. Rev. Mol. Cell Biol. 5, 343–354.

146. Hirota, K. and Semenza, G. L. 2005. Regulation of hypoxia-inducible factor 1 by prolyl and asparaginyl hydroxylases. Biochem. Biophys. Res. Commun. 338, 610–616.

147. Dunham-Snary, K. J., Wu, D., Sykes, E. A., Thakrar, A., Parlow, L. R. G., Mewburn, J. D., Parlow, J. L. and Archer, S. L. 2017. Hypoxic pulmonary vasoconstriction: From molecular mechanisms to medicine. Chest 151, 181–192.

148. Elks, P. M., van Eeden, F. J., Dixon, G., Wang, X., Reyes-Aldasoro, C. C., Ingham, P. W., Whyte, M. K., Walmsley, S. R. and Renshaw, S. A. 2011. Activation of hypoxia-inducible factor-1alpha (Hif-1alpha) delays inflammation resolution by reducing neutrophil apoptosis and reverse migration in a zebrafish inflammation model. Blood 118, 712–722.

149. McInturff, A. M., Cody, M. J., Elliott, E. A., Glenn, J. W., Rowley, J. W., Rondina, M. T. and

Yost, C. C. 2012. Mammalian target of rapamycin regulates neutrophil extracellular trap formation via induction of hypoxia-inducible factor 1 alpha. *Blood* 120, 3118–3125.

150. Minor, E. A., Court, B. L., Young, J. I. and Wang, G. 2013. Ascorbate induces Ten-eleven translocation (Tet) methylcytosine dioxygenase-mediated generation of 5-hydroxymethylcytosine. *J. Biol. Chem.* 288(19), 13669–13674.

151. Yin, R., Mao, S. Q., Zhao, B., Chong, Z., Yang, Y., Zhao, C., Zhang, D. et al. 2013. Ascorbic acid enhances Tet-mediated 5-methylcytosine oxidation and promotes DNA demethylation in mammals. *J. Am. Chem. Soc.* 135(28), 10396–10403.

152. Blaschke, K., Ebata, K. T., Karimi, M. M., Zepeda-Martinez, J. A., Goyal, P., Mahapatra, S., Tam, A. et al. 2013. Vitamin C induces Tet-dependent DNA demethylation and a blastocyst-like state in ES cells. *Nature* 500, 222–226.

153. Song, C. X. and He, C. 2013. Potential functional roles of DNA demethylation intermediates. *Trends Biochem. Sci.* 38, 480–484.

154. Camarena, V. and Wang, G. 2016. The epigenetic role of vitamin C in health and disease. *Cell Mol. Life Sci.* 73, 1645–1658.

155. Klose, R. J., Kallin, E. M. and Zhang, Y. 2006. JmjC-domain-containing proteins and histone demethylation. *Nat. Rev. Genet.* 7, 715–727.

156. Tsukada, Y., Fang, J., Erdjument-Bromage, H., Warren, M. E., Borchers, C. H., Tempst, P. and Zhang, Y. 2006. Histone demethylation by a family of JmjC domain-containing proteins. *Nature* 439, 811–816.

157. Wang, T., Chen, K., Zeng, X., Yang, J., Wu, Y., Shi, X., Qin, B. et al. 2011. The histone demethylases Jhdm1a/1b enhance somatic cell reprogramming in a vitamin-C-dependent manner. *Cell Stem Cell.* 9, 575–587.

158. Ebata, K. T., Mesh, K., Liu, S., Bilenky, M., Fekete, A., Acker, M. G., Hirst, M., Garcia, B. A. and Ramalho-Santos, M. 2017. Vitamin C induces specific demethylation of H3K9me2 in mouse embryonic stem cells via Kdm3a/b. *Epigenetics Chromatin.* 10, 36.

159. Manning, J., Mitchell, B., Appadurai, D. A., Shakya, A., Pierce, L. J., Wang, H., Nganga, V. et al. 2013. Vitamin C promotes maturation of T-cells. *Antioxid. Redox Signal.* 19, 2054–2067.

160. Sasidharan Nair, V., Song, M. H. and Oh, K. I. 2016. Vitamin C facilitates demethylation of the Foxp3 enhancer in a Tet-dependent manner. *J. Immunol.* 196, 2119–2131.

161. Song, M. H., Nair, V. S. and Oh, K. I. 2017. Vitamin C enhances the expression of IL17 in a Jmjd2-dependent manner. *BMB Reports* 50, 49–54.

162. Canali, R., Natarelli, L., Leoni, G., Azzini, E., Comitato, R., Sancak, O., Barella, L. and Virgili, F. 2014. Vitamin C supplementation modulates gene expression in peripheral blood mononuclear cells specifically upon an inflammatory stimulus: A pilot study in healthy subjects. *Genes. Nutr.* 9, 390.

163. Hemilä, H. and Chalker, E. 2013. Vitamin C for preventing and treating the common cold. *Cochrane Database Syst. Rev.* 1, CD000980.

164. Carr, A. C. 2018. Can a simple chemical help to both prevent and treat sepsis. *Crit. Care* 22, 247.

165. Travica, N., Ried, K., Sali, A., Scholey, A., Hudson, I. and Pipingas, A. 2017. Vitamin C status and cognitive function: A systematic review. *Nutrients* 9.

166. Kocot, J., Luchowska-Kocot, D., Kielczykowska, M., Musik, I. and Kurzepa, J. 2017. Does vitamin C influence neurodegenerative diseases and psychiatric disorders? *Nutrients* 9.

167. Hooper, M. H., Carr, A. and Marik, P. E. 2019. The adrenal-vitamin C axis: From fish to guinea pigs and primates. *Crit. Care* 23, 29.

168. Carr, A. C., Bozonet, S. M., Pullar, J. M., Simcock, J. W. and Vissers, M. C. 2013. Human skeletal muscle ascorbate is highly responsive to changes in vitamin C intake and plasma concentrations. *Am. J. Clin. Nutr.* 97, 800–807.

169. Carr, A. C., Vissers, M. C. M. and Cook, J. S. 2014. The effect of intravenous vitamin C on cancer- and chemotherapy-related fatigue and quality of life. *Front. Oncol.* 4, 1–7.

170. Burns, J. J., Mosbach, E. H. and Schulenberg, S. 1954. Ascorbic acid synthesis in normal and drug-treated rats, studied with L-ascorbic-1-C14 acid. *J. Biol. Chem.* 207, 679–687.

171. Conney, A. H., Bray, G. A., Evans, C. and Burns, J. J. 1961. Metabolic interactions between L-ascorbic acid and drugs. *Ann. N. Y. Acad. Sci.* 92, 115–127.

CHAPTER EIGHT

Vitamin C in Immune Cell Function

Abel Ang, Margreet C.M. Vissers, and Juliet M. Pullar

DOI: 10.1201/9780429442025-8

CONTENTS

INTRODUCTION

Humans have an absolute requirement for vitamin C (ascorbate) as part of their diet, with its absence resulting in the deficiency disease scurvy. Historically, scurvy was commonly observed in sailors on long voyages who were deprived of fresh fruit and vegetables for extended periods. Pneumonia was a well-recognized symptom of scurvy and often a cause of death for those with the illness [1]. More recently, there have been several Cochrane reviews that have methodically combined studies conducted to determine the effect of ascorbate supplementation on different respiratory infections [2,3]. These analyses have shown that ascorbate intake can reduce the incidence of the common cold and pneumonia in particular subgroups of the population, such as those under severe physical stress. There were also therapeutic effects of ascorbate on the duration and severity of these illnesses. Preclinical studies have also suggested that ascorbate decreases the

incidence and severity of infections [4]. Thus, the efficacy of ascorbate in treating infections as a concept is now gaining traction, particularly with recent clinical studies using substantial doses of ascorbate to supplement and treat patients with severe sepsis or septic shock [5,6]. Whereas many of these clinical effects of ascorbate supplementation may reflect its supportive functions on many biological processes, it is also possible that modulation of immune cell functions may affect the course of these acute infectious diseases.

There is widespread belief that vitamin C supports the immune system, but despite this, there remains considerable controversy surrounding its capacity to influence the immune and inflammatory response. A way to address this question is to consider the different facets of the immune system and the ability of ascorbate to modulate these activities. The core chemical property of ascorbate is its ability to act as an electron donor, with all of its known biological functions dependent on this activity, and hence its reputation as an antioxidant [7]. Through this reducing capacity, ascorbate functions as a cofactor to a number of copper (Cu)- and iron (Fe)-containing biosynthetic enzymes, exemplified by the Cu-containing dopamine β-hydroxylase, which converts dopamine to norepinephrine, and the Fe-dependent collagen prolyl and lysyl hydroxylases that form cross-links to stabilize the tertiary structure of collagen [8–11]. It has become apparent that the enzymes requiring ascorbate as a cofactor include other newly characterized hydroxylases that regulate gene transcription and cell signaling pathways [12,13]. These hydroxylases belong to the family of Fe-containing 2-oxoglutarate–dependent dioxygenases (2-OGDDs); members of this family are widespread throughout biology and include enzymes involved in biosynthesis, posttranslational protein modification, and the oxidative demethylation of methylcytosine and methylated histone residues [14–18]. Prominent examples of these enzymes include the DNA demethylases (ten-eleven translocases [TETs]), histone demethylases (Jumonji C domain-containing histone demethylases [JHMDs]), and the prolyl, lysyl, and asparagine hydroxylases (PHDs) that modify the α regulatory subunit of the hypoxia-inducible factors (HIFs) [15,18].

In this chapter, we consider the roles of ascorbate in regulating the immune system, particularly with respect to its cell signaling and gene regulatory cofactor activities. We examine the functional effects of ascorbate on cells of both the innate and adaptive immune responses and, as cancer has an inflammatory component [19], we include a discussion of the role of ascorbate on immune cells in a cancer setting. The contributions of ascorbate as an antioxidant in immune cells have been well reviewed by others [20–25] and will not be discussed here.

ASCORBATE LEVELS IN IMMUNE CELLS

One of the main drivers of the interest in the function of ascorbate in the immune system is that leukocytes accumulate the vitamin to high intracellular concentrations, signaling an important role for it in these cells [26–29]. The intracellular ascorbate concentrations in freshly isolated circulating lymphocytes, monocytes, and neutrophils have been reported to be ~3.5, ~3,

and ~1.5 mM, respectively, when plasma levels are at least 50 μM, the status in healthy individuals consuming ≥100 mg ascorbate daily [28,30]. However, in the general population without controlling for ascorbate intake, levels of freshly isolated circulating T and B cells were reported to be ~1 and 1.5 mM, respectively, while monocytes and neutrophils were similar [26].

When plasma levels fall below 20 μM, immune cell ascorbate content decreases, with intracellular concentrations at around 1.5, 1.2, and 0.5 mM in lymphocytes, monocytes, and neutrophils, respectively [28,30]. Plasma levels below 23 μM represent a state of hypovitaminosis C and are commonly seen in individuals with infrequent fresh fruit and vegetable intake [31–36]. In addition, there is substantial evidence that plasma and cellular ascorbate levels are depressed in patients with active inflammatory disease [37–40] and particularly in very ill patient populations [37,40,41]. Low ascorbate status has also been measured in patients with cancer, including hematological cancers [41–49]. Ascorbate loss during illness is thought to reflect increased turnover due to oxidative and metabolic stress [50,51]. These data suggest a significant degree of variability in immune cell ascorbate status as reflecting the range of plasma levels during normal health as seen in the general population, and also in illness [31–36].

It is also interesting to note that immune cells are capable of accumulating ascorbate beyond freshly isolated levels. When cultured in vitro in the presence of 100 μM ascorbate, immune cells can accumulate up to three times their initial ascorbate content with minimal loss over 24 hours [26]. This is equivalent to the 100 μM plasma levels achievable solely through dietary intake [52]. The highest resting ascorbate level measured under these in vitro conditions was in monocytes at 10 mM, while activated neutrophils accumulated levels up to 14 mM (Table 8.1) [26,29,53]. This variable availability of ascorbate in immune cells may modulate ascorbate-dependent enzyme reactions and thereby affect the functioning of the immune response.

The reasons for the high intracellular requirement for ascorbate in immune cells are suggestive of an essential role in these cells. The emerging understanding of the dependency of many critical functions of immune cells on enzymes that require ascorbate as a cofactor is likely to be pertinent

TABLE 8.1

Ascorbate levels in different immune cells from donors without diet restrictions (fresh isolated) or following in vitro *supplementation and accumulation*

Cell type	Fresh Isolated	In vitro Accumulation	References
Peripheral T cells	$\sim$1 mM	Up to 2.5 mM in 100 μM ascorbate	[26]
Peripheral B cells	$\sim$1.5 mM	Up to 2 mM in 100 μM ascorbate	[26]
Peripheral monocytes	$\sim$3 mM	Up to 10 mM in 100 μM ascorbate	[26]
Peripheral neutrophils	$\sim$1.5 mM	In excess of 5 mM in 50 μM ascorbate. Activated neutrophils take up $\sim$14 mM	[29,53,181]
Peritoneal macrophages	$\sim$2.5 mM	Up to 8 mM in 200 μM ascorbate. Phagocytosis depletes intracellular ascorbate	[182]

to this discussion. These functions include the differentiation of white cells in the bone marrow and, hence, the development of immune cells [54–58]. In addition, many inflammatory sites, including cancers, are known to be hypoxic environments [59–64], with activation of the hypoxic response being an integral component of inflammation. The interaction of ascorbate with these immune processes is discussed later.

HYPOXIA-INDUCIBLE FACTORS AND DEPENDENCY ON ASCORBATE

Immune cells undergo dramatic metabolic changes following activation, and increased aerobic glycolytic activity and fatty acid oxidation have been observed [65,66]. These metabolic changes, once thought to be a consequence of cell activation, are now being reexamined as a mechanism for phenotype switching, termed *metabolic reprogramming* (reviewed in [67]). Central to this switch are the HIFs, transcription factors that regulate the expression of hundreds of genes that not only upregulate the glycolytic machinery but also direct the inflammatory and immune response (reviewed in [66]) [62,63].

The hypoxia-induced transcription factors are expressed ubiquitously and are heterodimeric complexes of a constitutively expressed β subunit and a regulatory α subunit (isoforms HIF-1α, HIF-2α, HIF-3α). HIF activation is regulated by posttranslational modification of proline and asparagine residues on the HIF-α proteins [59–61,68]. This reaction is carried out by hydroxylases that are members of the 2-OGDD family [18,68,69]. Proline hydroxylation results in recruitment of von Hippel-Lindau protein and

targeting to the proteasome for protein degradation. Asparagine modification prevents the formation of an active transcription complex, and the combined hydroxylation events thereby provide a dual control mechanism to prevent inadvertent activation of HIF-mediated transcription [18,68,69]. Three proline hydroxylases, PHD1-3, and the asparagine hydroxylase known as factor inhibiting HIF (FIH) are grouped together as the HIF hydroxylases and represent a distinct subset of 2-OGGDs with unique oxygen sensing capacity [68,69]. These enzymes have been shown to require ascorbate for optimal activity [70,71]. This dependency has been demonstrated in cell free systems [68,71,72], and other reducing agents such as glutathione are much less effective as recyclers of the hydroxylase active site Fe^{2+} [71,73–76]. Depleted intracellular ascorbate levels have been shown to contribute to the upregulation of HIF activation, particularly under conditions of mild or moderate hypoxia [70,77].

FUNCTION OF THE HYPOXIC RESPONSE IN IMMUNE CELLS

The interaction between ascorbate and the HIFs is relevant to the function of immune cells in both inflammation and cancer. As mentioned previously, inflammatory sites are known to be under hypoxic stress, with the increased oxidative metabolism of inflammatory cells contributing to this [62–64]. Growing tumors are also well characterized as being hypoxic due to rapid proliferation and outgrowth of the established blood supply [78,79]. The resulting upregulation of the HIFs is instrumental in the activation of glycolysis, angiogenesis, resistance to chemotherapy, and promotion of a stem cell phenotype, thereby promoting tumor growth and metastasis [60,80,81]. At inflammatory sites and in tumor tissue, the hypoxic environment affects immune cell function and, given the interdependence between the activation of the HIFs and cellular ascorbate [70,82–87], we propose that many effects of ascorbate on immune cell function are likely to reflect the regulation of HIF-mediated functions.

Table 8.2 and Figure 8.1 show a summary of the potential interactions between immune cell functions and the hypoxic response. These interactions are discussed in more depth in the following sections.

Monocytes/Macrophages

The high ascorbate concentrations in monocytes [30] may, in part, be related to their dependency on HIF for many essential functions. HIF-1 has been shown to be activated in monocytes following activation with phorbol esters [88] and pathogenic stimuli [89–92], even under nonhypoxic conditions. That HIF activation is an integral part of monocyte function is indicated by the demonstrations that HIF-1α or HIF-2α deletion in myeloid cells caused profound impairment of cell aggregation, motility, invasiveness, and bacterial killing [90–92], resulting in decreased bacterial resistance and failure to restrict systemic spread of a localized infection [91–93]. HIF-1 appears to be important for monocyte-mediated host defense; HIF-1 activation has been shown to contribute to disease progression in colitis, and myeloid HIF-1α knockout shifts the balance to an anti-inflammatory phenotype resulting in a less severe inflammation [94]. The sepsis-related host immunosuppressive monocyte phenotype has also been shown to be mediated by chronic HIF-1α expression, resulting in suppressed pro-inflammatory cytokine expression and increased ability to induce Treg cell polarization [95].

In cancer, activation of HIF-1/2 in monocytes has been implicated in the development and phenotype of tumor-associated macrophages [93,96]. This is associated with an increased M2-like gene profile, increased expression of immunosuppressive and pro-tumor proteins such as arginase 1, iNOS, and VEGF, as well as induction of PD-L1 expression [93,96–98]. These changes lead to greater monocyte/macrophage tumor invasion [93] and tumor cytotoxic T-cell suppression [98,99]. Interestingly, a macrophage-targeted

HIF-1α and HIF-2α knockout resulted in delayed tumor progression in models of breast tumor, fibrosarcoma, and colitis-associated colon carcinoma [93,99,100].

The potential complexity of ascorbate engagement with immune cells in the hypoxic tumor microenvironment is well demonstrated by the observations that dendritic cells (DCs) treated with ascorbate secreted increased levels of interleukin (IL)-12p70 after activation with lipopolysaccharide (LPS) and induced more Th1 cytokine and interferon (IFN)-γ, but less Th2-cytokine, IL-5 expression in naïve T cells [101]. Ascorbate-treated DCs also increased the frequency of IFN-γ+ T cells when cocultured with both CD4+ and CD8+ T cells and demonstrated an improved antitumor effect [102].

Neutrophils

Neutrophils are short-lived cells that are the first responders to an inflammatory challenge. Their recruitment to, and clearance from, inflammatory sites is dependent on the regulation of cell death and survival pathways [103]. It appears that HIF-1 and ascorbate are intimately involved in determining neutrophil cell fate. Hypoxia has been shown to prolong neutrophil survival via activation of HIF-1 and its downstream pathways [104–106]. HIF-1 activation also enhanced neutrophil overall antibacterial function as demonstrated by increased susceptibility to bacterial keratitis in mice when HIF-1 was inhibited [107]. This was supported by findings of delayed rates of apoptosis and enhanced bacterial phagocytosis under normoxic conditions in neutrophils from patients with a monoallelic mutation of von Hippel-Lindau protein who exhibit a "partial hypoxic" phenotype [105]. These results suggest that a functional hypoxic response supports neutrophil function at hypoxic inflammatory sites in vivo. A similar antiapoptotic phenotype in ascorbate-deficient neutrophils was shown to be associated with HIF-1 activation under normoxic conditions [108]. Recognition of aged neutrophils by macrophages was also reported, and neutrophil clearance from an inflammatory site was delayed [108]. Interestingly, increasing neutrophil ascorbate content was found to inhibit Fas-induced neutrophil cell death [109] as well as the rates of neutrophil and monocyte apoptosis in patients with sepsis [110]. Also, in the ascorbate-dependent Gulo$^{-/-}$ mouse, a high

TABLE 8.2

Description of role of hypoxia-inducible factors (HIFs) in immune cells

Cell Type	Role of HIFs	References
Monocyte/ Macrophage	*Differentiation*	
	• Monocyte differentiation with PMA induces HIF-1α activation and downstream gene expression.	[88]
	• HIF-1α is essential for the regulation of glycolytic capacity in myeloid cells and subsequent acquisition of pro-inflammatory response toward bacterial and fungal infections.	[89,90]
	• HIF-1α and HIF-2α regulates the production of key immune effector molecules in response to bacterial infection.	[91–93]
	• High extracellular lactate induces M2-like polarization of macrophages mediated by HIF-1α activation.	[96]
	• Strong HIF-2α staining in TAM of clinical tumor sections.	[183]
	Chemotaxis and Effector Function	
	• HIF-1α-deficient myeloid cells in mice showed decreased bacterial resistance and failed to restrict systemic spread.	[91,92]
	• Myeloid HIF-1α deletion results in metabolic defects and profound impairment of myeloid cell aggregation, motility, invasiveness, and bacterial killing. Contrary report shows HIF pathways in myeloid cells are dispensable for myeloid cell tissue trafficking.	[90,184]
	• Hypoxia amplified macrophage production of M1 induced pro-inflammatory cytokines (IL-6, IL-12, IL-1β) in a HIF-2α-dependent manner.	[93]
	• HIF-2α promotes tumor-associated macrophage invasion into the tumor.	[93]
	Immunomodulation	
	• Hypoxic macrophage suppression of T-cell proliferation is HIF-1α dependent.	[99]
	• HIF-1α in myeloid cells contributes to the pro-inflammatory disease progression in colitis. Myeloid HIF-1α knockout appears to shift the balance to anti-inflammatory resulting in milder inflammation and less T_H17 cells.	[94]
	• Chronic HIF-1α expression mediates sepsis-related host immunosuppressive monocyte phenotype by suppressing pro-inflammatory cytokine expression and increased ability to induce Treg polarization.	[95]
	• HIF-1α expression in myeloid cells is correlated with suppression of tumor cytotoxic T-cell responsiveness.	[99]
	• HIF-1α binds to transcriptionally active HRE in the PD-L1 proximal promoter resulting in increased PL-L1 expression under hypoxia (0.1%).	[98]
	Angiogenesis	
	Myeloid HIF-1α stabilization enhanced neovascularization in matrigel plugs and significantly improved blood flow in a mouse model of hindlimb ischemia.	[185]
	Tumor Growth	
	Macrophage targeted HIF knockout reduced tumor growth in progressive models of breast cancer and fibrosarcoma in mice as well as compromised development of murine colitis associated cancer.	[93,99,100]
Neutrophil	*Cell Survival*	
	• Prolongs survival of neutrophils under hypoxia.	[104,106]
	• Patients with pVHL mutations show delayed neutrophil apoptosis.	[105]

(Continued)

Cell Type	Role of HIFs	References
	Effector Function	
	• Patients with pVHL mutations show enhanced neutrophil phagocytosis.	[105]
	• HIF-1α inhibition increases susceptibility to bacterial keratitis in mice. HIF-1α is essential for effective bacterial killing, apoptosis, and antimicrobial peptide production by PMNs.	[107]
T Cells	**Differentiation**	
	• HIF-1α enhances T_H17 development and attenuates Treg development. HIF-1α-deficient T cells were associated with diminished T_H17 and increased Treg cells in experimental autoimmune encephalitis in mice.	[124]
	• HIF-1α-dependent transcriptional program was important for mediating T-cell glycolytic activity. Blocking glycolysis inhibited T_H17 development while promoting Treg cell generation from CD4+ T cells.	[123]
	• Hypoxic stimulation increased Treg development in wild-type T cells, but HIF-1α knockout resulted in T_H17 instead.	[126]
	• Enhanced HIF-1α activity by CD8+ T cells led to a sustained effector state therefore inhibiting or delaying the terminal differentiation or memory formation of CD8+ effector cells. Glucose dependent.	[125]
	Effector Function	
	• HIF activity activated the expression of CTL genes responsible for effector molecules, costimulatory receptors, activation and inhibitory receptors, and key transcriptional regulators of effector and memory cell differentiation.	[125,129]
	• "Always-on" HIF signaling induced by Vhl deletion augmented the effector capacity of CTLs beyond the attenuated levels observed for wild-type cells during chronic infection and resulted in lethal immunopathology.	[125]
	• Lack of HIF-1α resulted in diminished T_H17 development but enhanced Treg cell differentiation and protected mice from autoimmune neuroinflammation and encephalitis.	[123,124]
	• Deletion of T-cell HIF-1α gene leads to higher levels of pro-inflammatory cytokines, stronger antibacterial effects with better survival of septic mice and more severe colitis-induced colonic inflammation than control mice with the upregulation of Th1 and Th17 cells.	[126,130]
	• Using CTL targeted HIF-1α deletion, HIF-1α was associated with expression of CD69 (marker of activated T cells) on CTLs in hypoxic regions of tumor.	[131]
	Tumor Growth	
	• VHL-deficient CTL recipients delayed growth of established tumors in mice. Moreover, 5 of 20 mice that received VHL-deficient CTLs had no detectable masses at the end of the study (50 days).	[125]
	• HIF-1α knockout CTLs accelerated tumor growth implants in mice, there was higher tumor VEGF-A content, fewer CTL infiltrates, and greater vasculature normalization.	[129]
B Cells	B-cell-specific HIF-1α deletions have reduced number of IL-10-producing B cells, which result in exacerbated collagen-induced arthritis and experimental autoimmune encephalomyelitis.	[186]

ABBREVIATIONS: CTL, cytotoxic T lymphocyte; HRE, hypoxia-response element; IL, interleukin; PMA, phorbol myristate acetate; PMN, poly-morphonuclear; TAM, tumor-associated macrophage; VEGF, vascular endothelial growth factor; VHL, von Hippel-Lindau.

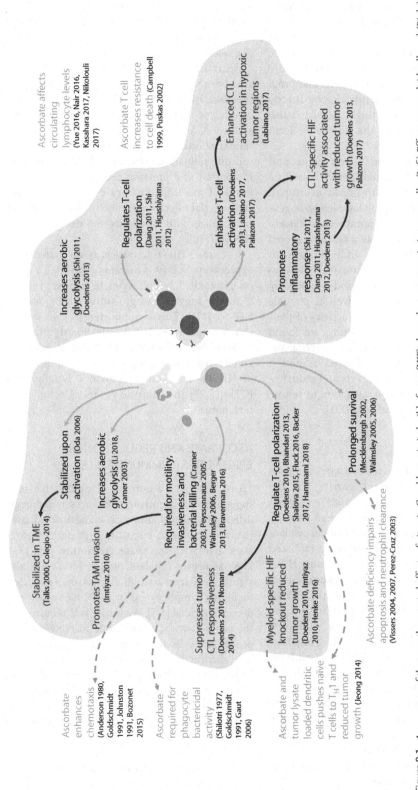

Figure 8.1. A summary of the recently reported effects of vitamin C and hypoxia-inducible factor (HIF)–dependent processes in immune cells. (Left) Effects on myeloid cells and (Right) lymphoid cells. Effects shown in black text represent a reported role of HIF, text in red indicates a reported effect of HIF in the context of cancer, and orange text indicates an effect of ascorbate on immune cells. The interrelationships between these are indicated by arrows. (References from figure: Anderson, 1980 [120]; Backer, 2017 [94]; Berger, 2013 [107]; Bhandari, 2013 [189]; Bozonet, 2015 [121]; Braverman, 2016 [92]; Campbell, 1999 [190]; Colegio, 2014 [96]; Cramer, 2003 [90]; Dang, 2011 [124]; Doedens, 2010, 2013 [99,125]; Fluck, 2016 [127]; Gaut, 2006 [115]; Goldschmidt, 1991 [114]; Hammami, 2018 [128]; Henke, 2016 [100]; Higashiyama, 2012 [126]; Imtiyaz, 2010 [93]; Jeong, 2014 [102]; Johnston, 1991 [191]; Kasahara, 2017 [175]; Labiano, 2017 [131]; Li, 2018 [89]; Mecklenburgh, 2002 [104]; Nair, 2016 [173]; Nikolouli, 2017 [174]; Norman, 2014 [98]; Oda, 2006 [88]; Palazon, 2017 [129]; Perez-Cruz, 2003 [109]; Peyssonnaux, 2005 [91]; Puskas, 2002 [192]; Shalova, 2015 [95]; Shi, 2011 [123]; Shilotri, 1977 [116]; Talks, 2000 [183]; Vissers, 2004, 2007 [108,193]; Walmsley, 2005, 2006 [105,106]; Yue, 2016 [172].)

ascorbate diet was found to increase circulating granulocyte and monocyte numbers [111]. More recently, ascorbate deficiency has been linked to increased generation of neutrophil extracellular traps (NETs) in the Gulo$^{-/-}$ mouse [112]. NETs are produced at the expense of neutrophil viability and are thought to immobilize bacteria and fungi and promote the resolution of infection by directly killing the pathogen. Interestingly, HIF-1 has also been postulated to regulate this alternative form of cell death [113].

Not all effects of ascorbate on neutrophil function will be HIF related. Severe ascorbate deficiency has been shown to impair neutrophil bactericidal ability toward phagocytosed pathogens following infection with actinomyces and *Klebsiella pneumoniae* [114,115], possibly as a result of altered oxidative capacity [116,117]. Individuals with genetic defects in neutrophil function have also shown decreased infectious episodes following vitamin C supplementation, as well as possible improvements in the bactericidal activity of their neutrophils [118–120]. In our study, we found neutrophils from individuals with suboptimal circulating ascorbate levels showed a modest increase in neutrophil chemotaxis and oxidative burst *ex vivo* following supplementation to restore vitamin C status to healthy levels [121].

T Cells

Differentiation of CD4 T cells dictates the type of inflammatory response occurring via the development of different T-helpers and iTreg subsets and their corresponding effector functions [66,122]. Therefore, depending on the nature of the insult or source of inflammation, the prevailing ratio and species of T cells could alter the outcome. HIF-1 appears to play an important, although unresolved, role in T-cell differentiation. For example, HIF-1α T-cell-targeted knockout protected mice from autoimmune neuro-inflammation and was associated with a shift from T_H17 to Treg response, possibly by increasing glycolysis [123–125], while the opposite was observed in irritable bowel disease where T-cell HIF-1α knockout increased T_H1 and T_H17 leading to severe colonic inflammation [126]. HIF-1α-mediated myeloid and DC-driven differentiation of T cells also greatly affected the inflammatory outcome; HIF-1α knockout in myeloid cells resulted in lesser T_H17 prevalence and decreased inflammation [94]. In DCs, HIF-1

knockout resulted in impaired Treg development and increased inflammation [127] and HIF-1-mediated events were reported to limit Th1 cell development by preventing IL-12 production and to exacerbate *Leishmania* infections [128].

Apart from T-cell differentiation, HIF-1 has also been shown to affect T-cell activation and function. HIF activation enhanced the expression of effector molecules, co-stimulatory receptors, activation and inhibitory receptors, and key transcriptional regulators of effector and memory cell differentiation [125,129]. However, this was in contrast to a previous report showing higher levels of pro-inflammatory cytokines, stronger antibacterial effects, and much better survival of septic mice with T-cell-targeted deletion of the HIF-1α [130].

In cancer, HIF-1 activation is associated with expression of CD69 (a marker of activated T cells) on cytotoxic T lymphocytes in hypoxic regions of tumor, suggesting a pro-tumor killing role for HIF-1α [131]. This is supported by two studies showing that T-cell HIF-1 activation significantly delayed tumor growth [125] and, conversely, accelerated tumor progression in the presence of HIF-1α knockout cytotoxic T lymphocytes (CTLs) [129] in a murine model of ectopic B16 melanoma.

ASCORBATE AND REGULATION OF EPIGENETICS IN IMMUNE CELLS

There have been a number of studies that have suggested that ascorbate influences lymphocyte differentiation, including early studies that indicated that an increase in circulating lymphocytes was associated with ascorbate availability [132,133]. High ascorbate supplementation for 1 year also significantly increased all circulating leukocytes, including lymphocytes, in SMP30KO ascorbate-dependent mice [111]. Ascorbate was required for the progression of mouse bone marrow–derived progenitor cells into functional T lymphocytes and also increased the natural killer (NK) cell population in vitro [134–136]. Many of these effects show a significant correlation with the regulation of the TET and Jumonji demethylases and epigenetic changes, rather than with the expression of HIF-1. This topic is discussed in the following sections.

The immune system consists of a diverse set of effector cells whose development is both strictly regulated as well as plastic in nature. This affords our immune system a nearly unlimited capacity to respond to environmental triggers in an appropriate manner. Epigenetics is now being recognized as an important player in the development and differentiation of our immune and inflammatory cells [137].

In mammals, one of the most widespread epigenetic modifications is DNA cytosine methylation, a modification that generally results in silencing of gene expression [138,139]. This modification can be actively reversed by the Tet enzymes that catalyze the oxidation of 5-methylcytosine (5mC) to 5-hydroxymethylcytosine (5hmC), potentially a stable epigenetic mark in itself [138], or initiate the generation of 5-formylcytosine (5fC) and 5-carboxylcytosine (5caC), which results in active regeneration of unmarked cytosine by excision-repair mechanisms [140]. Ascorbate availability has been shown to markedly enhance Tet activity [141,142] through its cofactor function, likely maintaining the active site Fe^{2+} of these dioxygenases [143]. Although other reducing agents could reduce Fe^{3+} and promote TET activity in a cell free system, ascorbate was shown to be the most efficient [143], and glutathione was incapable of increasing murine embryonic TET activity compared to equimolar ascorbate [141,142]. The Jumonji C domain-containing histone demethylases (JHDMs) are also members of the Fe- and 2-oxoglutarate dependent dioxygenase family, and similarly to TETs, full enzyme activity of JHDMs occurs when ascorbate is present [144,145]. The JHDMs are the third and largest class of demethylase enzymes, capable of removing all three histone lysine-methylation states through oxidative reactions [145].

At inflammatory sites and in tumor tissue, immune and inflammatory cells acquire specific phenotypes depending on the cues they receive. This, in part, involves the selective expression and repression or specific genes [146,147]. Given the interdependence between epigenetic regulation and cellular ascorbate, we propose that many effects of ascorbate on immune cell function could reflect the regulation of the DNA and histone demethylation enzymes, the TETs and JHDMs. Table 8.3 and Figure 8.2 show a summary of the interactions between epigenetic regulation and immune cell functions that are discussed in the following sections.

Monocytes/Macrophages

Epigenetic regulation plays an important role in macrophage differentiation, with rapid TET-dependent demethylation observed in colony stimulating factor 1–differentiated human monocytes [148,149]. TET2 transcription was further induced by LPS but not IL-4 stimulation [148]. The genes affected by TET-mediated demethylation are part of 10 consolidated pathways related to the regulation of the actin cytoskeleton, phagocytosis, and the innate immune system [149]. In macrophages, TET2 is thought to restrain the inflammatory response by upregulating expression of genes involved in dampening toll-like receptor 4 signaling [148]. This notion is supported by a report showing TET2 represses IL-6 production during LPS-induced inflammation and that TET2 knockout exacerbates the expression of macrophage pro-inflammatory molecules such as IL-6, MCP-1, MCP-3 in response to LPS stimulation, resulting in an enhanced inflammatory response [150].

The JHDM enzyme JMJD3 is expressed in monocytes/macrophages and is inducible by differentiating factors [151–153] as well as by pathogenic [154,155] and damage-associated molecules [156,157]. Although JMJD3 has been shown to affect gene expression in macrophages, the role of JMD3 in macrophage function is still unclear. For example, 70% of macrophage-LPS-inducible genes were found to be JMJD3 targets, but only a few hundred genes, including inducible inflammatory genes, were moderately affected by JMJD3 deletion [154]. However, Kruidenier et al. demonstrated a drastic drop in LPS-induced cytokine expression using a specific JMJD3 inhibitor and siRNA, among them TNF-α [158]. In contrast, Satoh et al. showed no effect in M1 cytokine secretion following LPS stimulation in JMJD3 knockout macrophages including TNF-α [151]. Contradictions aside, two studies looking at macrophage response to parasitic infection have associated JMJD3 demethylation activity to acquisition of an M2 phenotype, demonstrated by upregulation of M2 proteins such as arg1, Ym1, Fizz1, MR, and iNOS [151,156]. Two other studies have associated JMJD3 activity with an M1 macrophage phenotype following serum amyloid A stimulation [157] and in arthritis [152] resulting in induction of pro-inflammatory cytokines.

TABLE 8.3

Description of role of ten-eleven translocations (TETs) and Jumonji C domain-containing histone demethylases (JMJDs) in immune cells

Cell Type	Role of TETs and Jumonjis	References
Monocyte/ Macrophage	*Differentiation*	
	• TETs are highly expressed in murine macrophages differentiated by M-CSF. TET-2 transcription was further induced by LPS but not interleukin IL-4 stimulation.	[148]
	• Monocyte to macrophage differentiation by CSF-1 was shown to be followed by rapid demethylation (within 24 hours) mediated by TET enzymes. Identified affected genes are part of 10 consolidated pathways related to the regulation of actin cytoskeleton, phagocytosis, and innate immune system.	[149]
	• 70% of LPS-inducible genes were found to be JMJD3 targets. JMJD3 deletion showed moderately impaired Pol II recruitment and transcription. JMJD3 fine-tunes the transcriptional output of LPS-activated macrophages in an H3K27 demethylation-independent manner.	[154]
	• PMA stimulation increases expression of Jumonji proteins.	[158]
	• GM-CSF upregulates interferon regulatory factor 4 driven CCL17 production expression by enhancing JMJD3 demethylase activity in human monocytes and mouse macrophages.	[152]
	• JMJD3-mediated H3K27 demethylation is crucial for regulating M2 macrophage development in anti-helminth host responses; this is mediated by interferon regulatory factor 4.	[151,156]
	• IFN-γ and IL-4 stimulation increased JMJD3 expression in CSF-1 stimulated macrophages and contributes to gene expression relating to immune response and leukocyte activation.	[153]
	Cytokine Production	
	• TET-2 represses late phase expression of macrophage pro-inflammatory molecules such as IL-6, MCP-1, MCP-3 (by recruiting Hdac2) in response to LPS stimulation resulting in greater degree of inflammatory response in TET-2 knockout mice challenged with LPS and colitis.	[148,150]
	• JMJD3 knockout did not affect macrophage M1 (by LPS) cytokine secretion.	[154]
	• Inhibition of JMJD3 attenuated TNF-α and other pro-inflammatory cytokine production in primary human macrophages in response to LPS.	[158]
	• JMJD3 is highly inducible in serum amyloid A-stimulated macrophages and plays an important role in the induction of inflammatory cytokine genes (IL-23p19 and TREM-1).	[157]
Dendritic Cells	• Tet2 represses late phase expression of macrophage and dendritic cell pro-inflammatory molecules such as IL-6, MCP-1, MCP-3 in response to LPS stimulation resulting in greater degree of inflammatory response in Tet-2 knockout mice challenged with LPS and colitis.	[150]
	• DNA demethylation changes occur during development of monocytes to immature and mature dendritic cells.	[162]
	• KDM5B acts to repress type I IFN and other innate cytokines in DCs to promote an altered immune response following RSV infection that contributes to development of chronic disease.	[163]

(Continued)

TABLE 8.3 (*Continued*)
Description of role of ten-eleven translocations (TETs) and Jumonji C domain-containing histone demethylases (JMJDs) in immune cells

Cell Type	Role of TETs and Jumonjis	References
NK Cells	*Differentiation*	
	• TET-2 and TET-3 act together to control iNKT cell expansion and cell lineage specification. *Tet2/3* T-DKO mice showed an impressive expansion of iNKT cells even at very young ages.	[177]
	• UTX-mediated regulation of superenhancer accessibility was a key mechanism for commitment to the iNKT cell lineage *in vitro*.	[179]
	• CD4+ T-cell UTX and JMJD3 promote NKT cell development and are required for effective NKT function in mice.	[178]
	Effector Function	
	GSK-J4 (H3K27 demethylase inhibitor) reduced IFN-γ, TNFα, GM-CSF, and IL-10 levels in cytokine-stimulated NK cells while sparing their cytotoxic killing activity against cancer cells.	[180]
T Cells	*Differentiation*	
	• In a murine model of acute viral infection, TET-2 loss promotes early acquisition of a memory CD8+ T-cell fate. TET-2 is an important regulator of CD8+ T-cell fate decisions.	[166]
	• Widespread 5mC/5hmC remodeling during human CD4+ T-cell differentiation and is retained by CD4+ T memory cells at genes and cell-specific enhancers with known T-cell function.	[164]
	• Profound demethylation of H3K27 is observed by 1 day after activation in CD4+ T cells corresponds to pathways crucial to T-cell function, including T-cell activation and the JAK/STAT pathways.	[168,169]
	• JMJD3 deletion promotes T_H2 and T_H17 differentiation and inhibits T_H1 and Treg cell differentiation by mediating the methylation status of H3K27 and/or H3K4 in target genes and regulating target gene expression.	[170,171]
	Effector Function	
	• Prolonged antigen stimulation in peptide immunotherapy is associated with demethylation of conserved regions of PD-1 promoter, possibly via TETs leading to sustained PD-1 expression in CD4+ effector T cells.	[167]
	• Tet2 promotes DNA demethylation and activation of cytokine gene expression in T_H1 and T_H17 cells.	[165]
B Cells	Deletion of both the Tet2 and Tet3 genes in early B cells with *Mb1Cre* resulted in a developmental arrest at the pro-B to pre-B stage.	[187,188]

ABBREVIATIONS: CSF, colony-stimulating factor; DC, dendritic cell; GM-CSF, granulocyte-macrophage colony-stimulating factor; IFN, interferon; IL, interleukin; LPS, lipopolysaccharide; MCP, monocyte chemoattractant protein; NK, natural killer; PMA, phorbol myristate acetate; RSV, respiratory syncytial virus; T-DKO, TET2/3 double knock out; TNF, tumor necrosis factor.

Epigenetic processes regulated by the demethylases are associated with leukemogenesis, and ascorbate availability has been closely linked to this phenomenon. As mentioned previously, hematopoietic stem and progenitor cells (HSPCs) accumulate high intracellular concentrations of ascorbate, and this is essential for HSPC differentiation via support of TET2 activity [54].

TET2 inhibition in HSPCs by ascorbate depletion retards differentiation and increases HSPC frequency. TET2 mutations are also known to cooperate with FLT3[ITD] mutations to cause acute myeloid leukemia [54]. Ascorbate depletion coupled with FLT3[ITD] mutations was adequate for leukemogenesis [54]. It appears then, that ascorbate accumulation within HSCs promotes TET function

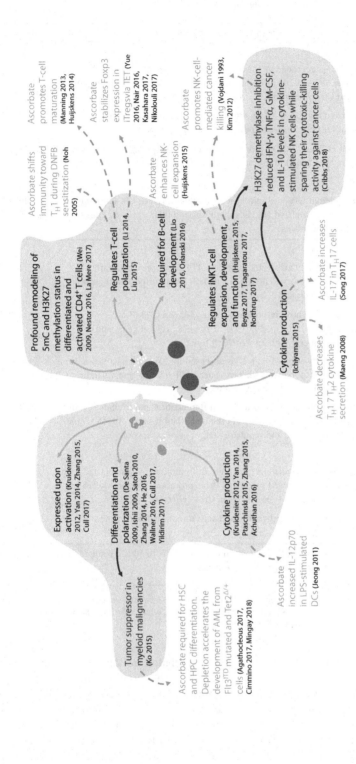

Figure 8.2. A summary of the recently reported effects of vitamin C, TET, and JMJD-dependent processes in immune cells. (Left) Effects on myeloid cells and (Right) lymphoid cells. Effects shown in black text represent a reported role of TET and JMJD, text in red indicates a reported effect of TET and JMJD, and orange text indicates an effect of ascorbate on immune cells. The interrelationships between these are indicated by arrows. (References from figure: Acuthan, 2016 [152]; Agathocleous, 2017 [54]; Cimmino, 2017 [159]; Cribbs, 2018 [180]; Cull, 2017 [148]; De Santa, 2009 [154]; He, 2016 [194]; Huijskens, 2015, 2014 [134,135]; Ichiyama, 2015 [165]; Ishii, 2009 [156]; Jeong, 2011 [101]; Kasahara, 2017 [175]; Kim, 2012 [195]; Ko, 2015 [146]; Kruidenier, 2012 [158]; La Mere, 2017 [168]; Li, 2014 [170]; Lio, 2016 [187]; Liu, 2015 [171]; Maeng, 2008 [196]; Manning, 2013 [136]; Mingay, 2018 [56]; Nair, 2016 [173]; Nestor, 2016 [164]; Nikolouli, 2017 [174]; Noh, 2005 [197]; Northrup, 2017 [178]; Orlanski, 2016 [188]; Ptaschinski, 2015 [163]; Satoh, 2010 [151]; Song, 2017 [176]; Tsagaratou, 2017 [177]; Vojdani, 1993 [198]; Wallner, 2016 [149]; Wei, 2009 [169]; Yan, 2014 [157]; Yıldırım, 2017 [153]; Yue, 2016 [172]; Zhang, 2014, 2015 [150,162].)

in vivo, limiting HSPC frequency and suppressing leukemogenesis. These findings were corroborated in part by another group that described the use of ascorbate as a combination therapy for treating leukemia [159]. Patients with leukemia often have low plasma ascorbate levels [41,46–48], and the capacity for ascorbate to influence the epigenetic drivers of some leukemias has led to conjecture that increased ascorbate supply may provide clinical benefit to some individuals with leukemia. Two recent publications have provided support for this hypothesis [160,161].

Dendritic Cells

DNA demethylation changes occur during development of monocytes into immature DCs and mature DCs [162]. TET2 represses late phase expression of DC pro-inflammatory molecules such as Il-6, MCP-1, and MCP-3 in response to LPS stimulation, and TET2 knockout results in a greater degree of inflammatory response in mice challenged with LPS and colitis [150]. KDM5B acts to repress type I IFN and other innate cytokines in DCs to promote an altered immune response following respiratory syncytial virus infection that contributes to development of chronic disease [163].

T Cells

Widespread DNA methylation remodeling has been reported at genes and cell-specific enhancers with known T-cell function during human CD4$^+$ T differentiation [164,165], and Tet2 was reported to be the critical DNA demethylase involved in the differentiation of T_H1 and T_H17 cells, leading to activation of effector cytokine gene expression [165]. TET2 has also been shown to regulate CD8$^+$ T-cell fate, particularly in formation of memory CD8$^+$ T cells [166]. Prolonged antigen stimulation in peptide immunotherapy is associated with demethylation of conserved regions of PD-1 promoter, possibly via TET, leading to sustained PD-1 expression in CD4$^+$ effector T cells [167].

Profound demethylation of histone H3K27 is observed after activation in CD4$^+$ T cells, and corresponds to pathways crucial to T-cell function, including T-cell activation and the regulation of the JAK/STAT pathways [168,169]. Deletion of the histone demethylase JMJD3 was found to regulate gene expression resulting in T_H2 and T_H17 differentiation and inhibiting T_H1 and Treg cell differentiation via altered methylation status of H3K27 and H3K4 [170,171].

Recent studies focusing on the role of ascorbate in T-cell differentiation and function suggest close alignment with epigenetic regulation and demethylase activity. Initial work showed ascorbate to be required for the progression of mouse bone marrow–derived progenitor cells into functional T lymphocytes in vitro and in vivo by a JmjC-mediated process [135,136]. Subsequent studies reported ascorbate-mediated stabilization of Foxp3 expression in TGF-β induced Tregs by TET enzymes [172,173]. Also, ascorbate enhanced alloantigen-induced Treg suppressive capacity in skin allograft and GVHD in mice was attributed to the stabilization of Foxp3 expression, presumably via demethylation of Foxp3 and other Treg-specific epigenetic genes [174,175]. Apart from Tregs, ascorbate has also been implicated in the maintenance of the T_H17 phenotype by increasing IL-17 expression in T_H17-differentiated T cells via reduced trimethylation of histone H3 lysine 9 (H3K9me3) in the regulatory elements of the IL17 locus [176].

NK Cells

A number of recent studies have demonstrated the impact of TET- and JHDM-mediated demethylation on NKT-cell development, proliferation, and function [177–179]. Interestingly, inhibition of the H3K27 demethylase reduced IFN-γ, TNFα, GM-CSF, and IL-10 levels in cytokine-stimulated NK cells while sparing their cytotoxic killing activity against cancer cells [180].

SUMMARY

The ability of ascorbate to influence the function of the many distinct cells of the immune system is becoming more established, alongside the mechanisms by which this might occur. Ascorbate availability will influence HIF activation and immune cell function in hypoxic inflammatory and tumor microenvironments. Evidence is mounting that DNA and histone demethylases, which are highly responsive to ascorbate, are intimately involved in the epigenetic remodeling of immune cells. However, there is still much to refine, owing both to our still evolving understanding of ascorbate biochemistry and to the very complexities of the immune system, with its capacity to defend against a diversity of pathogens as well as its proficiency in eliminating our own cells when required (cancer cells or dying/damaged cells). How this translates into an overall effect of ascorbate function on immunity is still emerging.

REFERENCES

1. Hess, A. F. 1932. Diet, nutrition and infection. N. Engl. J. Med. 207, 637–648.
2. Hemilä, H. and Chalker, E. 2013. Vitamin C for preventing and treating the common cold. Cochrane Database Syst. Rev. Jan 31;(1):CD000980.
3. Hemilä, H. and Louhiala, P. 2013. Vitamin C for preventing and treating pneumonia. Cochrane Database Syst. Rev. Aug 8;(8):CD005532.
4. Hemilä, H. 2017. Vitamin C and infections. Nutrients 9.
5. Marik, P. E., Khangoora, V., Rivera, R., Hooper, M. H. and Catravas, J. 2017. Hydrocortisone, vitamin C, and thiamine for the treatment of severe sepsis and septic shock: A retrospective before-after study. Chest 151, 1229–1238.
6. Fowler, A. A., 3rd, Syed, A. A., Knowlson, S., Sculthorpe, R., Farthing, D., DeWilde, C., Farthing, C. A. et al. 2014. Phase I safety trial of intravenous ascorbic acid in patients with severe sepsis. J. Transl. Med. 12, 32.
7. Du, J., Cullen, J. J. and Buettner, G. R. 2012. Ascorbic acid: Chemistry, biology and the treatment of cancer. Biochim. Biophys. Acta 1826, 443–457.
8. Peterkofsky, B. 1991. Ascorbate requirement for hydroxylation and secretion of procollagen: Relationship to inhibition of collagen synthesis in scurvy. Am. J. Clin. Nutr. 54, 1135s–1140s.
9. Pihlajaniemi, T., Myllyla, R. and Kivirikko, K. I. 1991. Prolyl 4-hydroxylase and its role in collagen synthesis. J. Hepatol. 13(Suppl 3), S2–7.
10. England, S. and Seifter, S. 1986. The biochemical functions of ascorbic acid. Annu. Rev. Nutr. 6, 365–406.
11. Levine, M., Morita, K., Heldman, E. and Pollard, H. B. 1985. Ascorbic acid regulation of norepinephrine biosynthesis in isolated chromaffin granules from bovine adrenal medulla. J. Biol. Chem. 260, 15598–15603.
12. Kuiper, C. and Vissers, M. C. 2014. Ascorbate as a co-factor for fe- and 2-oxoglutarate dependent dioxygenases: Physiological activity in tumor growth and progression. Front. Oncol. 4, 359.
13. Loenarz, C. and Schofield, C. J. 2011. Physiological and biochemical aspects of hydroxylations and demethylations catalyzed by human 2-oxoglutarate oxygenases. Trends Biochem. Sci. 36, 7–18.
14. Nelson, P. J., Pruitt, R. E., Henderson, L. L., Jenness, R. and Henderson, L. M. 1981. Effect of ascorbic acid deficiency on the in vivo synthesis of carnitine. Biochim. Biophys. Acta 672, 123–127.
15. Stolze, I. P., Mole, D. R. and Ratcliffe, P. J. 2006. Regulation of HIF: Prolyl hydroxylases. Novartis Found. Symp. 272, 15–25; discussion 25–36.
16. Young, J. I., Zuchner, S. and Wang, G. 2015. Regulation of the epigenome by vitamin C. Annu. Rev. Nutr. 35, 545–564.
17. Camarena, V. and Wang, G. 2016. The epigenetic role of vitamin C in health and disease. Cell. Mol. Life Sci. 73, 1645–1658.
18. Vissers, M. C., Kuiper, C. and Dachs, G. U. 2014. Regulation of the 2-oxoglutarate-dependent dioxygenases and implications for cancer. Biochem. Soc. Trans. 42, 945–951.
19. Diakos, C. I., Charles, K. A., McMillan, D. C. and Clarke, S. J. 2014. Cancer-related inflammation and treatment effectiveness. Lancet Oncol. 15, e493–503.
20. Bendich, A. 1990. Antioxidant vitamins and their functions in immune responses. Adv. Exp. Med. Biol. 262, 35–55.
21. Mangge, H., Becker, K., Fuchs, D. and Gostner, J. M. 2014. Antioxidants, inflammation and cardiovascular disease. World J. Cardiol. 6, 462–477.
22. Pohanka, M., Pejchal, J., Snopkova, S., Havlickova, K., Karasova, J. Z., Bostik, P. and Pikula, J. 2012. Ascorbic acid: An old player with a broad impact on body physiology including oxidative stress suppression and immunomodulation: A review. Mini Rev. Med. Chem. 12, 35–43.
23. Puertollano, M. A., Puertollano, E., de Cienfuegos, G. A. and de Pablo, M. A. 2011. Dietary antioxidants: Immunity and host defense. Curr. Top. Med. Chem. 11, 1752–1766.
24. Ginter, E., Simko, V. and Panakova, V. 2014. Antioxidants in health and disease. Bratisl. Lek. Listy 115, 603–606.
25. Maggini, S., Wenzlaff, S. and Hornig, D. 2010. Essential role of vitamin C and zinc in child immunity and health. J. Int. Med. Res. 38, 386–414.
26. Bergsten, P., Amitai, G., Kehrl, J., Dhariwal, K. R., Klein, H. G. and Levine, M. 1990. Millimolar concentrations of ascorbic acid in purified human mononuclear leukocytes. Depletion and reaccumulation. J. Biol. Chem. 265, 2584–2587.

27. Bergsten, P., Yu, R., Kehrl, J. and Levine, M. 1995. Ascorbic acid transport and distribution in human B lymphocytes. *Arch. Biochem. Biophys.* 317, 208–214.

28. Levine, M., Wang, Y., Padayatty, S. J. and Morrow, J. 2001. A new recommended dietary allowance of vitamin C for healthy young women. *Proc. Natl. Acad. Sci. USA* 98, 9842–9846.

29. Wang, Y., Russo, T. A., Kwon, O., Chanock, S., Rumsey, S. C. and Levine, M. 1997. Ascorbate recycling in human neutrophils: Induction by bacteria. *Proc. Natl. Acad. Sci. USA* 94, 13816–13819.

30. Levine, M., Conry-Cantilena, C., Wang, Y., Welch, R. W., Washko, P. W., Dhariwal, K. R., Park, J. B. et al. 1996. Vitamin C pharmacokinetics in healthy volunteers: Evidence for a recommended dietary allowance. *Proc. Natl. Acad. Sci. USA* 93, 3704–3709.

31. Schleicher, R. L., Carroll, M. D., Ford, E. S. and Lacher, D. A. 2009. Serum vitamin C and the prevalence of vitamin C deficiency in the United States: 2003–2004 National Health and Nutrition Examination Survey (NHANES). *Am. J. Clin. Nutr.* 90, 1252–1263.

32. Carr, A. C., Bozonet, S. M., Pullar, J. M., Simcock, J. W. and Vissers, M. C. 2013. A randomized steady-state bioavailability study of synthetic versus natural (kiwifruit-derived) vitamin C. *Nutrients* 5, 3684–3695.

33. Carr, A. C., Bozonet, S. M., Pullar, J. M., Simcock, J. W. and Vissers, M. C. 2013. Human skeletal muscle ascorbate is highly responsive to changes in vitamin C intake and plasma concentrations. *Am. J. Clin. Nutr.* 97, 800–807.

34. Carr, A. C., Bozonet, S. M. and Vissers, M. C. 2013. A randomised cross-over pharmacokinetic bioavailability study of synthetic versus kiwifruit-derived vitamin C. *Nutrients* 5, 4451–4461.

35. Carr, A. C. and Vissers, M. C. 2013. Synthetic or food-derived vitamin C—Are they equally bioavailable? *Nutrients* 5, 4284–4304.

36. Pullar, J. M., Bayer, S. and Carr, A. C. 2018. Appropriate handling, processing and analysis of blood samples is essential to avoid oxidation of vitamin C to dehydroascorbic acid. *Antioxidants* (Basel, Switzerland). 7.

37. Bonham, M. J., Abu-Zidan, F. M., Simovic, M. O., Sluis, K. B., Wilkinson, A., Winterbourn, C. C. and Windsor, J. A. 1999. Early ascorbic acid depletion is related to the severity of acute pancreatitis. *Br. J. Surg.* 86, 1296–1301.

38. Jiang, Q., Lykkesfeldt, J., Shigenaga, M. K., Shigeno, E. T., Christen, S. and Ames, B. N. 2002. Gamma-tocopherol supplementation inhibits protein nitration and ascorbate oxidation in rats with inflammation. *Free Radic. Biol. Med.* 33, 1534–1542.

39. Bakaev, V. V. and Duntau, A. P. 2004. Ascorbic acid in blood serum of patients with pulmonary tuberculosis and pneumonia. *Int. J. Tuberc. Lung Dis.* 8, 263–266.

40. Carr, A. C., Rosengrave, P. C., Bayer, S., Chambers, S., Mehrtens, J. and Shaw, G. M. 2017. Hypovitaminosis C and vitamin C deficiency in critically ill patients despite recommended enteral and parenteral intakes. *Crit. Care* (London, England). 21, 300.

41. Huijskens, M. J., Wodzig, W. K., Walczak, M., Germeraad, W. T. and Bos, G. M. 2016. Ascorbic acid serum levels are reduced in patients with hematological malignancies. *Results Immunol.* 6, 8–10.

42. Basu, T. K., Raven, R. W., Dickerson, J. W. and Williams, D. C. 1974. Leucocyte ascorbic acid and urinary hydroxyproline levels in patients bearing breast cancer with skeletal metastases. *Eur. J. Cancer* 10, 507–511.

43. Anthony, H. M. and Schorah, C. J. 1982. Severe hypovitaminosis C in lung-cancer patients: The utilization of vitamin C in surgical repair and lymphocyte-related host resistance. *Br. J. Cancer* 46, 354–367.

44. Mayland, C. R., Bennett, M. I. and Allan, K. 2005. Vitamin C deficiency in cancer patients. *Palliat. Med.* 19, 17–20.

45. Abou-Seif, M. A., Rabia, A. and Nasr, M. 2000. Antioxidant status, erythrocyte membrane lipid peroxidation and osmotic fragility in malignant lymphoma patients. *Clin. Chem. Lab. Med.* 38, 737–742.

46. Kennedy, D. D., Tucker, K. L., Ladas, E. D., Rheingold, S. R., Blumberg, J. and Kelly, K. M. 2004. Low antioxidant vitamin intakes are associated with increases in adverse effects of chemotherapy in children with acute lymphoblastic leukemia. *Am. J. Clin. Nutr.* 79, 1029–1036.

47. Nakagawa, K. 2000. Effect of chemotherapy on ascorbate and ascorbyl radical in cerebrospinal fluid and serum of acute lymphoblastic leukemia. *Cell. Mol. Biol.* (Noisy-Le-Grand, France). 46, 1375–1381.

48. Neyestani, T. R., Fereydouni, Z., Hejazi, S., Salehi-Nasab, F., Nateghifard, F., Maddah, M. and Karandish, M. 2007. Vitamin C status in Iranian children with acute lymphoblastic leukemia: Evidence for increased utilization. *J. Pediatr. Gastroenterol. Nutr.* 45, 141–144.

49. Sharma, M., Rajappa, M., Kumar, G. and Sharma, A. 2009. Oxidant-antioxidant status in Indian patients with carcinoma of posterior one-third of tongue. *Cancer Biomark.* 5, 253–260.

50. Evans-Olders, R., Eintracht, S. and Hoffer, L. J. 2010. Metabolic origin of hypovitaminosis C in acutely hospitalized patients. *Nutrition* (Burbank, Los Angeles County, CA). 26, 1070–1074.

51. Gan, R., Eintracht, S. and Hoffer, L. J. 2008. Vitamin C deficiency in a university teaching hospital. *J. Am. Coll. Nutr.* 27, 428–433.

52. Levine, M., Padayatty, S. J. and Espey, M. G. 2011. Vitamin C: A concentration-function approach yields pharmacology and therapeutic discoveries. *Adv. Nutr.* (Bethesda, MD). 2, 78–88.

53. Washko, P. W., Wang, Y. and Levine, M. 1993. Ascorbic acid recycling in human neutrophils. *J. Biol. Chem.* 268, 15531–15535.

54. Agathocleous, M., Meacham, C. E., Burgess, R. J., Piskounova, E., Zhao, Z., Crane, G. M., Cowin, B. L. et al. 2017. Ascorbate regulates haematopoietic stem cell function and leukaemogenesis. *Nature* 549, 476–481.

55. Mastrangelo, D., Pelosi, E., Castelli, G., Lo-Coco, F. and Testa, U. 2018. Mechanisms of anti-cancer effects of ascorbate: Cytotoxic activity and epigenetic modulation. *Blood Cells, Mol. Dis.* 69, 57–64.

56. Mingay, M., Chaturvedi, A., Bilenky, M., Cao, Q., Jackson, L., Hui, T., Moksa, M. et al. 2018. Vitamin C-induced epigenomic remodelling in IDH1 mutant acute myeloid leukaemia. *Leukemia* 32, 11–20.

57. Gillberg, L., Orskov, A. D., Liu, M., Harslof, L. B. S., Jones, P. A. and Gronbaek, K. 2018. Vitamin C—A new player in regulation of the cancer epigenome. *Semin. Cancer Biol.* 51, 59–67.

58. Vissers, M. C. M. and Das, A. B. 2018. Potential mechanisms of action for vitamin C in cancer: Reviewing the evidence. *Front. Physiol.* 9, 809.

59. Kaelin, W. G., Jr. and Ratcliffe, P. J. 2008. Oxygen sensing by metazoans: The central role of the HIF hydroxylase pathway. *Mol. Cell* 30, 393–402.

60. Ratcliffe, P. J. 2013. Oxygen sensing and hypoxia signalling pathways in animals: The implications of physiology for cancer. *J. Physiol.* 591, 2027–2042.

61. Semenza, G. L. 2010. HIF-1: Upstream and downstream of cancer metabolism. *Curr. Opin. Genet. Dev.* 20, 51–56.

62. Colgan, S. P., Campbell, E. L. and Kominsky, D. J. 2016. Hypoxia and mucosal inflammation. *Annu. Rev. Pathol.* 11, 77–100.

63. Whyte, M. K. and Walmsley, S. R. 2014. The regulation of pulmonary inflammation by the hypoxia-inducible factor-hydroxylase oxygen-sensing pathway. *Ann. Am. Thorac. Soc.* 11 (Suppl 5), S271–276.

64. Campbell, E. L., Bruyninckx, W. J., Kelly, C. J., Glover, L. E., McNamee, E. N., Bowers, B. E., Bayless, A. J. et al. 2014. Transmigrating neutrophils shape the mucosal microenvironment through localized oxygen depletion to influence resolution of inflammation. *Immunity* 40, 66–77.

65. Pearce, E. L. and Pearce, E. J. 2013. Metabolic pathways in immune cell activation and quiescence. *Immunity* 38, 633–643.

66. Corcoran, S. E. and O'Neill, L. A. 2016. HIF1alpha and metabolic reprogramming in inflammation. *J. Clin. Invest.* 126, 3699–3707.

67. Palsson-McDermott, E. M. and O'Neill, L. A. 2013. The Warburg effect then and now: From cancer to inflammatory diseases. *BioEssays: News and Reviews in Molecular, Cellular and Developmental Biology* 35, 965–973.

68. Ozer, A. and Bruick, R. K. 2007. Non-heme dioxygenases: Cellular sensors and regulators jelly rolled into one? *Nat. Chem. Biol.* 3, 144–153.

69. Islam, M. S., Leissing, T. M., Chowdhury, R., Hopkinson, R. J. and Schofield, C. J. 2018. 2-Oxoglutarate-Dependent Oxygenases. *Annu. Rev. Biochem.* 87, 585–620.

70. Kuiper, C., Dachs, G. U., Currie, M. J. and Vissers, M. C. 2014. Intracellular ascorbate enhances hypoxia-inducible factor (HIF)-hydroxylase activity and preferentially suppresses the HIF-1 transcriptional response. *Free Radic. Biol. Med.* 69, 308–317.

71. Flashman, E., Davies, S. L., Yeoh, K. K. and Schofield, C. J. 2010. Investigating the dependence of the hypoxia-inducible factor hydroxylases (factor inhibiting HIF and prolyl hydroxylase domain 2) on ascorbate and other reducing agents. *Biochem. J.* 427, 135–142.

72. Jaakkola, P., Mole, D. R., Tian, Y. M., Wilson, M. I., Gielbert, J., Gaskell, S. J., von Kriegsheim, A. et al. 2001. Targeting of HIF-alpha to the von Hippel-Lindau ubiquitylation complex by O2-regulated prolyl hydroxylation. *Science* (New York, NY). 292, 468–472.

73. Hirsilä, M., Koivunen, P., Gunzler, V., Kivirikko, K. I. and Myllyharju, J. 2003. Characterization of the human prolyl 4-hydroxylases that modify the hypoxia-inducible factor. *J. Biol. Chem.* 278, 30772–30780.

74. Nytko, K. J., Maeda, N., Schlafli, P., Spielmann, P., Wenger, R. H. and Stiehl, D. P. 2011. Vitamin C is dispensable for oxygen sensing in vivo. *Blood* 117, 5485–5493.

75. Nytko, K. J., Spielmann, P., Camenisch, G., Wenger, R. H. and Stiehl, D. P. 2007. Regulated function of the prolyl-4-hydroxylase domain (PHD) oxygen sensor proteins. *Antioxid Redox Signal.* 9, 1329–1338.

76. Osipyants, A. I., Poloznikov, A. A., Smirnova, N. A., Hushpulian, D. M., Khristichenko, A. Y., Chubar, T. A., Zakhariants, A. A. et al. 2018. L-ascorbic acid: A true substrate for HIF prolyl hydroxylase? *Biochimie.* 147, 46–54.

77. Campbell, E. J., Vissers, M. C. and Dachs, G. U. 2016. Ascorbate availability affects tumor implantation-take rate and increases tumor rejection in Gulo(-/-) mice. *Hypoxia (Auckland, N.Z.).* 4, 41–52.

78. Adam, J., Yang, M., Soga, T. and Pollard, P. J. 2014. Rare insights into cancer biology. *Oncogene* 33, 2547–2556.

79. Brahimi-Horn, M. C., Chiche, J. and Pouyssegur, J. 2007. Hypoxia and cancer. *J. Mol. Med. (Berl.)* 85, 1301–1307.

80. Potter, C. and Harris, A. L. 2004. Hypoxia inducible carbonic anhydrase IX, marker of tumour hypoxia, survival pathway and therapy target. *Cell Cycle* 3, 164–167.

81. Semenza, G. L. 2015. Regulation of the breast cancer stem cell phenotype by hypoxia-inducible factors. *Clin. Sci. (Lond.)* 129, 1037–1045.

82. Karaczyn, A., Ivanov, S., Reynolds, M., Zhitkovich, A., Kasprzak, K. S. and Salnikow, K. 2006. Ascorbate depletion mediates up-regulation of hypoxia-associated proteins by cell density and nickel. *J. Cell. Biochem.* 97, 1025–1035.

83. Knowles, H. J., Mole, D. R., Ratcliffe, P. J. and Harris, A. L. 2006. Normoxic stabilization of hypoxia-inducible factor-1alpha by modulation of the labile iron pool in differentiating U937 macrophages: Effect of natural resistance-associated macrophage protein 1. *Cancer Res.* 66, 2600–2607.

84. Knowles, H. J., Raval, R. R., Harris, A. L. and Ratcliffe, P. J. 2003. Effect of ascorbate on the activity of hypoxia-inducible factor in cancer cells. *Cancer Res.* 63, 1764–1768.

85. Kuiper, C., Vissers, M. C. and Hicks, K. O. 2014. Pharmacokinetic modeling of ascorbate diffusion through normal and tumor tissue. *Free Radic. Biol. Med.* 77, 340–352.

86. Kuiper, C., Dachs, G. U., Munn, D., Currie, M. J., Robinson, B. A., Pearson, J. F. and Vissers, M. C. 2014. Increased tumor ascorbate is associated with extended disease-free survival and decreased hypoxia-inducible factor-1 activation in human colorectal cancer. *Front. Oncol.* 4, 10.

87. Kuiper, C., Molenaar, I. G., Dachs, G. U., Currie, M. J., Sykes, P. H. and Vissers, M. C. 2010. Low ascorbate levels are associated with increased hypoxia-inducible factor-1 activity and an aggressive tumor phenotype in endometrial cancer. *Cancer Res.* 70, 5749–5758.

88. Oda, T., Hirota, K., Nishi, K., Takabuchi, S., Oda, S., Yamada, H., Arai, T. et al. 2006. Activation of hypoxia-inducible factor 1 during macrophage differentiation. *Am. J. Physiology. Cell Physiology* 291, C104–113.

89. Li, C., Wang, Y., Li, Y., Yu, Q., Jin, X., Wang, X., Jia, A. et al. 2018. HIF1alpha-dependent glycolysis promotes macrophage functional activities in protecting against bacterial and fungal infection. *Sci. Rep.* 8, 3603.

90. Cramer, T., Yamanishi, Y., Clausen, B. E., Forster, I., Pawlinski, R., Mackman, N., Haase, V. H. et al. 2003. HIF-1alpha is essential for myeloid cell-mediated inflammation. *Cell* 112, 645–657.

91. Peyssonnaux, C., Datta, V., Cramer, T., Doedens, A., Theodorakis, E. A., Gallo, R. L., Hurtado-Ziola, N., Nizet, V. and Johnson, R. S. 2005. HIF-1alpha expression regulates the bactericidal capacity of phagocytes. *J. Clin. Invest.* 115, 1806–1815.

92. Braverman, J., Sogi, K. M., Benjamin, D., Nomura, D. K. and Stanley, S. A. 2016. HIF-1alpha is an essential mediator of IFN-gamma-dependent immunity to *Mycobacterium tuberculosis*. *J. Immunol.* (Baltimore, MD: 1950). 197, 1287–1297.

93. Imtiyaz, H. Z., Williams, E. P., Hickey, M. M., Patel, S. A., Durham, A. C., Yuan, L. J., Hammond, R., Gimotty, P. A., Keith, B. and Simon, M. C. 2010. Hypoxia-inducible factor 2alpha regulates macrophage function in mouse models of acute and tumor inflammation. *J. Clin. Invest.* 120, 2699–2714.

94. Backer, V., Cheung, F. Y., Siveke, J. T., Fandrey, J. and Winning, S. 2017. Knockdown of myeloid

cell hypoxia-inducible factor-1alpha ameliorates the acute pathology in DSS-induced colitis. *PLOS ONE* 12, e0190074.

95. Shalova, I. N., Lim, J. Y., Chittezhath, M., Zinkernagel, A. S., Beasley, F., Hernandez-Jimenez, E., Toledano, V. et al. 2015. Human monocytes undergo functional re-programming during sepsis mediated by hypoxia-inducible factor-1alpha. *Immunity* 42, 484–498.

96. Colegio, O. R., Chu, N. Q., Szabo, A. L., Chu, T., Rhebergen, A. M., Jairam, V., Cyrus, N. et al. 2014. Functional polarization of tumour-associated macrophages by tumour-derived lactic acid. *Nature* 513, 559–563.

97. Corzo, C. A., Condamine, T., Lu, L., Cotter, M. J., Youn, J. I., Cheng, P., Cho, H. I. et al. 2010. HIF-1alpha regulates function and differentiation of myeloid-derived suppressor cells in the tumor microenvironment. *J. Exp. Med.* 207, 2439–2453.

98. Noman, M. Z., Desantis, G., Janji, B., Hasmim, M., Karray, S., Dessen, P., Bronte, V. and Chouaib, S. 2014. PD-L1 is a novel direct target of HIF-1alpha, and its blockade under hypoxia enhanced MDSC-mediated T cell activation. *J. Exp. Med.* 211, 781–790.

99. Doedens, A. L., Stockmann, C., Rubinstein, M. P., Liao, D., Zhang, N., DeNardo, D. G., Coussens, L. M., Karin, M., Goldrath, A. W. and Johnson, R. S. 2010. Macrophage expression of hypoxia-inducible factor-1 alpha suppresses T-cell function and promotes tumor progression. *Cancer Res.* 70, 7465–7475.

100. Henke, N., Ferreiros, N., Geisslinger, G., Ding, M. G., Essler, S., Fuhrmann, D. C., Geis, T., Namgaladze, D., Dehne, N. and Brune, B. 2016. Loss of HIF-1alpha in macrophages attenuates AhR/ARNT-mediated tumorigenesis in a PAH-driven tumor model. *Oncotarget* 7, 25915–25929.

101. Jeong, Y. J., Hong, S. W., Kim, J. H., Jin, D. H., Kang, J. S., Lee, W. J. and Hwang, Y. I. 2011. Vitamin C-treated murine bone marrow-derived dendritic cells preferentially drive naive T cells into Th1 cells by increased IL-12 secretions. *Cell. Immunol.* 266, 192–199.

102. Jeong, Y. J., Kim, J. H., Hong, J. M., Kang, J. S., Kim, H. R., Lee, W. J. and Hwang, Y. I. 2014. Vitamin C treatment of mouse bone marrow-derived dendritic cells enhanced CD8(+) memory T cell production capacity of these cells in vivo. *Immunobiology* 219, 554–564.

103. Maianski, N. A., Maianski, A. N., Kuijpers, T. W. and Roos, D. 2004. Apoptosis of neutrophils. *Acta Haematol.* 111, 56–66.

104. Mecklenburgh, K. I., Walmsley, S. R., Cowburn, A. S., Wiesener, M., Reed, B. J., Upton, P. D., Deighton, J., Greening, A. P. and Chilvers, E. R. 2002. Involvement of a ferroprotein sensor in hypoxia-mediated inhibition of neutrophil apoptosis. *Blood* 100, 3008–3016.

105. Walmsley, S. R., Cowburn, A. S., Clatworthy, M. R., Morrell, N. W., Roper, E. C., Singleton, V., Maxwell, P., Whyte, M. K. and Chilvers, E. R. 2006. Neutrophils from patients with heterozygous germline mutations in the von Hippel Lindau protein (pVHL) display delayed apoptosis and enhanced bacterial phagocytosis. *Blood* 108, 3176–3178.

106. Walmsley, S. R., Print, C., Farahi, N., Peyssonnaux, C., Johnson, R. S., Cramer, T., Sobolewski, A. et al. 2005. Hypoxia-induced neutrophil survival is mediated by HIF-1alpha-dependent NF-kappaB activity. *J. Exp. Med.* 201, 105–115.

107. Berger, E. A., McClellan, S. A., Vistisen, K. S. and Hazlett, L. D. 2013. HIF-1alpha is essential for effective PMN bacterial killing, antimicrobial peptide production and apoptosis in Pseudomonas aeruginosa keratitis. *PLoS Pathog.* 9, e1003457.

108. Vissers, M. C. M. and Wilkie, R. P. 2007. Ascorbate deficiency results in impaired neutrophil apoptosis and clearance and is associated with up-regulation of hypoxia-inducible factor 1alpha. *J. Leukoc. Biol.* 81, 1236–1244.

109. Perez-Cruz, I., Carcamo, J. M. and Golde, D. W. 2003. Vitamin C inhibits FAS-induced apoptosis in monocytes and U937 cells. *Blood* 102, 336–343.

110. Ferron-Celma, I., Mansilla, A., Hassan, L., Garcia-Navarro, A., Comino, A. M., Bueno, P. and Ferron, J. A. 2009. Effect of vitamin C administration on neutrophil apoptosis in septic patients after abdominal surgery. *J. Surg. Res.* 153, 224–230.

111. Uchio, R., Hirose, Y., Murosaki, S., Yamamoto, Y. and Ishigami, A. 2015. High dietary intake of vitamin C suppresses age-related thymic atrophy and contributes to the maintenance of immune cells in vitamin C-deficient senescence marker protein-30 knockout mice. *Br. J. Nutr.* 113, 603–609.

112. Mohammed, B. M., Fisher, B. J., Kraskauskas, D., Farkas, D., Brophy, D. F., Fowler, A. A., 3rd and Natarajan, R. 2013. Vitamin C: A novel regulator of neutrophil extracellular trap formation. Nutrients 5, 3131–3151.

113. McInturff, A. M., Cody, M. J., Elliott, E. A., Glenn, J. W., Rowley, J. W., Rondina, M. T. and Yost, C. C. 2012. Mammalian target of rapamycin regulates neutrophil extracellular trap formation via induction of hypoxia-inducible factor 1 alpha. Blood 120, 3118–3125.

114. Goldschmidt, M. C. 1991. Reduced bactericidal activity in neutrophils from scorbutic animals and the effect of ascorbic acid on these target bacteria in vivo and in vitro. Am. J. Clin. Nutr. 54, 1214s–1220s.

115. Gaut, J. P., Belaaouaj, A., Byun, J., Roberts, L. J., 2nd, Maeda, N., Frei, B. and Heinecke, J. W. 2006. Vitamin C fails to protect amino acids and lipids from oxidation during acute inflammation. Free Radic. Biol. Med. 40, 1494–1501.

116. Shilotri, P. G. 1977. Phagocytosis and leukocyte enzymes in ascorbic acid deficient guinea pigs. J. Nutr. 107, 1513–1516.

117. Shilotri, P. G. 1977. Glycolytic, hexose monophosphate shunt and bactericidal activities of leukocytes in ascorbic acid deficient guinea pigs. J. Nutr. 107, 1507–1512.

118. Weening, R. S., Schoorel, E. P., Roos, D., van Schaik, M. L., Voetman, A. A., Bot, A. A., Batenburg-Plenter, A. M., Willems, C., Zeijlemaker, W. P. and Astaldi, A. 1981. Effect of ascorbate on abnormal neutrophil, platelet and lymphocytic function in a patient with the Chediak-Higashi syndrome. Blood 57, 856–865.

119. Patrone, F., Dallegri, F., Bonvini, E., Minervini, F. and Sacchetti, C. 1982. Effects of ascorbic acid on neutrophil function. Studies on normal and chronic granulomatous disease neutrophils. Acta vitaminologica et enzymologica 4, 163–168.

120. Anderson, R. 1981. Assessment of oral ascorbate in three children with chronic granulomatous disease and defective neutrophil motility over a 2-year period. Clin. Exp. Immunol. 43, 180–188.

121. Bozonet, S. M., Carr, A. C., Pullar, J. M. and Vissers, M. C. 2015. Enhanced human neutrophil vitamin C status, chemotaxis and oxidant generation following dietary supplementation with vitamin C-rich SunGold kiwifruit. Nutrients 7, 2574–2588.

122. Groux, H. and Powrie, F. 1999. Regulatory T cells and inflammatory bowel disease. Immunol. Today 20, 442–445.

123. Shi, L. Z., Wang, R., Huang, G., Vogel, P., Neale, G., Green, D. R. and Chi, H. 2011. HIF1alpha-dependent glycolytic pathway orchestrates a metabolic checkpoint for the differentiation of TH17 and Treg cells. J. Exp. Med. 208, 1367–1376.

124. Dang, E. V., Barbi, J., Yang, H. Y., Jinasena, D., Yu, H., Zheng, Y., Bordman, Z. et al. 2011. Control of T(H)17/T(reg) balance by hypoxia-inducible factor 1. Cell 146, 772–784.

125. Doedens, A. L., Phan, A. T., Stradner, M. H., Fujimoto, J. K., Nguyen, J. V., Yang, E., Johnson, R. S. and Goldrath, A. W. 2013. Hypoxia-inducible factors enhance the effector responses of CD8(+) T cells to persistent antigen. Nat. Immunol. 14, 1173–1182.

126. Higashiyama, M., Hokari, R., Hozumi, H., Kurihara, C., Ueda, T., Watanabe, C., Tomita, K. et al. 2012. HIF-1 in T cells ameliorated dextran sodium sulfate-induced murine colitis. J. Leukoc. Biol. 91, 901–909.

127. Fluck, K., Breves, G., Fandrey, J. and Winning, S. 2016. Hypoxia-inducible factor 1 in dendritic cells is crucial for the activation of protective regulatory T cells in murine colitis. Mucosal Immunol. 9, 379–390.

128. Hammami, A., Abidin, B. M., Heinonen, K. M. and Stager, S. 2018. HIF-1alpha hampers dendritic cell function and Th1 generation during chronic visceral leishmaniasis. Sci. Rep. 8, 3500.

129. Palazon, A., Tyrakis, P. A., Macias, D., Velica, P., Rundqvist, H., Fitzpatrick, S., Vojnovic, N. et al. 2017. An HIF-1alpha/VEGF-A axis in cytotoxic T cells regulates tumor progression. Cancer Cell 32, 669–683.e665.

130. Thiel, M., Caldwell, C. C., Kreth, S., Kuboki, S., Chen, P., Smith, P., Ohta, A., Lentsch, A. B., Lukashev, D. and Sitkovsky, M. V. 2007. Targeted deletion of HIF-1alpha gene in T cells prevents their inhibition in hypoxic inflamed tissues and improves septic mice survival. PLOS ONE 2, e853.

131. Labiano, S., Melendez-Rodriguez, F., Palazon, A., Teijeira, A., Garasa, S., Etxeberria, I., Aznar, M. A. et al. 2017. CD69 is a direct HIF-1alpha target gene in hypoxia as a mechanism enhancing expression on tumor-infiltrating T lymphocytes. Oncoimmunology 6, e1283468.

132. Fraser, R. C., Pavlovic, S., Kurahara, C. G., Murata, A., Peterson, N. S., Taylor, K. B. and Feigen, G. A. 1980. The effect of variations in vitamin C intake on the cellular immune response of guinea pigs. Am. J. Clin. Nutr. 33, 839–847.

133. Kennes, B., Dumont, I., Brohee, D., Hubert, C. and Neve, P. 1983. Effect of vitamin C supplements on cell-mediated immunity in old people. *Gerontology* 29, 305–310.

134. Huijskens, M. J., Walczak, M., Koller, N., Briede, J. J., Senden-Gijsbers, B. L., Schnijderberg, M. C., Bos, G. M. and Germeraad, W. T. 2014. Technical advance: Ascorbic acid induces development of double-positive T cells from human hematopoietic stem cells in the absence of stromal cells. *J. Leukoc. Biol.* 96, 1165–1175.

135. Huijskens, M. J., Walczak, M., Sarkar, S., Atrafi, F., Senden-Gijsbers, B. L., Tilanus, M. G., Bos, G. M., Wieten, L. and Germeraad, W. T. 2015. Ascorbic acid promotes proliferation of natural killer cell populations in culture systems applicable for natural killer cell therapy. *Cytotherapy* 17, 613–620.

136. Manning, J., Mitchell, B., Appadurai, D. A., Shakya, A., Pierce, L. J., Wang, H., Nganga, V., Swanson, P. C., May, J. M., Tantin, D. and Spangrude, G. J. 2013. Vitamin C promotes maturation of T-cells. *Antioxid Redox Signal.* 19, 2054–2067.

137. Busslinger, M. and Tarakhovsky, A. 2014. Epigenetic control of immunity. *Cold Spring Harbor Perspect. Biol.* 6.

138. Delatte, B., Deplus, R. and Fuks, F. 2014. Playing TETris with DNA modifications. *EMBO J.* 33, 1198–1211.

139. Shen, L., Song, C. X., He, C. and Zhang, Y. 2014. Mechanism and function of oxidative reversal of DNA and RNA methylation. *Annu. Rev. Biochem.* 83, 585–614.

140. Wu, X. and Zhang, Y. 2017. TET-mediated active DNA demethylation: Mechanism, function and beyond. *Nat. Rev. Genet.* 18, 517–534.

141. Blaschke, K., Ebata, K. T., Karimi, M. M., Zepeda-Martinez, J. A., Goyal, P., Mahapatra, S., Tam, A. et al. 2013. Vitamin C induces Tet-dependent DNA demethylation and a blastocyst-like state in ES cells. *Nature* 500, 222–226.

142. Minor, E. A., Court, B. L., Young, J. I. and Wang, G. 2013. Ascorbate induces ten-eleven translocation (Tet) methylcytosine dioxygenase-mediated generation of 5-hydroxymethylcytosine. *J. Biol. Chem.* 288, 13669–13674.

143. Hore, T. A. 2017. Modulating epigenetic memory through vitamins and TET: Implications for regenerative medicine and cancer treatment. *Epigenomics* 9, 863–871.

144. Tsukada, Y., Fang, J., Erdjument-Bromage, H., Warren, M. E., Borchers, C. H., Tempst, P. and Zhang, Y. 2006. Histone demethylation by a family of JmjC domain-containing proteins. *Nature* 439, 811–816.

145. Klose, R. J., Kallin, E. M. and Zhang, Y. 2006. JmjC-domain-containing proteins and histone demethylation. *Nat. Rev. Genet.* 7, 715–727.

146. Ko, M., An, J. and Rao, A. 2015. DNA methylation and hydroxymethylation in hematologic differentiation and transformation. *Curr. Opin. Cell Biol.* 37, 91–101.

147. Ma, L., Qi, T., Wang, S., Hao, M., Sakhawat, A., Liang, T., Zhang, L., Cong, X. and Huang, Y. 2019. Tet methylcytosine dioxygenase 1 promotes hypoxic gene induction and cell migration in colon cancer. *J. Cell. Physiol.* 234, 6286–6297.

148. Cull, A. H., Snetsinger, B., Buckstein, R., Wells, R. A. and Rauh, M. J. 2017. Tet2 restrains inflammatory gene expression in macrophages. *Exp. Hematol.* 55, 56–70.e13.

149. Wallner, S., Schroder, C., Leitao, E., Berulava, T., Haak, C., Beisser, D., Rahmann, S. et al. 2016. Epigenetic dynamics of monocyte-to-macrophage differentiation. *Epigenetics & Chromatin* 9, 33.

150. Zhang, Q., Zhao, K., Shen, Q., Han, Y., Gu, Y., Li, X., Zhao, D. et al. 2015. Tet2 is required to resolve inflammation by recruiting Hdac2 to specifically repress IL-6. *Nature* 525, 389–393.

151. Satoh, T., Takeuchi, O., Vandenbon, A., Yasuda, K., Tanaka, Y., Kumagai, Y., Miyake, T. et al. 2010. The Jmjd3-Irf4 axis regulates M2 macrophage polarization and host responses against helminth infection. *Nat. Immunol.* 11, 936–944.

152. Achuthan, A., Cook, A. D., Lee, M. C., Saleh, R., Khiew, H. W., Chang, M. W., Louis, C. et al. 2016. Granulocyte macrophage colony-stimulating factor induces CCL17 production via IRF4 to mediate inflammation. *J. Clin. Invest.* 126, 3453–3466.

153. Yildirim-Buharalioglu, G., Bond, M., Sala-Newby, G. B., Hindmarch, C. C. and Newby, A. C. 2017. Regulation of epigenetic modifiers, including KDM6B, by interferon-gamma and interleukin-4 in human macrophages. *Front. Immunol.* 8, 92.

154. De Santa, F., Narang, V., Yap, Z. H., Tusi, B. K., Burgold, T., Austenaa, L., Bucci, G. et al. 2009. Jmjd3 contributes to the control of gene expression in LPS-activated macrophages. *EMBO J.* 28, 3341–3352.

155. De Santa, F., Totaro, M. G., Prosperini, E., Notarbartolo, S., Testa, G. and Natoli, G. 2007.

The histone H3 lysine-27 demethylase Jmjd3 links inflammation to inhibition of polycomb-mediated gene silencing. *Cell* 130, 1083–1094.

156. Ishii, M., Wen, H., Corsa, C. A., Liu, T., Coelho, A. L., Allen, R. M., Carson, W. F. T. et al. 2009. Epigenetic regulation of the alternatively activated macrophage phenotype. *Blood* 114, 3244–3254.

157. Yan, Q., Sun, L., Zhu, Z., Wang, L., Li, S. and Ye, R. D. 2014. Jmjd3-mediated epigenetic regulation of inflammatory cytokine gene expression in serum amyloid A-stimulated macrophages. *Cell. Signal.* 26, 1783–1791.

158. Kruidenier, L., Chung, C. W., Cheng, Z., Liddle, J., Che, K., Joberty, G., Bantscheff, M. et al. 2012. A selective Jumonji H3K27 demethylase inhibitor modulates the proinflammatory macrophage response. *Nature* 488, 404–408.

159. Cimmino, L., Dolgalev, I., Wang, Y., Yoshimi, A., Martin, G. H., Wang, J., Ng, V. et al. 2017. Restoration of TET2 function blocks aberrant self-renewal and leukemia progression. *Cell* 170, 1079–1095.e1020.

160. Foster, M. N., Carr, A. C., Antony, A., Peng, S. and Fitzpatrick, M. G. 2018. Intravenous vitamin C administration improved blood cell counts and health-related quality of life of patient with history of relapsed acute myeloid leukaemia. *Antioxidants (Basel, Switzerland).* 7.

161. Zhao, H., Zhu, H., Huang, J., Zhu, Y., Hong, M., Zhu, H., Zhang, J. et al. 2018. The synergy of vitamin C with decitabine activates TET2 in leukemic cells and significantly improves overall survival in elderly patients with acute myeloid leukemia. *Leuk. Res.* 66, 1–7.

162. Zhang, X., Ulm, A., Somineni, H. K., Oh, S., Weirauch, M. T., Zhang, H. X., Chen, X., Lehn, M. A., Janssen, E. M. and Ji, H. 2014. DNA methylation dynamics during ex vivo differentiation and maturation of human dendritic cells. *Epigenetics & Chromatin* 7, 21.

163. Ptaschinski, C., Mukherjee, S., Moore, M. L., Albert, M., Helin, K., Kunkel, S. L. and Lukacs, N. W. 2015. RSV-Induced H3K4 demethylase KDM5B leads to regulation of dendritic cell-derived innate cytokines and exacerbates pathogenesis in vivo. *PLoS Pathog.* 11, e1004978.

164. Nestor, C. E., Lentini, A., Hagg Nilsson, C., Gawel, D. R., Gustafsson, M., Mattson, L., Wang, H. et al. 2016. 5-Hydroxymethylcytosine remodeling precedes lineage specification during differentiation of human CD4(+) T cells. *Cell Rep.* 16, 559–570.

165. Ichiyama, K., Chen, T., Wang, X., Yan, X., Kim, B. S., Tanaka, S., Ndiaye-Lobry, D. et al. 2015. The methylcytosine dioxygenase Tet2 promotes DNA demethylation and activation of cytokine gene expression in T cells. *Immunity* 42, 613–626.

166. Carty, S. A., Gohil, M., Banks, L. B., Cotton, R. M., Johnson, M. E., Stelekati, E., Wells, A. D., Wherry, E. J., Koretzky, G. A. and Jordan, M. S. 2018. The loss of TET2 promotes CD8(+) T cell memory differentiation. *J. Immunol. (Baltimore, MD: 1950).* 200, 82–91.

167. McPherson, R. C., Konkel, J. E., Prendergast, C. T., Thomson, J. P., Ottaviano, R., Leech, M. D., Kay, O. et al. 2014. Epigenetic modification of the PD-1 (Pdcd1) promoter in effector CD4(+) T cells tolerized by peptide immunotherapy. *eLife* 3.

168. LaMere, S. A., Thompson, R. C., Meng, X., Komori, H. K., Mark, A. and Salomon, D. R. 2017. H3K27 Methylation dynamics during CD4 T cell activation: regulation of JAK/STAT and IL12RB2 expression by JMJD3. *J. Immunol. (Baltimore, MD: 1950).* 199, 3158–3175.

169. Wei, G., Wei, L., Zhu, J., Zang, C., Hu-Li, J., Yao, Z., Cui, K. et al. 2009. Global mapping of H3K4me3 and H3K27me3 reveals specificity and plasticity in lineage fate determination of differentiating CD4+ T cells. *Immunity* 30, 155–167.

170. Li, Q., Zou, J., Wang, M., Ding, X., Chepelev, I., Zhou, X., Zhao, W. et al. 2014. Critical role of histone demethylase Jmjd3 in the regulation of CD4+ T-cell differentiation. *Nat. Commun.* 5, 5780.

171. Liu, Z., Cao, W., Xu, L., Chen, X., Zhan, Y., Yang, Q., Liu, S. et al. 2015. The histone H3 lysine-27 demethylase Jmjd3 plays a critical role in specific regulation of Th17 cell differentiation. *J. Mol. Cell Biol.* 7, 505–516.

172. Yue, X., Trifari, S., Aijo, T., Tsagaratou, A., Pastor, W. A., Zepeda-Martinez, J. A., Lio, C. W. et al. 2016. Control of Foxp3 stability through modulation of TET activity. *J. Exp. Med.* 213, 377–397.

173. Sasidharan Nair, V., Song, M. H. and Oh, K. I. 2016. Vitamin C facilitates demethylation of the Foxp3 enhancer in a TET-dependent manner. *J. Immunol. (Baltimore, MD: 1950).* 196, 2119–2131.

174. Nikolouli, E., Hardtke-Wolenski, M., Hapke, M., Beckstette, M., Geffers, R., Floess, S., Jaeckel, E. and Huehn, J. 2017. Alloantigen-induced

regulatory T cells generated in presence of vitamin C display enhanced stability of Foxp3 expression and promote skin allograft acceptance. *Front.Immunol.* 8, 748.

175. Kasahara, H., Kondo, T., Nakatsukasa, H., Chikuma, S., Ito, M., Ando, M., Kurebayashi, Y. et al. 2017. Generation of allo-antigen-specific induced Treg stabilized by vitamin C treatment and its application for prevention of acute graft versus host disease model. *Int. Immunol.* 29, 457–469.

176. Song, M. H., Nair, V. S. and Oh, K. I. 2017. Vitamin C enhances the expression of IL17 in a Jmjd2-dependent manner. *BMB Rep.* 50, 49–54.

177. Tsagaratou, A., Gonzalez-Avalos, E., Rautio, S., Scott-Browne, J. P., Togher, S., Pastor, W. A., Rothenberg, E. V., Chavez, L., Lahdesmaki, H. and Rao, A. 2017. TET proteins regulate the lineage specification and TCR-mediated expansion of iNKT cells. *Nat. Immunol.* 18, 45–53.

178. Northrup, D., Yagi, R., Cui, K., Proctor, W. R., Wang, C., Placek, K., Pohl, L. R. et al. 2017. Histone demethylases UTX and JMJD3 are required for NKT cell development in mice. *Cell Biosci.* 7, 25.

179. Beyaz, S., Kim, J. H., Pinello, L., Xifaras, M. E., Hu, Y., Huang, J., Kerenyi, M. A. et al. 2017. The histone demethylase UTX regulates the lineage-specific epigenetic program of invariant natural killer T cells. *Nat. Immunol.* 18, 184–195.

180. Cribbs, A., Hookway, E. S., Wells, G., Lindow, M., Obad, S., Oerum, H., Prinjha, R. K. et al. 2018. Inhibition of histone H3K27 demethylases selectively modulates inflammatory phenotypes of natural killer cells. *J. Biol. Chem.* 293, 2422–2437.

181. Washko, P., Rotrosen, D. and Levine, M. 1991. Ascorbic acid in human neutrophils. *Am. J. Clin. Nutr.* 54, 1221s–1227s.

182. May, J. M., Li, L., Qu, Z. C. and Huang, J. 2005. Ascorbate uptake and antioxidant function in peritoneal macrophages. *Arch. Biochem. Biophys.* 440, 165–172.

183. Talks, K. L., Turley, H., Gatter, K. C., Maxwell, P. H., Pugh, C. W., Ratcliffe, P. J. and Harris, A. L. 2000. The expression and distribution of the hypoxia-inducible factors HIF-1alpha and HIF-2alpha in normal human tissues, cancers, and tumor-associated macrophages. *Am. J. Pathol.* 157, 411–421.

184. Gardner, P. J., Liyanage, S. E., Cristante, E., Sampson, R. D., Dick, A. D., Ali, R. R. and Bainbridge, J. W. 2017. Hypoxia inducible factors are dispensable for myeloid cell migration into the inflamed mouse eye. *Sci. Rep.* 7, 40830.

185. Ahn, G. O., Seita, J., Hong, B. J., Kim, Y. E., Bok, S., Lee, C. J., Kim, K. S. et al. 2014. Transcriptional activation of hypoxia-inducible factor-1 (HIF-1) in myeloid cells promotes angiogenesis through VEGF and S100A8. *Proc. Natl. Acad. Sci. USA* 111, 2698–2703.

186. Meng, X., Grotsch, B., Luo, Y., Knaup, K. X., Wiesener, M. S., Chen, X. X., Jantsch, J., Fillatreau, S., Schett, G. and Bozec, A. 2018. Hypoxia-inducible factor-1alpha is a critical transcription factor for IL-10-producing B cells in autoimmune disease. *Nat. Commun.* 9, 251.

187. Lio, C. W., Zhang, J., Gonzalez-Avalos, E., Hogan, P. G., Chang, X. and Rao, A. 2016. Tet2 and Tet3 cooperate with B-lineage transcription factors to regulate DNA modification and chromatin accessibility. *eLife* 5.

188. Orlanski, S., Labi, V., Reizel, Y., Spiro, A., Lichtenstein, M., Levin-Klein, R., Koralov, S. B. et al. 2016. Tissue-specific DNA demethylation is required for proper B-cell differentiation and function. *Proc. Natl. Acad. Sci. USA* 113, 5018–5023.

189. Bhandari, T., Olson, J., Johnson, R. S. and Nizet, V. 2013. HIF-1alpha influences myeloid cell antigen presentation and response to subcutaneous OVA. vaccination. *J. Mol. Med* 91, 1199–1205.

190. Campbell, J. D., Cole, M., Bunditrutavorn, B. and Vella, A. T. 1999. Ascorbic acid is a potent inhibitor of various forms of T cell apoptosis. *Cell. Immunol* 194, 1–5.

191. Johnston, C. S. and Huang, S. N. 1991. Effect of ascorbic acid nutriture on blood histamine and neutrophil chemotaxis in guinea pigs. *J. Nutr* 121, 126–130.

192. Puskas, F., Gergely, P., Niland, B., Banki, K. and Perl, A. 2002. Differential regulation of hydrogen peroxide and Fas-dependent apoptosis pathways by dehydroascorbate, the oxidized form of vitamin C. *Antioxid. Redox Signal.* 4, 357–369.

193. Vissers, M. C. and Hampton, M. B. 2004. The role of oxidants and vitamin C on neutrophil apoptosis and clearance. *Biochem. Soc. Trans* 32, 499–501.

194. He, C., Larson-Casey, J. L., Gu, L., Ryan, A. J., Murthy, S. and Carter, A. B. 2016. Cu,Zn-superoxide dismutase-mediated redox regulation of Jumonji domain containing 3 modulates macrophage polarization and pulmonary fibrosis. *Am. J. Respir. Cell Mol. Biol* 55, 58–71.

195. Kim, J. E., Cho, H. S., Yang, H. S., Jung, D. J., Hong, S. W., Hung, C. F., Lee, W. J. and Kim, D. 2012. Depletion of ascorbic acid impairs NK cell activity against ovarian cancer in a mouse model. *Immunobiology* 217, 873–881.

196. Maeng, H. G., Lim, H., Jeong, Y. J., Woo, A., Kang, J. S., Lee, W. J. and Hwang, Y. I. 2009. Vitamin C enters mouse T cells as dehydroascorbic acid in vitro and does not recapitulate *in vivo* vitamin C effects. *Immunobiology* 214, 311–320.

197. Noh, K., Lim, H., Moon, S. K., Kang, J. S., Lee, W. J., Lee, D. and Hwang, Y. I. 2005. Mega-dose Vitamin C modulates T cell functions in Balb/c mice only when administered during T cell activation. *Immunol. Lett* 98, 63–72.

198. Vojdani, A. and Ghoneum, M. 1993. In vivo effect of ascorbic acid on enhancement of human natural killer cell activity. *Nutr. Res* 13, 753–764.

CHAPTER NINE

Role of Vitamin C in Chronic Wound Healing

Juliet M. Pullar and Margreet C.M. Vissers

DOI: 10.1201/9780429442025-9

CONTENTS

INTRODUCTION

The skin plays a vital role in maintaining homeostasis of the human body and in protecting us from insults or threats from the external environment, including pathogens. To fulfill many of its functions, reviewed in [1], skin must maintain its integrity. Wound healing is the physiologic process by which skin is repaired in response to injury. In this review, the focus is on skin; however, the same wound healing process occurs in most other organ systems when they become damaged.

Vitamin C (ascorbate) is an essential micronutrient with a multitude of functions in the body largely related to its role as a cofactor for various biosynthetic and gene regulatory enzymes and its antioxidant capacity. There are a number of biological activities of vitamin C that may impact on wound healing. There is evidence to suggest that prolonged and severe vitamin C deficiency dramatically impairs wound healing. In this review, we discuss the biology of cutaneous wound healing and consider the evidence for the use of vitamin C to promote the healing of chronic wounds.

STRUCTURE OF SKIN

The skin is composed of two main layers—the outermost epidermis and the underlying dermis, with the differing structures of the two layers clearly reflecting their distinct functions. The epidermis is a stratified squamous epithelium

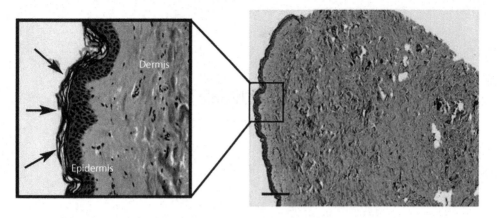

Figure 9.1. Micrograph of a human breast skin sample. The view on the right shows the thickness of the dermis (pink staining) in comparison to the much thinner epidermis (purple staining). The scale bar indicates 200 μm. A zoomed-in image is shown within the box. The stratified nature of the cells of the epidermis is evident. The outermost layer of the epidermis, the stratum corneum, is indicated by the arrows, with its characteristic basket-weave structure. Within the dermis, the collagen bundles are very clear, as are the scattered purple-stained fibroblasts. (Originally published in Pullar, J. M. et al. 2017. *Nutrients* 9.)

predominantly made up of keratinocytes. The keratinocytes closest to the dermis, termed the *basal cells*, divide, and cells move outward toward the surface, differentiating as they do so (Figure 9.1). During differentiation, keratinocytes lose almost all of their organelles and much of their cytoplasm, they increase production of specialized structural proteins, their cell membrane is replaced by a cellular envelope of cross-linked proteins, and they secrete lipids [2,3]. These changes allow the formation of the outermost epidermal layer, the stratum corneum, which consists of flattened, metabolically inactive cells, now termed *corneocytes*, sealed together with lipid-rich intercellular domains forming a nearly water-impermeable barrier (Figure 9.1). The stratum corneum interacts with the external environment and fulfills almost all of the physical barrier function of the skin [4]. The epidermis is continually renewing; it takes 30–60 days for new keratinocytes to reach the skin surface [1].

In comparison, the thicker dermal layer consists mainly of a complex extracellular matrix that provides strength and flexibility and gives structural support to the epidermis. It is also home to the vascular, lymphatic, and neuronal systems of the skin. The dermis is relatively acellular (Figure 9.1). Fibroblasts are the main cell type present and are responsible for synthesizing many components of the extracellular matrix [5]. The majority of the dermis is composed of collagen fibers that account for 70%–80% of its dry weight. They are arranged

in bundles in a random basket-weave pattern and provide the tensile strength of the skin, preventing it from tearing when stretched. Elastin fibers are also present in the dermis, although they are much less abundant than collagen. Elastin provides the skin with recoil or resilience, allowing it to return to an unstretched state. Both of these fibers are embedded into a proteoglycan matrix that, in conjunction with polymeric hyaluronic acid, is crucial for maintaining skin hydration, lubricating between collagen and elastic fiber networks during skin movement, and providing resistance to compression [1]. The dermis differs from the epidermis in that it is not continually replacing itself. Many of the extracellular matrix molecules are extremely long lived, with the half-life of skin collagen estimated at 15 years [6]. Lying between the two layers is the dermal-epidermal junction, a specialized basement membrane that fixes the epidermis to the dermis below.

WOUND HEALING IN ACUTE VERSUS CHRONIC WOUNDS

An acute wound in a healthy individual follows a reproducible trajectory of wound healing that can be described in three overlapping phases—inflammation, proliferation, and tissue remodeling. The process relies on a complex interplay of many different cell types, their mediators and the extracellular matrix, and is reviewed in detail in [7–9]. The first stage of wound repair is the inflammatory phase, which begins immediately after injury when vasoconstriction and clot formation combine to stem bleeding. The fibrin clot becomes a scaffold for the infiltrating cells required during healing, with neutrophils among the first cells recruited to the wound site. The recruitment of leukocytes to the wound depends on signals including platelet degranulation, activation of complement, products of bacterial or extracellular matrix degradation, and injured or activated cells [7,10]. The influx of neutrophils is followed 1–2 days later by the infiltration of monocytes, which subsequently differentiate into macrophages [11], although there are also resident macrophages present in the skin. Neutrophils and macrophages, as phagocytic cells, clear the wound of infectious material and any damaged tissue or debris. Macrophages also remove apoptotic neutrophils and are critically involved in orchestrating the next phases of the healing process, providing signals for reepithelialization, angiogenesis, and dermal repair [11–13]. Neutrophils are also thought to secrete cytokines, growth factors, and other mediators involved in activating endothelial cells, keratinocytes, fibroblasts, and inflammatory cells present in the wound [14]. In a normal acute wound, the inflammatory stage takes 2–5 days [9].

The next stage in the healing process is the proliferative phase, which is characterized by reepithelialization of the wound, restoration of the vascular system, and formation of granulation tissue. Reepithelialization occurs via a combination of migration and proliferation of the keratinocytes that reside closest to the leading edge of the wound. Thus soon after injury, keratinocytes at the wound edge begin to migrate over the wound [7,9]. Several days later, basal keratinocytes and nearby stem cells start to proliferate so as to provide enough cells to cover the wound [9]. Fibroblasts from a number of sources infiltrate the wound area where they proliferate and synthesize extracellular matrix components including collagen, fibronectin, proteoglycans, and hyaluronic acid [9]. These cells also secrete proteinases that remove the fibrin clot, replacing it with a more stable fibronectin-rich matrix that allows subsequent collagen deposition [15,16]. At the same time, angiogenesis is initiated via growth factor production by macrophages, keratinocytes, and fibroblasts leading neighboring endothelial

cells to become activated and to migrate into the wound. This response is enhanced in hypoxia, which leads to activation of hypoxia-inducible factor (HIF) [17]. The entire process results in seen granulation tissue being formed, made up of blood vessels, fibroblasts, inflammatory cells, endothelial cells, and new extracellular matrix of collagen, glycoprotein, fibronectin, and hyaluronic acid [18]. Granulation tissue forms a scaffold filling the wound bed and is required for complete reepithelialization of thicker wounds so that migrating keratinocytes have viable tissue to migrate across [18]. Alongside these events, fibroblasts also differentiate into myofibroblasts in response to stimulation by macrophages [19]. The myofibroblasts at the edge of the wound, which now contain actin, then contract, decreasing the size of the wound and also contribute to the deposition of the extracellular matrix.

The final stage of repair is remodeling, which starts 2–3 weeks after the original injury and can last a year or more [8]. The dermis is cleared of cells that are no longer needed. The loose extracellular matrix of the wound that has allowed migration and repair to occur is slowly replaced by a more dense matrix composed predominantly of collagen. Over time, much of the type III collagen, laid down during the proliferative phase, is replaced by the stronger type I collagen [20]. This requires the action of matrix metalloproteinases released from macrophages, keratinocytes, endothelial cells, and fibroblasts. The collagen fibers, which were originally arranged in a random basket-weave pattern in undamaged skin, become realigned parallel to the epithelium. Gradually the tensile strength of the wound increases; however, the strength of the skin at the repair site is never as great as the uninjured skin [20]. Scar tissue is also less functional and may not have the normal skin appendages, for example, sweat glands, depending on the thickness of the injury.

If something goes wrong with the skin repair process, and it is impaired or halted, then the wound moves from an acute to a chronic healing process. While the impairment can occur at any stage, it most commonly gets halted in the inflammatory phase [9,13]. The resultant protracted or excessive inflammation is likely to be detrimental with elevated levels of proteases, breakdown of cytokines and extracellular matrix proteins, and increased levels of oxidative species observed in chronic wounds. One cell that is at least partially responsible for the damage observed is the neutrophil, which has been shown to be present in chronic wounds for extended periods of time [14]. Evidence is mounting that the transition from the inflammatory to the proliferative phase is compromised in chronic wounds [9]. A recent hypothesis suggests that macrophages may be critical, with depletion of these cells resulting in defective healing [21]. In particular, their switch from the classical pro-inflammatory M1 phenotype to an anti-inflammatory M2 phenotype may be important for the transition to the proliferative phase [12,13,18]. M1 macrophages are associated with a phagocytic phenotype and in the production of pro-inflammatory mediators, whereas M2 macrophages express anti-inflammatory mediators, promoting fibroblast proliferation, extracellular matrix synthesis, angiogenesis, and reepithelialization [18]. The removal of apoptotic neutrophils by macrophages also seems to be important for this transition [22]. Other possible mechanisms implicated in the pathogenesis of chronic wounds include the sustained presence of bacterial infection leading to prolonged inflammation, dysregulation of protease production and activation, depletion of stem cells required for epidermal and blood vessel regeneration, and inadequate local angiogenesis [18].

PROBLEMS WITH CHRONIC WOUNDS

Chronic wounds are defined as wounds that have failed to heal in an orderly and timely manner or those that have proceeded through the repair process without reaching anatomical and functional integrity [23]. They are often a significant burden to an individual, causing pain, reduced mobility and capacity to work, social isolation, and poor quality of life [24,25]. They are also a growing problem worldwide, with rates increasing due to aging populations, and increasing rates of obesity, vascular disease, and diabetes, which are all associated with impaired wound healing.

The costs associated with the care and management of wounds, especially chronic wounds, is substantial. For example, recent economic modeling estimated the direct healthcare-related costs of chronic wounds in Australia at A\$3 billion, which equates to approximately 2% of the national healthcare budget [26]. The vast majority (~90%) of chronic wounds comprise only three types: venous leg ulcers, diabetic foot ulcers, and pressure wounds [27]. There are a range of underlying causes of chronic wounds; these include venous or arterial insufficiency, diabetes mellitus, ischemia, advanced age, and local pressure effects. Nutritional deficiencies are also thought to contribute to defects in wound healing. This review considers the effect of vitamin C on chronic wounds only.

VITAMIN C

Vitamin C is essential for both plants and animals [28]. However, humans, along with a few other species, have lost the ability to synthesize it from glucose and are therefore reliant on dietary sources. Following ingestion, vitamin C is absorbed via the small intestine into the bloodstream, and then actively accumulated into the tissues via specialized transporters [29,30]. Typical plasma levels are 20–80 μmol/L, while concentrations in cells can reach low millimolar levels. Unusually, vitamin C is differentially distributed throughout the body with some tissues having higher concentrations than others [29]. This is thought to be due to differing requirements for the vitamin in those tissues. In particular, the brain has substantially higher concentrations than many other tissues [31], indicating a vital role in this tissue.

Vitamin C (ascorbate) is best described as a water-soluble electron donor. This reducing capacity forms the basis for all of its known biological activities. It can consecutively undergo two one-electron oxidations to form ascorbate radical and dehydroascorbate, which can both be reduced back to ascorbate, either enzymatically or by glutathione [32]. The oxidized form of vitamin C, dehydroascorbic acid, has a half-life of about 15 minutes at 37°C due to irreversible hydrolysis of the ring in which 2,3-diketogulonic acid is formed [33]. This means that as vitamin C is metabolized, some of it is lost, and so a regular dietary source is required to maintain body levels, or the deficiency disease scurvy will result.

Among its most significant activities, ascorbate functions as an enzyme cofactor for a number of monooxygenases and dioxygenases for which it provides electrons to maintain their catalytic metals in the reduced forms [32,34]. Prominent examples include the iron (Fe)-dependent prolyl and lysyl hydroxylases responsible for stabilization of the triple helix structure of collagen and the copper (Cu)-containing dopamine β-hydroxylase, which converts dopamine to norephineprine in the chromaffin granules of adrenal medulla [35]. However, of late it has become apparent that ascorbate is also a cofactor for other newly described hydroxylases that regulate gene transcription and cell signaling pathways—all members of the Fe-containing 2-oxoglutarate–dependent dioxygenases (2-OGDDs) family [36,37]. These are the prolyl, lysyl, and arginine hydroxylases that modify the α regulatory subunit of the hypoxia-inducible factors (HIFs) [38,39], and the DNA demethylases (ten-eleven translocases [TETs]) and histone demethylases (Jumonji C domain-containing histone demethylases [JHMDs]), which are involved in epigenetic regulation [40]. These later findings suggest a broader role for vitamin C in health and disease than was previously appreciated. Another major function of vitamin C is as an antioxidant in which it can scavenge radicals and oxidants and thereby prevent them from damaging important biomolecules such as protein, nucleic acids, or lipids [41,42]. It is also thought to interact with and recycle other antioxidants such as the lipid-soluble α-tocopherol and tetrahydrobiopterin, thereby maintaining their antioxidant activity [42].

Vitamin C Content of Skin

The vitamin C content of human skin has been measured in several studies [43–45]. Skin concentrations are very similar to other tissues, likely in the millimolar range, and well above plasma levels. They are also comparable to the levels of other water-soluble antioxidants in skin such as glutathione [43–46]. Skin is a difficult tissue in which to measure vitamin C as it is very resilient to solubilization, and also quite sensitive to oxidation, particularly the extremely thin epidermal layer [44]. In the two studies that distinguished between the different layers, the epidermis consistently had higher levels of vitamin C than the dermis, with differences of two- to five-fold being reported [43,44]. This most likely reflects the differing cellularity of the epidermis and dermis, as the majority of the vitamin C in the skin is believed to be contained within intracellular compartments. While there is some uncertainty in the published values due to the considerable variation observed and the small number of studies, the levels measured in the epidermis are among the highest concentrations of vitamin C that have been found in any tissue [47].

The concentration of ascorbate found in the dermis, with a lower estimate of ~3 mg/100 mg wet weight, is similar to skeletal muscle and just a little lower than many other tissues such as kidney or liver [47]. Most extracellular fluids are thought to contain similar concentrations of ascorbate to plasma—that is, 20–80 μmol/L [48]—and as far as we know, this is the case for skin. Given these findings, and because the dermis is relatively acellular (see Figure 9.2), it is likely that the fibroblasts present have exceptionally high intracellular ascorbate concentrations. It is plausible that fibroblasts would have high ascorbate levels, as a primary function of these cells is to regenerate extracellular matrix, the great majority of which is collagen, and collagen synthesis is enhanced in the presence of ascorbate [49–53].

Vitamin C is transported into the skin from the blood vessels present in the dermal layer. Uptake from the plasma and delivery to the skin layers is mediated by specific sodium-dependent vitamin C transporters (SVCTs) that are present throughout the body. Interestingly, cells in the epidermis express both SVCT1 and SVCT2 [54]. This contrasts with many other tissues and cell types in the body, which mostly express SVCT2 [30,54,55]. SVCT1 expression is largely confined to the epithelial cells in the small intestine and the kidney, where it is associated with whole-body

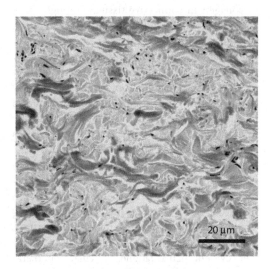

Figure 9.2. Hematoxylin-eosin stain of the dermis of human skin. The purple nuclear staining shows the sparse presence of fibroblasts in the dermis, with the collagen fibers staining pink. Thus, the great majority of the dermis is extracellular matrix. (Originally published in Pullar, J. M. et al. 2017. *Nutrients* 9.)

homeostasis of the vitamin [30,56,57]. Localization of SVCT1 to the epidermis is interesting due to the lack of vasculature in this tissue and raises the possibility that the combined expression of both transporters could assist in ensuring effective uptake and intracellular accumulation of the vitamin. Together with the purported high levels of vitamin C measured in the epidermal layer, the dual expression of the SVCTs suggests a high dependency on vitamin C in this tissue.

EFFECT OF VITAMIN C ON WOUND HEALING

Wound Healing in Scurvy

In the absence of vitamin C, the deficiency disease scurvy may develop. It is not known precisely at what plasma ascorbate concentration individuals actually develop scurvy [58]; however, those with plasma levels of less than 11 μmol/L are considered at severe risk of developing the disease. At these concentrations, it seems likely that many cells and tissues within the body will become depleted of the vitamin. Scurvy is associated with a range of clinical features including fatigue, bleeding, anemia, gum abnormalities, and skin and musculoskeletal manifestations [59,60] and can ultimately prove fatal if left untreated. Many of the symptoms of scurvy are thought to arise from a defect in the synthesis and maturation of collagen in connective tissue, although other cofactor activities also likely contribute.

Prominent among the many signs of scurvy is poor wound healing and wound dehiscence. Early in the 1940s, John Crandon investigated the effect of vitamin C deficiency on wound healing by experimenting on himself [61]. He was depleted of vitamin C over 6 months, and periodically his ability to recover from an acute wound was determined. After 3 months of depletion, a time at which he had no measurable vitamin C in both his plasma and white blood cells, although lacking other symptoms of scurvy, there was no defect in his capacity for wound healing; complete healing was observed 10 days after an incision. In contrast, at 6 months, when noticeable symptoms of scurvy were evident such as fatigue, follicular hyperkeratosis, and perifollicular hemorrhages, there was a substantial defect in healing. Ten days after an incision was made on his back, despite reepithelialization of the wound, a biopsy showed no healing to have occurred underneath this layer and a disorganized blood clot filling the space. There was a lack of extracellular matrix and capillary formation [61]. At the same time, older previously healed wounds also began to break down. Immediately after the second biopsy, he received daily intravenous vitamin C doses of 1000 mg. A repeat biopsy, 10 days after the vitamin C therapy was initiated, found good healing of the original wound and restoration of the collagen matrix. It was suggested that repletion of the vitamin was able to resolve the healing defect and strongly implicated lack of collagen deposition as one of the major impacts of vitamin C deficiency on wound healing.

Other studies have also suggested that an impairment of wound healing occurs in scurvy. However, it is interesting to consider whether it is an earlier or later manifestation of the disease. In a 1969 publication in which a less severe form of scurvy was induced in four to six prisoners, no defect in the healing of a surgical incision was observed among any of the participants, despite other signs of scurvy being present such as fatigue, follicular hyperkeratosis, muscle aches, and bleeding and swelling of the gums [59]. A study of conscientious objectors in World War II, which provided a detailed timeline of induced scurvy, showed that while hyperkeratosis of hair follicles was evident after 120–180 days of deficiency, abnormalities with wound healing did not occur until weeks later at approximately 190 days [62]. This was just after the first appearance of hemorrhagic follicles and at a similar time to swelling and hemorrhage of the gums. There was a reduced tendency to heal in the totally vitamin C–deprived group, and also, wounds made earlier in the study that had been exhibiting normal healing became red and livid. In a second study group, supplementation with as little as 10 mg of vitamin C daily was enough to prevent any obvious defects in wound healing, or indeed, any symptoms of scurvy altogether. Findings from these studies indicate that poor wound healing is not amongst the very earliest signs of vitamin C deficiency, but rather it occurs slightly later, probably concurrent with many of the other common symptoms of scurvy.

Animal Studies and Wound Healing

The effect of severe ascorbate deficiency on wounds has also been investigated in animal studies [63–67]. Light and electron microscopy of wounds induced in naturally vitamin C–deficient guinea pigs fed a scorbutic diet and confirmed as scorbutic, provided further support for collagen synthesis being impaired in severe vitamin C deficiency [63]. There was little collagen in the extracellular space of the wounds of the scorbutic animals, with levels greatly reduced compared to control animals. The authors also observed intracellular lipid deposits in the fibroblasts of the scorbutic wounds and an altered endoplasmic reticulum structure in these cells; they proposed these findings were related to an abnormality of protein synthesis. Interestingly, the scorbutic animals had consistently elevated numbers of macrophages compared to controls, at what was presumably the later stages of the inflammatory phase. Capillary formation was also delayed in the scorbutic animals.

Another animal model of vitamin C deficiency is the Gulo$^{-/-}$ knockout mouse, which lacks gulonolactone oxidase, the terminal enzyme in the synthetic pathway, and displays a phenotype similar to human scurvy when the vitamin is withdrawn from its diet [68,69]. Using this model, Mohammed and coworkers also showed significantly decreased collagen deposition at days 7 and 14 following full-thickness excisional wounding of vitamin C–deficient animals [70]. Granulation tissue was much less dense; the dermis was more cellular, and the basement membrane not as distinct. It should be noted these animals were depleted of vitamin C but not completely deficient in the vitamin with plasma levels approximately 32% of sufficient animals at day 7. Deficiency also increased mRNA expression of pro-inflammatory signals such as interleukin-1β and decreased expression of various wound healing mediators including vascular endothelial growth factor (VEGF). These results contrast with an earlier study in which vitamin C deficiency did not affect the formation of collagen in the skin of severely vitamin C–deficient Gulo$^{-/-}$ mice [69]. However, in that study, hydroxyproline levels were used to assess the amount of collagen. Surprisingly, the authors observed an increase in skin hydroxyproline during deficiency and suggested this may be artefactual due to a loss in tissue hydration. In addition, this was not a wound model, rather hydroxyproline was measured in skin generally, and given the long half-life of collagen, the short time course of deficiency is unlikely to have affected skin hydroxyproline levels markedly over this time.

The effect of vitamin C deficiency on skin tensile strength after wounding has also been investigated. Mohammed et al. [70] reported a decrease in healed skin stiffness with vitamin C deficiency in the Gulo mouse. Repletion of ascorbate by intraperitoneal infusion was unable to restore the tensile strength of the wound despite it improving other healing outcomes such as deposition of collagen. The reason for this discrepancy is unclear, but it may be related to the extent of collagen deposition and its state. In a small study of guinea pigs in which moderate and high dietary doses of vitamin C were compared, a small, nonsignificant increase in wound tensile strength was observed at the higher vitamin C doses [71], with the authors suggesting there was greater wound integrity in the higher-dose group. Finally, supplementation with ascorbate was also able to increase the tensile strength of an incisional wound in a diabetic rat model [67].

Studies in replete guinea pigs and rats have shown that wound skin vitamin C levels are decreased during wounding even as late as 14 days after injury [65,72]. In rats, a species that can make its own ascorbate, topical vitamin C treatment of wounds resulted in less inflammation, with a lower number of macrophages and more granulation tissue [73]. Collagen fibers were thicker in the treated group, and the complete contraction of wounds occurred earlier in this group. These findings imply that vitamin C is consumed during the healing process, and that physiologic doses of vitamin C may not be sufficient to restore or maintain levels.

It is well established that scurvy affects wound healing rates and causes wound dehiscence [61,62,74]. The findings from the preclinical studies detailed above also suggest that vitamin C appears to be essential for wound healing, with functions seemingly beyond its role in collagen synthesis and maturation, although these remain important consequences. One of the questions this review seeks to answer is whether having low, suboptimal levels of vitamin C could also affect rates of healing and healing outcomes, particularly for chronic wounds. There is some support for this proposition from the Gulo mouse study in which mice were depleted of vitamin C but were not completely deficient [70]. There were a number of effects on wound repair and the suggestion that the wounds of the deficient animals persisted in a pro-inflammatory state, possibly akin to that of a chronic wound.

Biological Roles of Vitamin C in Wound Repair

There are a number of biological functions of vitamin C that may explain its ability to modulate the wound healing process. It is implicated in all three major phases of healing—the inflammatory, proliferation, and remodeling phases.

Promotion of Collagen Formation

Foremost is the effect of vitamin C on the synthesis and maturation of collagen. During wound repair, there is an increased demand for collagen, the synthesis of which is mainly carried out by fibroblasts in the dermis. Ascorbate is a cofactor for the proline and lysine hydroxylases responsible for the posttranslational hydroxylation of procollagen chains [34] that strengthens and stabilizes the structure of mature collagen, allowing its proper assembly and export. Collagen chains that contain insufficient hydroxyproline residues cannot fold into triple helical structures that are stable at body temperature and are either degraded in the cell or gradually secreted as nonfunctional proteins [75]. In addition to cross-linking the collagen molecule by hydroxylation, vitamin C stimulates collagen mRNA production, and protein expression by fibroblasts, particularly type I and III collagen, both of which are critical for wound healing [49–53]. At present, we do not understand whether hydroxylation or synthesis is the major effect of ascorbate on collagen metabolism and whether the two are linked [34,51].

Ability to Regulate Inflammatory Responses

Ascorbate supports the local inflammatory response that occurs early in the wound healing process. The initial responder cell, the neutrophil, requires vitamin C for optimizing many of its functions. We have previously shown that neutrophils from Gulo$^{-/-}$ mice severely deficient in vitamin C fail to undergo spontaneous apoptosis, eventually dying by necrosis, an effect that was associated with HIF-1α activation [76]. The ability of macrophages to phagocytose and clear aged ascorbate-deficient neutrophils was also impaired. Mohammed and coworkers [70] also found attenuated apoptosis in vitamin C–deficient neutrophils from Gulo$^{-/-}$ mice. In chronic wounds, it is hypothesized that the formation of apoptotic neutrophils and their removal is a critical step for the resolution of the

inflammatory phase and progression of wound healing [14], implying this is one mechanism by which vitamin C could modulate healing.

As part of the repair response, neutrophils must migrate to the wound site and phagocytose and kill any invading pathogens. There is some suggestion that severe ascorbate deficiency impairs neutrophil chemotaxis [77,78] and that ascorbate supplementation may enhance the chemotactic ability of human neutrophils [79–82]. Ascorbate deficiency may also impair neutrophil bactericidal ability [77,83,84], possibly by attenuating the oxidative burst.

Neutrophil extracellular traps (NETs) are web-like structures composed of granular proteins and nuclear material that are released from dying neutrophils via a process termed NETosis. They are postulated to immobilize bacteria and fungi and mediate the extracellular killing of pathogens [85,86]. A model of experimental sepsis in the Gulo$^{-/-}$ mouse found that vitamin C deficiency increased the generation of NETs in vivo, with vitamin C supplementation able to attenuate this response [70]. Experiments conducted with human neutrophils activated to release NETs also showed that preloading of cells with ascorbate reduced the formation of NETs [70]. Recent work in diabetic patients and animals models suggests NETs may have a role in delayed healing of wounds associated with diabetes [87,88]. Neutrophils from diabetics were primed to release NETs, overexpressing PAD4, an enzyme important in chromatin decondensation during NET release [87]. In mouse models of diabetes, pharmacologic inhibition or genetic knockout of NET formation accelerated wound healing. In humans, proteomic analysis showed NET proteins were elevated in the wounds of nonhealing diabetic foot ulcers compared to those that exhibited rapid healing, and biomarkers of NETs were elevated in the circulation of those with diabetic foot ulcers [88]. Interestingly, HIF-1α can regulate NET formation in response to lipopolysaccharide [89].

Vitamin C is also an antioxidant that can dispose of free radicals and other oxidants, which are likely to be generated at wound sites due to the activation of phagocytic cells during the inflammatory phase. The capacity of these cells to generate and release a large amount of oxidative species is substantial [90]. However, vitamin C is only one player in the antioxidant defense mounted against such species; there are many others. Vitamin C is particularly effective at reducing oxidative damage to the skin when it is used in conjunction with vitamin E [91,92]. It is unclear whether the antioxidant activity of vitamin C is important during the process of wound healing.

Interaction with Cell Signaling Pathways

Cell culture studies demonstrate that vitamin C can promote the differentiation of keratinocytes, which is critical for reepithelialization in wound healing [93]. In an organotypic culture model, vitamin C enhanced differentiation of rat epidermal keratinocytes, increasing filaggrin expression, a differentiation marker, and the number of keratohyalin granules [94]. It also improved the ultrastructural organization of the stratum corneum, enhancing the characteristic basket-weave pattern of this layer, and promoting extracellular lipid deposition. Finally, vitamin C markedly increased the barrier function of the epidermis, as demonstrated by reductions in transepidermal water loss and permeation rates of corticosterone and mannitol. In support of these findings, others have also shown that vitamin C promotes the synthesis and organization of barrier lipids, increases cornified envelope formation during differentiation, and promotes barrier function [95–99]. More specifically, vitamin C has been shown to enhance ceramide synthesis, which is the major lipid component of the stratum corneum [95,97,99]. It particularly increases formation of ceramides 6 and 7 [95], which are hydroxylated derivatives. The increased synthesis seems to be related to upregulation of ceramide synthetic enzymes including serine palmitoyltransferase and ceramide synthase [97,99].

Vitamin C also increases proliferation and migration of dermal fibroblasts [50,100–102]. These cells are vital for effective wound healing, synthesizing much of the required extracellular matrix components including collagen. Normally fairly quiescent, fibroblasts are stimulated to rapidly proliferate during wounding. As stated, vitamin C can enhance the proliferative capacity of these cells. The underlying signaling mechanism driving this activity is not yet known; however, induction of large numbers of genes involved in DNA replication and regulation of the cell cycle were observed after prolonged incubation with a vitamin C derivative [100]. Similarly, fibroblasts must also migrate into the wound, a characteristic

that can be promoted by vitamin C, again likely by effects on gene expression [100].

There is also evidence that vitamin C can promote the formation of the dermal-epidermal junction [101,103], via effects on gene expression of both keratinocytes and fibroblasts [101]. Vitamin C accelerated deposition of type IV and VII collagens, procollagens I and III, laminin 10/11, nidogen, tenascin C, and fibrillin-1 [101], at least some of which were mediated by increased mRNA expression. Augmented glycosaminoglycan synthesis, including substantial increases in heparin sulfate, chrondroitin sulfate, and hyaluronic acid, has been demonstrated with vitamin C treatment of human skin fibroblasts [104]. Hyaluronic acid, which is a major component of the extracellular matrix, was recently shown to be affected by vitamin C during keratinocyte differentiation [105]. The authors suggest that vitamin C stabilized hyaluronic acid by supressing both synthesis and degradation of hyaluronic acid and that this contributed to normal epidermal differentiation. Vitamin C has been shown to both stimulate [106] and inhibit elastin synthesis in cultured fibroblasts [107]. Finally, vitamin C can modulate the expression of antioxidant enzymes including those involved in DNA repair [100] and, as such, can help repair oxidized bases.

Modulation of Hypoxia-Inducible Factor Signaling

Thus, there is substantial evidence that vitamin C can interact with cell signaling pathways to modulate gene expression and cell outcomes. One obvious mechanism for this is the stabilization and activation of the HIFs, metabolic sensors that respond to hypoxia and control the expression of hundreds of genes [108]. Wound sites are often under hypoxic stress, at least initially, due to the disruption of the vasculature around the wound and the increased demand for oxygen by the infiltrating inflammatory cells. This leads to an upregulation of the HIFs, which are thought to be intimately involved in determining healing outcomes, contributing to many stages of the repair process including cell migration and proliferation, growth factor production, extracellular matrix synthesis remodeling, and angiogenesis [17]. In inflammatory cells, for example, HIF-1α or HIF-2α deletion demonstrated that these proteins are essential regulators of energy metabolism and the switch to glycolysis, motility, invasiveness, and bactericidal activity [109–111]. HIF-1α activation is also a vital stimulator of angiogenesis, increasing expression of genes such as VEGF [112]. Similarly, stabilization of HIF-1α dramatically enhanced fibroblast proliferation and improved time to healing in a mouse model of diabetes [113]. HIFs also upregulate the expression of many other genes involved in wound repair, including transforming growth factor-β, collagens, fibronectin, and matrix metalloproteinases [17]. HIF transcription factors are regulated via posttranslational modification by the HIF hydroxylases, which require ascorbate for optimal activity [36,114–116]. Thus, vitamin C status can modulate HIF-1 expression, with depletion leading to upregulation of HIF activation, particularly under mild or moderate hypoxia [115,117]. Interestingly, HIF-1 deficiency has been shown to correlate with impaired healing in diabetes [118]. There are also substantial research efforts underway investigating modulation of the HIF pathway for the development of new therapies for the treatment of chronic wounds [17].

Vitamin C is also an epigenetic modifier, acting as a cofactor for the TET family of enzymes that catalyze the removal of methylated cytosine through its hydroxylation to 5-hydroxymethylcytosine [119–121]. Because TETs have a specific requirement for vitamin C to maintain enzyme activity [36], this provides a further mechanism by which the vitamin may affect gene expression and cell function, particularly differentiation. The ability of vitamin C to regulate the activity of the histone demethylases, JHMDs which require ascorbate for full enzyme activity, also affects this. However, whether epigenetic modifications generally are important for wound repair is still unclear [122]. There is some suggestion that myofibroblast differentiation and collagen synthesis can be regulated by epigenetic enzymes, although not the vitamin C–requiring ones described earlier [123]. Keratinocyte proliferation, migration, and differentiation during wound repair may also be subject to epigenetic regulation [122].

Chronic Wounds and Vitamin C: Human Studies

Results from human and animal studies suggest that severe and prolonged vitamin C deficiency causes impaired wound healing. A question for this review, however, is what happens to wound healing when the vitamin C deficiency is less severe? Does suboptimal vitamin C status impact

on the poor healing often observed in those with chronic wounds? The recent study in the Gulo mouse [70] in which substantial defects in healing were observed with suboptimal but not extreme vitamin C deficiency provides some support for this premise. Thus, it seems possible that there may be effects on the biological processes involved in wound healing when body vitamin C status is below optimal, although to a lesser extent than that observed in scurvy. Some functions may be unaffected until severe deficiency.

Vitamin C Status of Those with Chronic Wounds

If vitamin C deficiency were to affect the healing of particular chronic wounds, it might be expected that those who have such wounds had low vitamin C levels. There have been a number of studies investigating the vitamin C status of individuals with chronic wounds (see Table 9.1). Normally, fasting plasma ascorbate concentrations are used to assess a person's vitamin C status [124,125]. Historically, white blood cell vitamin C concentrations have also been used to assess vitamin C status, although this is less common today. The studies described suggest low vitamin C status is widespread among those with chronic wounds, including in pressure ulcers, diabetic foot ulcers, and some types of leg ulcers. However, the studies are generally small, and vitamin C has not always been well sampled and measured [126]. One of the more interesting investigations was conducted by Balaji and Mosley [127] in which the vitamin C status of individuals with large leg ulcers was measured using a saturation test. While many of those with leg ulcers had vitamin C deficiency, this was particularly the case for participants who had neither an arterial nor venous pathology associated with their leg ulcer, with seven of the eight individuals deficient in vitamin C. These results imply there may be particular chronic wound groups for which a nutritional deficiency in vitamin C may be more prevalent. These are the wounds that should be targeted for interventional studies, as the deficiency may contribute to the delayed healing. In addition, supplementation of replete individuals is pointless [125]. A 1992 study of elderly patients with femoral neck fractures who were at risk of pressure ulcers showed that the vitamin C status was lower in those who developed pressure ulcers than those who did not $(p < .001)$ [128].

TABLE 9.1
Studies providing baseline vitamin C status of patients with chronic wounds

Wound Type	Participants	Vitamin C Method	Baseline Vitamin C Status	References
Adult trauma patients with wounds unclosed or persistent secretion at 10 days postsurgery or trauma.	N = 20 Average age ~45 years	Plasma ascorbate by HPLC method; sample preparation OK, fasting sample.	15/20 patients had ≤25 μmol/L ascorbate. Split into two groups: Median (quartiles) = 16.7 (9.5, 21.9) μmol/L and 27.1 (13.1, 37.7) μmol/L in the other group.	[134]
Pressure ulcers in surgical patients	N = 20 Average age 74.5 years	WBC ascorbate— spectrophotometric method, not described.	WBC ascorbate 22–24 μg/10^8 cells at baseline. Just above the lower normal limit of 18 μg/10^8 cells used in the laboratory at the time.	[131]
Pressure ulcers in patients from nursing homes/hospital.	N = 88	Plasma ascorbate by fluorimetric assay. Nonfasting samples taken immediately after breakfast.	Average ~20 μmol/L plasma ascorbate.	[132]
Patients with large leg ulcers (SA > 100 cm²)	N = 50 Average age of 76 years	Vitamin C deficiency was assessed by saturation test—i.e., urinary excretion after loading dose of vitamin C.	60% of participants were vitamin C deficient. 5/17 of arterial leg ulcer patients. 18/25 of venous leg ulcer patients. 7/8 of those with ulcers that were neither venous nor arterial.	[127]
Cohort study of elderly patients with femoral neck fracture– pressure ulcers	N = 21 Aged ≥75 years	Spectrophotometric method of white blood cell ascorbate; fasting sample.	WBC vitamin C levels were lower in those who developed pressure sores than those who did not. 6.3 μg/10^8 cells versus 12.8 μg/10^8 cells; $p < .001$. Lab winter reference range lower limit of 15 μg/10^8 cells.	[128]
Older patients with chronic venous leg ulcers of at least 3 months' duration (wound clinic)	N = 12 Average age of 64 years	Self-reported nutritional data—FFQ.	Vitamin C intake on average 60 mg/day by FFQ. Lower then recommended intake in United States of 75–90 mg.	[135]
Women aged >60 years treated for venous leg ulcers for 6 months or more	N = 9 Average age of 80 years	Self-reported nutritional data—food diaries for 7 days. Assessing vitamin C intake.	Vitamin C intake on average 43 mg/day. Lower then recommended Swedish intake.	[136]
Overweight and obese patients with venous leg ulcers from wound clinic	N = 8 All >50 years Average age of 67 years	Serum vitamin C assayed— spectrophotometrically. Not clear how samples handled or whether fasting, but at least one value very high at 144 μmol/L. Dietary vitamin C intake also assessed—3-day dietary recall.	Average serum ascorbate of 47 μmol/L, or 34 μmol/L with high value excluded. Median dietary vitamin C intake of 87 mg. Close to recommended intake.	[137]

(Continued)

TABLE 9.1 (*Continued*)

Studies providing baseline vitamin C status of patients with chronic wounds

Wound Type	Participants	Vitamin C Method	Baseline Vitamin C Status	References
Diabetics with lack of healing of diabetic foot ulcer or with low fruit and vegetable intake (diabetic clinic)	N = 11 Average age of 55 years	Serum vitamin C measured by HPLC. Unclear how samples handled and whether fasting. Definition of deficiency not given.	Median serum ascorbate of 19 μmol/L all cohort. Median serum ascorbate of 17 μmol/L those with ulcers. 6/7 with ulcers were vitamin C deficient, and the one who was not was zinc deficient. Only 1/4 without ulcers were vitamin C deficient.	[133]

ABBREVIATIONS: FFQ, food frequency questionnaire; HPLC, high-performance liquid chromatography; WBC, white blood cell.

NOTE: Plasma vitamin C status classification; levels of ≥ 70 μmol/L are high or saturating, 50–70 μmol/L are healthy, 23–50 μmol/L are suboptimal, <23 μmol/L are deficient, with <11 μmol/L plasma ascorbate severely deficient and the individuals considered at risk of scurvy.

Interventional Studies of Chronic Wounds

Early work suggested an effect of vitamin C on wound healing [66,74,129]. Not long after his self-experimentation on wound healing in scurvy, Crandon also examined the poor healing of surgical wounds in those who were low in vitamin C through a case series design [74]. He found effects of vitamin C on the occurrence of wound disruption, likely related to the low tensile strength of the wound, but he was mainly focused on confirming the relationship in those with scurvy-like levels of vitamin C.

However, despite the widely held belief that vitamin C is important for wound healing [130], causal relationship data are lacking. There are very few clinical studies, particularly randomized controlled trials, that have investigated whether oral vitamin C supplementation improves the healing of chronic wounds. The key studies can be found in Table 9.2. In one study, supplementation with oral vitamin C of 1000 mg/d improved the healing of pressure wounds [131]. However, a much larger study in a similar group found no effect of supplementation on healing rates or outcomes [132]. It is unclear whether this relates to differences in the size, healing rate, and duration of the wounds in the two studies, or whether the larger study achieved the "correct" outcome. More research is needed, especially a sufficiently powered study in those who are actually vitamin C deficient. The recent study by Christie-David and Gunton provides good pilot evidence for low vitamin C status in some diabetic foot ulcer patients,

with restoration of levels very rapidly improving healing [133]. Finally, a study of individuals with large leg ulcers provides very preliminary evidence that vitamin C supplementation can improve healing [127]. Certainly many of the participants were vitamin C deficient, particularly those with venous leg ulcers and leg ulcers that were neither venous nor arterial.

SUMMARY

Chronic wounds are characterized by persistent inflammation, with a failure to enter the proliferative phase and increased proteolytic activity preventing adequate deposition of extracellular matrix components and granulation tissue [17]. It is widely recognized that scorbutic individuals exhibit impaired or delayed healing. In accounting for this, attention has been focused on vitamin C's well-known effects on collagen synthesis and maturation. However, evidence is also emerging for other activities. There is considerable data suggesting that vitamin C can decrease the inflammatory response and thus, possibly, aid in the transition to the proliferative phase. Vitamin C also enhances formation of the skin barrier, promoting ceramide synthesis and differentiation of keratinocytes. The vitamin C–responsive transcription factor HIF is an important determinant of healing outcomes, with its role in cell survival, cell migration, cell division, growth factor release, and extracellular matrix synthesis.

TABLE 9.2

Interventional studies providing evidence of vitamin C effects on chronic wounds

Wound Type	Design	Intervention	Key Findings	Strengths/Limitations	References
Pressure ulcers in surgical patients	Double-blind RCT. N = 20. Average age 74.5 years. Wound assessed by clinician, serial photography and ulcer tracings weekly. Average ulcer size ~16 cm².	Oral vitamin C—1000 mg/d for 12 weeks. Placebo tablet. Measured vitamin C in WBCs— 65.6 µg/10⁸ cells versus placebo 25.8 µg/10⁸ cells after 1-month treatment.	Accelerated healing rate and reduced pressure ulcer area. At 1 month, reduction in area 42.7% in placebo versus 84% with ascorbate, $p < .005$. Six healed completely in ascorbate group and three in placebo group over the 12 weeks. Average healing rate of 2.47 cm²/week versus 1.45 cm²/week.	Measured both vitamin C status and wound healing over time. Limited by small study size. Large ulcers.	[131]
Pressure ulcers in patients from 11 nursing homes and 1 hospital	Multicenter RCT. N = 88. Included partial-thickness skin loss or worse. Average ulcer size of 1.4 cm².	Oral vitamin C 1000 mg/day for 12 weeks. Control group received 20 mg vitamin C/day. Large increase in plasma vitamin C from 20 to 84 µmol/L in vitamin C group. Placebo group increased from 22 to 27.3 µmol/L over 2 weeks.	No difference between healing outcomes of the two groups. Healing rate 0.21 and 0.27 cm²/week in intervention versus control group.	Larger study. Placebo has 20 mg vitamin C—not true placebo. Ulcers small—maybe new? Chronic? Quite low healing rates. Plasma vitamin C not fasting. But large increase in treated group.	[132]
Patients with large leg ulcers, includes venous, arterial, and other leg ulcers	Case series N = 50. Average age of 76 years. Not interventional trial, but they mention supplementing patients who were vitamin C deficient.	60% of participants were vitamin C deficient. 1500 mg vitamin C daily and skin graft in those vitamin C deficient.	The authors state that everyone that was deficient healed quickly with skin graft and restoration of deficiency, although this was not substantiated with actual data.	Large ulcers—average area of 169 cm². Small numbers in each of the three ulcer types. Very limited in that healing data not provided so unclear whether benefit. Also supplemented those who were zinc deficient with zinc.	[127]
Diabetics with lack of healing of diabetic foot ulcer or with low fruit and vegetable intake	Retrospective cohort study that also supplemented those deficient in vitamin C who had an ulcer. N = 11 altogether, but only 7 of 11 had ulcers. Six of the seven with ulcers had vitamin C deficiency.	500 or 1000 mg/d vitamin C. For those with diabetic foot ulcers who were also vitamin C deficient.	Pilot evidence for vitamin C intervention in diabetic foot ulcer patients (with low vitamin C) rapidly improving healing. In five of the six individuals with ulcers who were vitamin C deficient, healing occurred within 2–3 weeks of starting supplementing with vitamin C. The ulcer in the other person also healed but recurred intermittently.	Very small pilot study. Ulcers had been present 3–22 months, so chronic. Unclear how long supplemented. One person with ulcer was zinc deficient (not vitamin C deficient), and healed with zinc supplementation.	[133]

ABBREVIATIONS: RCT, randomized controlled trial; WBC, white blood cell.

NOTE: Studies in which vitamin C is combined with other agents as an intervention do not allow determination of the causative agent. They have not been included in this analysis.

While we know that severe vitamin C deficiency can have an impact on wound healing, it is less well understood whether suboptimal vitamin C status can contribute to the poor healing outcomes observed in those with chronic wounds. There are a number of small studies investigating the vitamin C status of individuals with chronic wounds, which suggest that many of those with chronic wounds may be vitamin C deficient. Whether this is due to increased vitamin C consumption during wound healing is not clear. The extent to which low vitamin C status may contribute to the clinical burden of chronic wounds is unknown, due to the few clinical trials in this area. More research is urgently needed, as providing vitamin C to patients with nonhealing wounds to rectify any deficiency would be a simple intervention with potentially substantial health outcome benefits.

REFERENCES

1. Weller, R. P., Hunter, J. A., Savin, J. A. and Dahl, M. V. 2008. The function and structure of skin. In *Clinical Dermatology*, 4th Edition, Wiley-Blackwell, Massachusetts, MA, USA.

2. Marks, R. 2004. The stratum corneum barrier: The final frontier. *J. Nutr.* 134, 2017s–2021s.

3. Wickert, R. R. and Visscher, M. O. 2006. Structure and function of the epidermal barrier. *Am. J. Infect. Control* 34, S98–S110.

4. Proksch, E., Brandner, J. M. and Jensen, J. M. 2008. The skin: An indispensable barrier. *Exper. Dermatol.* 17, 1063–1072.

5. Tracy, L. E., Minasian, R. A. and Caterson, E. J. 2016. Extracellular matrix and dermal fibroblast function in the healing wound. *Adv. Wound Care (New Rochelle)* 5, 119–136.

6. Verzijl, N., DeGroot, J., Thorpe, S. R., Bank, R. A., Shaw, J. N., Lyons, T. J., Bijlsma, J. W., Lafeber, F. P., Baynes, J. W. and TeKoppele, J. M. 2000. Effect of collagen turnover on the accumulation of advanced glycation end products. *J. Biol. Chem.* 275, 39027–39031.

7. Singer, A. J. and Clark, R. A. 1999. Cutaneous wound healing. *N. Engl. J. Med.* 341, 738–746.

8. Gurtner, G. C., Werner, S., Barrandon, Y. and Longaker, M. T. 2008. Wound repair and regeneration. *Nature* 453, 314–321.

9. Landen, N. X., Li, D. and Stahle, M. 2016. Transition from inflammation to proliferation: A critical step during wound healing. *Cell Mol. Life Sci.* 73, 3861–3885.

10. Golebiewska, E. M. and Poole, A. W. 2015. Platelet secretion: From haemostasis to wound healing and beyond. *Blood Rev.* 29, 153–162.

11. Rodrigues, M. and Gurtner, G. 2017. Black, white, and gray: Macrophages in skin repair and disease. *Curr. Pathobiol. Rep.* 5, 333–342.

12. Rodero, M. P. and Khosrotehrani, K. 2010. Skin wound healing modulation by macrophages. *Int. J. Clin. Exp. Pathol.* 3, 643–653.

13. Krzyszczyk, P., Schloss, R., Palmer, A. and Berthiaume, F. 2018. The role of macrophages in acute and chronic wound healing and interventions to promote pro-wound healing phenotypes. *Front. Physiol.* 9, 419.

14. Wilgus, T. A., Roy, S. and McDaniel, J. C. 2013. Neutrophils and wound repair: Positive actions and negative reactions. *Adv. Wound Care (New Rochelle)* 2, 379–388.

15. Barker, T. H. and Engler, A. J. 2017. The provisional matrix: Setting the stage for tissue repair outcomes. *Matrix Biol.* 60–61, 1–4.

16. Sottile, J. and Hocking, D. C. 2002. Fibronectin polymerization regulates the composition and stability of extracellular matrix fibrils and cell-matrix adhesions. *Mol. Biol. Cell* 13, 3546–3559.

17. Hong, W. X., Hu, M. S., Esquivel, M., Liang, G. Y., Rennert, R. C., McArdle, A., Paik, K. J. et al. 2014. The role of hypoxia-inducible factor in wound healing. *Adv. Wound. Care (New Rochelle)* 3, 390–399.

18. Sorg, H., Tilkorn, D. J., Hager, S., Hauser, J. and Mirastschijski, U. 2017. Skin wound healing: An update on the current knowledge and concepts. *Eur. Surg. Res.* 58, 81–94.

19. Evans, R. A., Tian, Y. C., Steadman, R. and Phillips, A. O. 2003. TGF-beta1-mediated fibroblast-myofibroblast terminal differentiation-the role of Smad proteins. *Exp. Cell Res.* 282, 90–100.

20. Xue, M. and Jackson, C. J. 2015. Extracellular matrix reorganization during wound healing and its impact on abnormal scarring. *Adv. Wound Care (New Rochelle)* 4, 119–136.

21. Lucas, T., Waisman, A., Ranjan, R., Roes, J., Krieg, T., Muller, W., Roers, A. and Eming, S. A. 2010. Differential roles of macrophages in diverse phases of skin repair. *J. Immunol.* 184, 3964–3977.

22. Fadok, V. A., Bratton, D. L., Konowal, A., Freed, P. W., Westcott, J. Y. and Henson, P. M. 1998. Macrophages that have ingested apoptotic cells in vitro inhibit proinflammatory cytokine

production through autocrine/paracrine mechanisms involving TGF-beta, PGE2, and PAF. *J. Clin. Invest.* 101, 890–898.

23. Lazarus, G. S., Cooper, D. M., Knighton, D. R., Margolis, D. J., Pecoraro, R. E., Rodeheaver, G. and Robson, M. C. 1994. Definitions and guidelines for assessment of wounds and evaluation of healing. *Arch. Dermatol.* 130, 489–493.

24. European Wound Management Association (EWMA). 2008. *Position Document: Hard-to-HealWwounds: A Holistic Approach.* MEP Ltd, London.

25. Heinen, M. M., Persoon, A., van de Kerkhof, P., Otero, M. and van Achterberg, T. 2007. Ulcer-related problems and health care needs in patients with venous leg ulceration: A descriptive, cross-sectional study. *Int. J. Nurs. Stud.* 44, 1296–1303.

26. Australian Centre for Health Services Innovation (AusHSI). 2017. *Chronic Wounds in Australia.* Australia.

27. Mustoe, T. A., O'Shaughnessy, K. and Kloeters, O. 2006. Chronic wound pathogenesis and current treatment strategies: A unifying hypothesis. *Plast. Reconstr. Surg.* 117, 35S–41S.

28. Smirnoff, N. 2018. Ascorbic acid metabolism and functions: A comparison of plants and mammals. *Free Radic. Biol. Med.* 122, 116–129.

29. Lindblad, M., Tveden-Nyborg, P. and Lykkesfeldt, J. 2013. Regulation of vitamin C homeostasis during deficiency. *Nutrients* 5, 2860–2879.

30. Wohlrab, C., Phillips, E. and Dachs, G. U. 2017. Vitamin C transporters in cancer: Current understanding and gaps in knowledge. *Front. Oncol.* 7, 74.

31. Harrison, F. E. and May, J. M. 2009. Vitamin C function in the brain: Vital role of the ascorbate transporter SVCT2. *Free Radic. Biol. Med.* 46, 719–730.

32. Du, J., Cullen, J. J. and Buettner, G. R. 2012. Ascorbic acid: Chemistry, biology and the treatment of cancer. *Biochim. Biophys. Acta* 1826, 443–457.

33. Bode, A. M., Cunningham, L. and Rose, R. C. 1990. Spontaneous decay of oxidized ascorbic acid (dehydro-L-ascorbic acid) evaluated by high-pressure liquid chromatography. *Clin. Chem.* 36, 1807–1809.

34. Englard, S. and Seifter, S. 1986. The biochemical functions of ascorbic acid. *Annu. Rev. Nutr.* 6, 365–406.

35. Levine, M., Morita, K., Heldman, E. and Pollard, H. B. 1985. Ascorbic acid regulation of norepinephrine biosynthesis in isolated chromaffin granules from bovine adrenal medulla. *J. Biol. Chem.* 260, 15598–15603.

36. Kuiper, C. and Vissers, M. C. 2014. Ascorbate as a co-factor for Fe- and 2-oxoglutarate dependent dioxygenases: Physiological activity in tumor growth and progression. *Front. Oncol.* 4, 359.

37. Loenarz, C. and Schofield, C. J. 2011. Physiological and biochemical aspects of hydroxylations and demethylations catalyzed by human 2-oxoglutarate oxygenases. *Trends Biochem. Sci.* 36, 7–18.

38. Vissers, M. C., Kuiper, C. and Dachs, G. U. 2014. Regulation of the 2-oxoglutarate-dependent dioxygenases and implications for cancer. *Biochem. Soc. Trans.* 42, 945–951.

39. Stolze, I. P., Mole, D. R. and Ratcliffe, P. J. 2006. Regulation of HIF: Prolyl hydroxylases. *Novartis Found Symp.* 272, 15–25; discussion 25–36.

40. Young, J. I., Zuchner, S. and Wang, G. 2015. Regulation of the epigenome by vitamin C. *Annu. Rev. Nutr.* 35, 545–564.

41. Carr, A. and Frei, B. 1999. Does vitamin C act as a pro-oxidant under physiological conditions? *FASEB J.* 13, 1007–1024.

42. Carr, A. C., Zhu, B. Z. and Frei, B. 2000. Potential antiatherogenic mechanisms of ascorbate (vitamin C) and alpha-tocopherol (vitamin E). *Circ. Res.* 87, 349–354.

43. Rhie, G., Shin, M. H., Seo, J. Y., Choi, W. W., Cho, K. H., Kim, K. H., Park, K. C., Eun, H. C. and Chung, J. H. 2001. Aging- and photoaging-dependent changes of enzymic and nonenzymic antioxidants in the epidermis and dermis of human skin in vivo. *J. Invest. Dermatol.* 117, 1212–1217.

44. Shindo, Y., Witt, E., Han, D., Epstein, W. and Packer, L. 1994. Enzymic and non-enzymic antioxidants in epidermis and dermis of human skin. *J. Invest. Dermatol.* 102, 122–124.

45. McArdle, F., Rhodes, L. E., Parslew, R., Jack, C. I., Friedmann, P. S. and Jackson, M. J. 2002. UVR-induced oxidative stress in human skin in vivo: Effects of oral vitamin C supplementation. *Free Radic. Biol. Med.* 33, 1355–1362.

46. Wheeler, L. A., Aswad, A., Connor, M. J. and Lowe, N. 1986. Depletion of cutaneous glutathione and the induction of inflammation by 8-methoxypsoralen plus UVA radiation. *J. Invest. Dermatol.* 87, 658–662.

47. Pullar, J. M., Carr, A. C. and Vissers, M. C. M. 2017. The roles of vitamin C in skin health. *Nutrients* 9.

48. Wilson, J. X. 2005. Regulation of vitamin C transport. *Annu. Rev. Nutr.* 25, 105–125.

49. Kishimoto, Y., Saito, N., Kurita, K., Shimokado, K., Maruyama, N. and Ishigami, A. 2013. Ascorbic acid enhances the expression of type 1 and type 4 collagen and SVCT2 in cultured human skin fibroblasts. *Biochem. Biophys. Res. Commun.* 430, 579–584.

50. Phillips, C. L., Combs, S. B. and Pinnell, S. R. 1994. Effects of ascorbic acid on proliferation and collagen synthesis in relation to the donor age of human dermal fibroblasts. *J. Invest. Dermatol.* 103, 228–232.

51. Geesin, J. C., Darr, D., Kaufman, R., Murad, S. and Pinnell, S. R. 1988. Ascorbic acid specifically increases type I and type III procollagen messenger RNA levels in human skin fibroblast. *J. Invest. Dermatol.* 90, 420–424.

52. Tajima, S. and Pinnell, S. R. 1996. Ascorbic acid preferentially enhances type I and III collagen gene transcription in human skin fibroblasts. *J. Dermatol. Sci.* 11, 250–253.

53. May, J. M. and Qu, Z. C. 2005. Transport and intracellular accumulation of vitamin C in endothelial cells: Relevance to collagen synthesis. *Arch. Biochem. Biophys.* 434, 178–186.

54. Steiling, H., Longet, K., Moodycliffe, A., Mansourian, R., Bertschy, E., Smola, H., Mauch, C. and Williamson, G. 2007. Sodium-dependent vitamin C transporter isoforms in skin: Distribution, kinetics, and effect of UVB-induced oxidative stress. *Free Radic. Biol. Med.* 43, 752–762.

55. Michels, A. J., Hagen, T. M. and Frei, B. 2013. Human genetic variation influences vitamin C homeostasis by altering vitamin C transport and antioxidant enzyme function. *Annu. Rev. Nutr.* 33, 45–70.

56. May, J. M. 2011. The SLC23 family of ascorbate transporters: Ensuring that you get and keep your daily dose of vitamin C. *Br. J. Pharmacol.* 164, 1793–1801.

57. Savini, I., Rossi, A., Pierro, C., Avigliano, L. and Catani, M. V. 2008. SVCT1 and SVCT2: Key proteins for vitamin C uptake. *Amino. Acids.* 34, 347–355.

58. Padayatty, S. J. and Levine, M. 2016. Vitamin C: The known and the unknown and goldilocks. *Oral Dis.* 22, 463–493.

59. Hodges, R. E., Baker, E. M., Hood, J., Sauberlich, H. E. and March, S. C. 1969. Experimental scurvy in man. *Am. J. Clin. Nutr.* 22, 535–548.

60. Hodges, R. E., Hood, J., Canham, J. E., Sauberlich, H. E. and Baker, E. M. 1971. Clinical manifestations of ascorbic acid deficiency in man. *Am. J. Clin. Nutr.* 24, 432–443.

61. Crandon, J. H., Lund, C. C. and Dill, D. B. 1940. Experimental human scurvy. *N. Engl. J. Med.* 223, 353–369.

62. Medical Research Council. 1948. Vitamin C requirement of human adults. Experimental study of vitamin C deprivation in man. *Lancet* 251, 853–858.

63. Ross, R. and Benditt, E. P. 1962. Wound healing and collagen formation. II. Fine structure in experimental scurvy. *J. Cell Biol.* 12, 533–551.

64. Wolbach, S. B. 1933. Controlled formation of collagen and reticulum. A study of the source of intercellular substance in recovery from experimental scorbutus. *Am. J. Pathol.* 9, 689–700, 685.

65. Kim, M., Otsuka, M., Yu, R., Kurata, T. and Arakawa, N. 1993. The distribution of ascorbic acid and dehydroascorbic acid during tissue regeneration in wounded dorsal skin of guinea pigs. *Internat. J. Vit. Nutr. Res.* 63, 56–59.

66. Lanman, T. H. and Ingalls, T. H. 1937. Vitamin C deficiency and wound healing: An experimental and clinical study. *Ann. Surg.* 105, 616–625.

67. Kamer, E., Recai Unalp, H., Gundogan, O., Diniz, G., Ortac, R., Olukman, M., Derici, H. and Ali Onal, M. 2010. Effect of ascorbic Acid on incisional wound healing in streptozotocin-induced diabetic rats. *Wounds* 22, 27–31.

68. Maeda, N., Hagihara, H., Nakata, Y., Hiller, S., Wilder, J. and Reddick, R. 2000. Aortic wall damage in mice unable to synthesize ascorbic acid. *Proc. Natl. Acad. Sci. USA* 97, 841–846.

69. Parsons, K. K., Maeda, N., Yamauchi, M., Banes, A. J. and Koller, B. H. 2006. Ascorbic acid-independent synthesis of collagen in mice. *Am. J. Physiol. Endocrinol. Metab.* 290, E1131–1139.

70. Mohammed, B. M., Fisher, B. J., Kraskauskas, D., Ward, S., Wayne, J. S., Brophy, D. F., Fowler, A. A., 3rd, Yager, D. R. and Natarajan, R. 2016. Vitamin C promotes wound healing through novel pleiotropic mechanisms. *Int. Wound. J.* 13, 572–584.

71. Silverstein, R. J. and Landsman, A. S. 1999. The effects of a moderate and high dose of vitamin C on wound healing in a controlled guinea pig model. *J. Foot Ankle Surg.* 38, 333–338.

72. Shukla, A., Rasik, A. M. and Patnaik, G. K. 1997. Depletion of reduced glutathione, ascorbic acid,

vitamin E and antioxidant defence enzymes in a healing cutaneous wound. *Free Radic. Res.* 26, 93–101.

73. Lima, C. C., Pereira, A. P., Silva, J. R., Oliveira, L. S., Resck, M. C., Grechi, C. O., Bernardes, M. T. et al. 2009. Ascorbic acid for the healing of skin wounds in rats. *Braz. J. Biol.* 69, 1195–1201.

74. Lund, C. C. and Crandon, J. H. 1941. Ascorbic acid and human wound healing. *Ann. Surg.* 114, 776–790.

75. Pihlajaniemi, T., Myllyla, R. and Kivirikko, K. I. 1991. Prolyl 4-hydroxylase and its role in collagen synthesis. *J. Hepatol.* 13 (Suppl 3), S2–7.

76. Vissers, M. C. and Wilkie, R. P. 2007. Ascorbate deficiency results in impaired neutrophil apoptosis and clearance and is associated with up-regulation of hypoxia-inducible factor 1alpha. *J. Leukoc. Biol.* 81, 1236–1244.

77. Goldschmidt, M. C. 1991. Reduced bactericidal activity in neutrophils from scorbutic animals and the effect of ascorbic acid on these target bacteria in vivo and in vitro. *Am. J. Clin. Nutr.* 54, 1214S–1220S.

78. Johnston, C. S. and Huang, S. N. 1991. Effect of ascorbic acid nutriture on blood histamine and neutrophil chemotaxis in guinea pigs. *J. Nutr.* 121, 126–130.

79. Bozonet, S. M., Carr, A. C., Pullar, J. M. and Vissers, M. C. 2015. Enhanced human neutrophil vitamin C status, chemotaxis and oxidant generation following dietary supplementation with vitamin C-rich SunGold kiwifruit. *Nutrients* 7, 2574–2588.

80. Anderson, R., Oosthuizen, R., Maritz, R., Theron, A. and Van Rensburg, A. J. 1980. The effects of increasing weekly doses of ascorbate on certain cellular and humoral immune functions in normal volunteers. *Am. J. Clin. Nutr.* 33, 71–76.

81. Vohra, K., Khan, A. J., Telang, V., Rosenfeld, W. and Evans, H. E. 1990. Improvement of neutrophil migration by systemic vitamin C in neonates. *J. Perinatol.* 10, 134–136.

82. Goetzl, E. J., Wasserman, S. I., Gigli, I. and Austen, K. F. 1974. Enhancement of random migration and chemotactic response of human leukocytes by ascorbic acid. *J. Clin. Invest.* 53, 813–818.

83. Shilotri, P. G. 1977. Phagocytosis and leukocyte enzymes in ascorbic acid deficient guinea pigs. *J. Nutr.* 107, 1513–1516.

84. Shilotri, P. G. 1977. Glycolytic, hexose monophosphate shunt and bactericidal activities of leukocytes in ascorbic acid deficient guinea pigs. *J. Nutr.* 107, 1507–1512.

85. Boeltz, S., Amini, P., Anders, H. J., Andrade, F., Bilyy, R., Chatfield, S., Cichon, I. et al. 2019. To NET or not to NET: Current opinions and state of the science regarding the formation of neutrophil extracellular traps. *Cell Death Differ* 26, 395–408.

86. Parker, H., Albrett, A. M., Kettle, A. J. and Winterbourn, C. C. 2012. Myeloperoxidase associated with neutrophil extracellular traps is active and mediates bacterial killing in the presence of hydrogen peroxide. *J. Leukoc. Biol.* 91, 369–376.

87. Wong, S. L., Demers, M., Martinod, K., Gallant, M., Wang, Y., Goldfine, A. B., Kahn, C. R. and Wagner, D. D. 2015. Diabetes primes neutrophils to undergo NETosis, which impairs wound healing. *Nat. Med.* 21, 815–819.

88. Fadini, G. P., Menegazzo, L., Rigato, M., Scattolini, V., Poncina, N., Bruttocao, A., Ciciliot, S. et al. 2016. NETosis delays diabetic wound healing in mice and humans. *Diabetes* 65, 1061–1071.

89. McInturff, A. M., Cody, M. J., Elliott, E. A., Glenn, J. W., Rowley, J. W., Rondina, M. T. and Yost, C. C. 2012. Mammalian target of rapamycin regulates neutrophil extracellular trap formation via induction of hypoxia-inducible factor 1 alpha. *Blood* 120, 3118–3125.

90. Winterbourn, C. C., Kettle, A. J. and Hampton, M. B. 2016. Reactive oxygen species and neutrophil function. *Annu. Rev. Biochem.* 85, 765–792.

91. Steenvoorden, D. P. and van Henegouwen, G. M. 1997. The use of endogenous antioxidants to improve photoprotection. *J. Photochem. Photobiol. B.* 41, 1–10.

92. Lin, J. Y., Selim, M. A., Shea, C. R., Grichnik, J. M., Omar, M. M., Monteiro-Riviere, N. A. and Pinnell, S. R. 2003. UV photoprotection by combination topical antioxidants vitamin C and vitamin E. *J. Am. Acad. Dermatol.* 48, 866–874.

93. Wikramanayake, T. C., Stojadinovic, O. and Tomic-Canic, M. 2014. Epidermal differentiation in barrier maintenance and wound healing. *Adv. Wound Care (New Rochelle)* 3, 272–280.

94. Pasonen-Seppanen, S., Suhonen, T. M., Kirjavainen, M., Suihko, E., Urtti, A., Miettinen, M., Hyttinen, M., Tammi, M. and Tammi, R. 2001. Vitamin C enhances differentiation of a continuous keratinocyte cell line (REK) into epidermis with normal stratum corneum

ultrastructure and functional permeability barrier. *Histochem. Cell biol.* 116, 287–297.

95. Ponec, M., Weerheim, A., Kempenaar, J., Mulder, A., Gooris, G. S., Bouwstra, J. and Mommaas, A. M. 1997. The formation of competent barrier lipids in reconstructed human epidermis requires the presence of vitamin C. *J. Invest. Dermatol.* 109, 348–355.

96. Kim, S. W., Lee, I. W., Cho, H. J., Cho, K. H., Kim, K. H., Chung, J. H., Song, P. I. and Park, K. C. 2002. Fibroblasts and ascorbate regulate epidermalization in reconstructed human epidermis. *J. Dermatol. Sci.* 30, 215–223.

97. Katsuyama, Y., Taira, N., Tsuboi, T., Yoshioka, M., Masaki, H. and Muraoka, O. 2017. 3-O-Laurylglyceryl ascorbate reinforces skin barrier function through not only the reduction of oxidative stress but also the activation of ceramide synthesis. *Int. J. Cosmet. Sci.* 39, 49–55.

98. Savini, I., Catani, M. V., Rossi, A., Duranti, G., Melino, G. and Avigliano, L. 2002. Characterization of keratinocyte differentiation induced by ascorbic acid: Protein kinase C involvement and vitamin C homeostasis. *J. Invest. Dermatol.* 118, 372–379.

99. Kim, K. P., Shin, K. O., Park, K., Yun, H. J., Mann, S., Lee, Y. M. and Cho, Y. 2015. Vitamin C stimulates epidermal ceramide production by regulating its metabolic enzymes. *Biomol. Ther.* 23, 525–530.

100. Duarte, T. L., Cooke, M. S. and Jones, G. D. 2009. Gene expression profiling reveals new protective roles for vitamin C in human skin cells. *Free Radic. Biol. Med.* 46, 78–87.

101. Marionnet, C., Vioux-Chagnoleau, C., Pierrard, C., Sok, J., Asselineau, D. and Bernerd, F. 2006. Morphogenesis of dermal-epidermal junction in a model of reconstructed skin: Beneficial effects of vitamin C. *Exp. Dermatol.* 15, 625–633.

102. Hata, R., Sunada, H., Arai, K., Sato, T., Ninomiya, Y., Nagai, Y. and Senoo, H. 1988. Regulation of collagen metabolism and cell growth by epidermal growth factor and ascorbate in cultured human skin fibroblasts. *Eur. J. Biochem.* 173, 261–267.

103. Boyce, S. T., Supp, A. P., Swope, V. B. and Warden, G. D. 2002. Vitamin C regulates keratinocyte viability, epidermal barrier, and basement membrane in vitro, and reduces wound contraction after grafting of cultured skin substitutes. *J. Invest. Dermatol.* 118, 565–572.

104. Kao, J., Huey, G., Kao, R. and Stern, R. 1990. Ascorbic acid stimulates production of glycosaminoglycans in cultured fibroblasts. *Exp. Mol. Pathol.* 53, 1–10.

105. Hamalainen, L., Karkkainen, E., Takabe, P., Rauhala, L., Bart, G., Karna, R., Pasonen-Seppanen, S., Oikari, S., Tammi, M. I. and Tammi, R. H. 2018. Hyaluronan metabolism enhanced during epidermal differentiation is suppressed by vitamin C. *Br. J. Dermatol.* 179, 651–661.

106. Hinek, A., Kim, H. J., Wang, Y., Wang, A. and Mitts, T. F. 2014. Sodium L-ascorbate enhances elastic fibers deposition by fibroblasts from normal and pathologic human skin. *J. Dermatol. Sci.* 75, 173–182.

107. Davidson, J. M., LuValle, P. A., Zoia, O., Quaglino, D., Jr. and Giro, M. 1997. Ascorbate differentially regulates elastin and collagen biosynthesis in vascular smooth muscle cells and skin fibroblasts by pretranslational mechanisms. *J. Biol. Chem.* 272, 345–352.

108. Ratcliffe, P. J. 2013. Oxygen sensing and hypoxia signalling pathways in animals: The implications of physiology for cancer. *J. Physiol.* 591, 2027–2042.

109. Peyssonnaux, C., Datta, V., Cramer, T., Doedens, A., Theodorakis, E. A., Gallo, R. L., Hurtado-Ziola, N., Nizet, V. and Johnson, R. S. 2005. HIF-1alpha expression regulates the bactericidal capacity of phagocytes. *J. Clin. Invest.* 115, 1806–1815.

110. Braverman, J., Sogi, K. M., Benjamin, D., Nomura, D. K. and Stanley, S. A. 2016. HIF-1alpha is an essential mediator of IFN-gamma-dependent immunity to mycobacterium tuberculosis. *J. Immunol.* 197, 1287–1297.

111. Imtiyaz, H. Z., Williams, E. P., Hickey, M. M., Patel, S. A., Durham, A. C., Yuan, L. J., Hammond, R., Gimotty, P. A., Keith, B. and Simon, M. C. 2010. Hypoxia-inducible factor 2alpha regulates macrophage function in mouse models of acute and tumor inflammation. *J. Clin. Invest.* 120, 2699–2714.

112. Ahluwalia, A. and Tarnawski, A. S. 2012. Critical role of hypoxia sensor—HIF-1alpha in VEGF gene activation. Implications for angiogenesis and tissue injury healing. *Curr. Med. Chem.* 19, 90–97.

113. Zhang, X., Yan, X., Cheng, L., Dai, J., Wang, C., Han, P. and Chai, Y. 2013. Wound healing improvement with PHD-2 silenced fibroblasts in diabetic mice. *PLOS ONE* 8, e84548.

114. Flashman, E., Davies, S. L., Yeoh, K. K. and Schofield, C. J. 2010. Investigating the dependence of the hypoxia-inducible factor hydroxylases (factor inhibiting HIF and prolyl

hydroxylase domain 2) on ascorbate and other reducing agents. *Biochem. J.* 427, 135–142.

115. Kuiper, C., Dachs, G. U., Currie, M. J. and Vissers, M. C. 2014. Intracellular ascorbate enhances hypoxia-inducible factor (HIF)-hydroxylase activity and preferentially suppresses the HIF-1 transcriptional response. *Free Radic. Biol. Med.* 69, 308–317.

116. Wohlrab, C., Vissers, M. C. M., Phillips, E., Morrin, H., Robinson, B. A. and Dachs, G. U. 2018. The association between ascorbate and the hypoxia-inducible factors in human renal cell carcinoma requires a functional von hippel-lindau protein. *Front. Oncol.* 8, 574.

117. Campbell, E. J., Vissers, M. C. and Dachs, G. U. 2016. Ascorbate availability affects tumor implantation-take rate and increases tumor rejection in Gulo(−/−) mice. *Hypoxia (Auckl)* 4, 41–52.

118. Thangarajah, H., Yao, D., Chang, E. I., Shi, Y., Jazayeri, L., Vial, I. N., Galiano, R. D. et al. 2009. The molecular basis for impaired hypoxia-induced VEGF expression in diabetic tissues. *Proc. Natl. Acad. Sci. USA* 106, 13505–13510.

119. Minor, E. A., Court, B. L., Young, J. I. and Wang, G. 2013. Ascorbate induces ten-eleven translocation (Tet) methylcytosine dioxygenase-mediated generation of 5-hydroxymethylcytosine. *J. Biol. Chem.* 288, 13669–13674.

120. Blaschke, K., Ebata, K. T., Karimi, M. M., Zepeda-Martinez, J. A., Goyal, P., Mahapatra, S., Tam, A. et al. 2013. Vitamin C induces Tet-dependent DNA demethylation and a blastocyst-like state in ES cells. *Nature* 500, 222–226.

121. Yin, R., Mao, S. Q., Zhao, B., Chong, Z., Yang, Y., Zhao, C., Zhang, D. et al. 2013. Ascorbic acid enhances Tet-mediated 5-methylcytosine oxidation and promotes DNA demethylation in mammals. *J. Am. Chem. Soc.* 135, 10396–10403.

122. Lewis, C. J., Mardaryev, A. N., Sharov, A. A., Fessing, M. Y. and Botchkarev, V. A. 2014. The epigenetic regulation of wound healing. *Adv. Wound Care (New Rochelle)* 3, 468–475.

123. Glenisson, W., Castronovo, V. and Waltregny, D. 2007. Histone deacetylase 4 is required for TGFbeta1-induced myofibroblastic differentiation. *Biochim. Biophys. Acta* 1773, 1572–1582.

124. Jacob, R. A. 1990. Assessment of human vitamin C status. *J. Nutr.* 120 (Suppl 11), 1480–1485.

125. Lykkesfeldt, J. and Poulsen, H. E. 2010. Is vitamin C supplementation beneficial? Lessons learned from randomised controlled trials. *Br. J. Nutr.* 103, 1251–1259.

126. Pullar, J. M., Bayer, S. and Carr, A. C. 2018. Appropriate handling, processing and analysis of blood samples is essential to avoid oxidation of vitamin C to dehydroascorbic acid. *Antioxidants (Basel, Switzerland)* 7.

127. Balaji, P. and Mosley, J. G. 1995. Evaluation of vascular and metabolic deficiency in patients with large leg ulcers. *Ann. R. Coll. Surg. Engl.* 77, 270–272.

128. Goode, H. F., Burns, E. and Walker, B. E. 1992. Vitamin C depletion and pressure sores in elderly patients with femoral neck fracture. *BMJ* 305, 925–927.

129. Wolfer, J. A., Hoebel, F. C. 1939. The significance of cevitamic acid deficiency in surgical patients. *Surg. Gynec. Obstet.* 69, 745–755.

130. Molnar, J. A., Underdown, M. J. and Clark, W. A. 2014. Nutrition and chronic wounds. *Adv. Wound Care (New Rochelle)* 3, 663–681.

131. Taylor, T. V., Rimmer, S., Day, B., Butcher, J. and Dymock, I. W. 1974. Ascorbic acid supplementation in the treatment of pressure-sores. *Lancet* 2, 544–546.

132. ter Riet, G., Kessels, A. G. and Knipschild, P. G. 1995. Randomized clinical trial of ascorbic acid in the treatment of pressure ulcers. *J. Clin Epidemiol.* 48, 1453–1460.

133. Christie-David, D. J. and Gunton, J. E. 2017. Vitamin C deficiency and diabetes mellitus—Easily missed? *Diabet. Med.* 34, 294–296.

134. Blass, S. C., Goost, H., Tolba, R. H., Stoffel-Wagner, B., Kabir, K., Burger, C., Stehle, P. and Ellinger, S. 2012. Time to wound closure in trauma patients with disorders in wound healing is shortened by supplements containing antioxidant micronutrients and glutamine: A PRCT. *Clin. Nutr.* 31, 469–475.

135. McDaniel, J. C., Kemmner, K. G. and Rusnak, S. 2015. Nutritional profile of older adults with chronic venous leg ulcers: A pilot study. *Geriatr. Nurs.* 36, 381–386.

136. Wissing, U., Unosson, M., Lennernas, M. A. and Ek, A. C. 1997. Nutritional intake and physical activity in leg ulcer patients. *J. Adv. Nurs.* 25, 571–578.

137. Tobon, J., Whitney, J. D. and Jarrett, M. 2008. Nutritional status and wound severity of overweight and obese patients with venous leg ulcers: A pilot study. *J. Vasc. Nurs.* 26, 43–52.

PART FOUR

Vitamin C and Neurological Function

CHAPTER TEN

Vitamin C in Neurological Function and Neurodegenerative Disease

Shilpy Dixit, David C. Consoli, Krista C. Paffenroth,
Jordyn M. Wilcox, and Fiona E. Harrison

DOI: 10.1201/9780429442025-10

CONTENTS

INTRODUCTION

The brain has the highest concentration of vitamin C (ascorbic acid) compared to other organs in the body, except for the adrenal glands [1]. During dietary deficiency, efficient transport and recycling properties conserve vitamin C in the brain, even at the expense of other organs [2,3]. Preservation of adequate vitamin C in the brain is supported by the two-step, sodium- and energy-dependent transport system via the sodium-dependent vitamin C transporter type 2 (SVCT2), which is expressed in the choroid plexus and in neurons [4,5]. Retention is supported by efficient recycling by astrocytes from twice-oxidized dehydroascorbic acid, and the once-oxidized ascorbate free radical. Nevertheless, we have recently come to understand multiple, specific roles for vitamin C in the brain in healthy functioning, and for its deficiency in neurological disease.

The brain has sufficient plasticity that it can withstand relatively large amounts of damage and yet still function adequately in most areas, even if not optimally. Many degenerative diseases actually begin their course years or decades prior to clinical presentation of cognitive or motor symptoms, offering a long period during which genetic and environmental influences may interact to hasten or modify disease pathogenesis. The β-amyloid plaques characteristic of Alzheimer disease (AD) are detectable in many individuals from the third decade of life and are also common in nondemented elder individuals [6–9]. The neurodegenerative conditions described herein each have a strong genetic or other molecular basis for the disorder; however, vitamin C has many specific roles in the brain that render its deficiency likely to manifest in faster decline or stronger symptomology. Each of these degenerative diseases is associated with elevations in general oxidative stress. It is therefore highly likely that there is increased demand for vitamin C in affected brains with accelerating pathologic accumulation depleting supplies. Chronic vitamin C deficiency could, in turn, accelerate disease progress by promoting oxidative stress conditions. Recent research has highlighted a number of additional roles for vitamin C in the brain, and compromise of each of these may also lead to specific disease-relevant changes in neural tissues. Data supporting the involvement of these mechanisms are described for each disease, where relevant, and the manners in which vitamin C deficiency in each of these areas may have an impact on or be impacted by disease are discussed.

Vitamin C–Dependent Mechanisms in the Brain

Vitamin C is a powerful antioxidant. It is soluble in aqueous solutions and readily donates an electron to neutralize other radicals. Importantly, vitamin C can also recycle other antioxidants such as vitamin E to further impact oxidative homeostasis within the cell. Its low reduction potential allows the molecule to easily donate an electron to oxidized molecules and is the basis for many of the roles that are described later. Previous detailed reviews have summarized the large amount of data in this area [10,11], and we therefore have chosen to focus on underlying mechanisms behind the posited relationships, including global oxidative stress, mitochondrial dysfunction, and glutamatergic excitotoxicity (see Figure 10.1).

Role of Vitamin C in Oxidative Stress

Free radicals are atoms (or groups of atoms) that exist with unpaired electrons, making them highly reactive. Production of radical species is a normal part of cellular function, and the by-products are often required for multiple roles within the cell, including cell signaling, gene transcription, regulation of cerebral blood flow, among others. Oxidative stress in the brain reflects an imbalance between the production of free radicals and antioxidant and repair capacity (reviewed in [12,13]). The brain is proposed to be particularly sensitive to production of reactive oxygen species given that it has very high-energy demands, utilizing an estimated 20% of all oxygen consumed in the body. Significant production of excess electrons is an inevitable part of mitochondrial function, and effective repair mechanisms are in place to mitigate potential damage. Each of the diseases described in the following sections involves impaired mitochondrial function [14–17] and increased oxidative stress markers [11,18–22]. Partial deletion of the SVCT2, which has been localized to the mitochondrial membrane [23,24], decreased mitochondrial function resulting in decreased energy production and increased local oxidative stress [25], supporting the role of vitamin C in mitochondrial health. Further, each disease is also associated with inflammatory response, excitotoxicity, protein aggregates, and disruptions

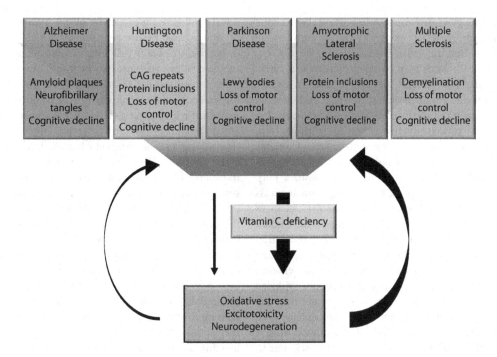

Alzheimer Disease	Huntington Disease	Parkinson Disease	Amyotrophic Lateral Sclerosis	Multiple Sclerosis
Amyloid plaques Neurofibrillary tangles Cognitive decline	CAG repeats Protein inclusions Loss of motor control Cognitive decline	Lewy bodies Loss of motor control Cognitive decline	Protein inclusions Loss of motor control Cognitive decline	Demyelination Loss of motor control Cognitive decline

Vitamin C deficiency

Oxidative stress
Excitotoxicity
Neurodegeneration

Figure 10.1. Overall hypothesis: Vitamin C deficiency can exacerbate multiple underlying pathologic pathways mediating neurodegenerative disease development.

in the utilization of transition metals. Each of these factors can contribute to cytotoxicity and to self-perpetuating cycles of damage. The maintenance of appropriate levels of the major antioxidant vitamin C is therefore a critical preventative barrier to accelerating oxidative damage.

Vitamin C as an Enzyme Cofactor

The ability of vitamin C to readily donate electrons is critical for its role in the synthesis of collagen and carnitine, as well as neuropeptide amidation and catecholamine biosynthesis. For the latter, vitamin C recycles tetrahydrobiopterin, which is required for synthesis of l-3,4-dyhydroxyphenylalanine (L-dopa) by the enzyme tyrosine hydroxylase. Vitamin C is also the major electron donor for dopamine β-hydroxylase, which is required to generate norepinephrine from dopamine [26–28]. Accordingly, dopamine and each of its breakdown products (DOPAC, HVA, 3MT) as well as the serotonin derivative 5-HIAA were all decreased in the cerebral cortex of ascorbate-deficient mice [29]. Interestingly, neurotransmitters dopamine, norepinephrine, and serotonin are also auto-oxidizable and can generate radical species under conditions of

excess, potentially contributing to an oxidative environment under conditions of enhanced stress or imbalance.

Vitamin C also serves as a cofactor for ten-eleven translocation (TET) dioxygenases, which are required for catalysis of the oxidation of 5-methylcytosine (5mC) into 5-hydroxymethylcytosine (5hmC) [30]. Some histone demethylases further require vitamin C as a cofactor for demethylation of histones. Through these roles, vitamin C is emerging as a critical determinant for epigenetic regulation [31–33], but the potential importance of this in degenerative disorders has not yet been established. Given the late onset of diagnosable phenotypes, and long time course for disease development, this role offers a great potential for diet to impact disease course.

Glutamate Clearance and Excitotoxicity

Neurotransmitter signaling in the brain is sensitive to both oxidative stress in general and vitamin C levels specifically, and it is also a potential contributor to damage. Glutamate is the primary excitotoxic neurotransmitter of the brain and becomes neurotoxic when not cleared efficiently from the synapse. Without tight regulation of release and uptake, glutamate can spread beyond the synaptic cleft and propagate excitatory signals and damage in surrounding cells. Overexcitation of postsynaptic cells leads to prolonged increases in intracellular calcium and sodium, ultimately leading to impaired mitochondrial function and elevated production of nitric oxide and other radicals that lead to antioxidant imbalance and ultimately to cell death. The interaction with vitamin C release and glutamatergic stimulation has been demonstrated in whole brain and synaptosomes [34]. Glutamate clearance via the GLT-1 glutamate transporter on the astrocyte leads to osmolytic changes in the astrocyte, and cell swelling permits efflux of vitamin C through volume-regulated anion channels [34,35]. This relationship was originally termed a "hetero-exchange," although evidence for the reverse relationship of vitamin C–induced changes in glutamate clearance does not have the same support. Disruption of this process has been observed in mouse models of Huntington disease (HD) [36,37] and is also implicated in seizures in a mouse model of AD with low vitamin C [38].

Methodological Considerations and Clinical Research Challenges

There are two key areas that bear closer examination when examining the literature concerning human studies of vitamin C levels and neurodegenerative disorders. First, in clinical or population samples, there is often no direct measurement of plasma or tissue levels of vitamin C. Second, there are many other factors that may affect vitamin C levels independent of disease. Mutations in the vitamin C transporters SVCT1, which are primarily involved in gut absorption, are strongly related to altered blood vitamin C levels [39,40] and can mitigate the link between dietary intake and circulating levels. A link between single nucleotide polymorphisms (SNPs) in these transporters and neurodegenerative disorders has not yet been reported. Nutritional intake may be assessed using dietary intake questionnaires about typical or recent intake, which can give a good general idea of dietary habits, but they are prone to errors of recall or estimates of portion size. Performing these studies in an aging and potentially cognitively impaired population may lead to greater estimation errors due to altered cognition (memory for foods and attention to recall tasks) [41]. Such questionnaires may also not always be representative of lifetime intake levels. Nevertheless, this type of

dietary information can allow global stratification of groups based on dietary patterns. It is also very important to obtain full information about any supplement use. Such use may be more common in older participants [42,43] and can significantly bias results. For example, multivitamin use is sometimes permitted without being classed as supplement use, and the amount of vitamin C in a typical supplement (60–120 mg) may be sufficient when combined with even a moderate dietary intake to saturate intestinal transport [44]. In fact, supplement use may be a stronger predictor of vitamin C level than dietary intake [45]. Such classification therefore decreases the percent of vitamin-deficient individuals in the comparison groups. With even the most accurate intake information, individual differences in disease status, obesity, smoking status, and even local pollution levels may impact the utilization of intracellular vitamin C [45]. While these factors by no means invalidate a substantial body of literature, they should be considered when interpreting the literature concerning vitamin C levels and disease prevalence.

Additional challenges are found in the use of rodent models and investigation of vitamin C and the brain. The guinea pig is, like humans, unable to synthesize its own vitamin C [46] and is therefore a good model for deficiency research. However, cost and methodological challenges (e.g., identifying good antibodies and adapting behavioral studies), render this work challenging in guinea pigs and require specialized laboratories and equipment [47–49]. Rats and mice are typically able to synthesize vitamin C and so cannot be rendered deficient since physiologic stress can lead to increased production in the liver. A range of models exist to combat this challenge, including mutations relating to synthesis (osteogenic disorder Shionogi rats [50], Gulo$^{-/-}$ [51]) and transport (SVCT1$^{-/-}$ [52], SVCT2$^{+/-}$ [53], SVCT2-Tg [54]) (see Table 10.1), but they are not always employed in the majority of studies. Modifications to brain vitamin C level through a combination of genetic manipulation and diet have shown that decreased vitamin C is associated with increased oxidative stress and earlier disease onset [22,55].

VITAMIN C AND ALZHEIMER DISEASE

Background, Presentation, Etiology

Alzheimer disease is a progressive neurodegenerative disorder and the most common cause for

TABLE 10.1
Common rodent models used in nutritional studies of vitamin C

Rodent Model	Description	Advantage for Vitamin C Studies	References
Guinea pigs	Innate lack of functional L-gulono-γ-lactone oxidase (Gulo) gene and therefore are dependent on an exogenous vitamin C source.	Validated animal model for vitamin C deprivation studies by allowing resemblance of human deficiency without genetic interference.	[48,56,57]
Gulo$^{-/-}$ mice	Knockout of gene encoding L-gulono-1,4-lactone oxidase (Gulo) in mice, the enzyme responsible for the final step in synthesizing vitamin C.	Allows for deficiency studies in mice using varied vitamin C supplementation. When severely vitamin C deficient, mice have decreased blood glucose, increased oxidative damage to lipids and proteins in the cortex.	[29,51,55]
SMP30/GNL mice	Senescence marker protein-30 (SMP30) is the lactone-hydrolyzing enzyme gluconolactonases (GNLs) necessary for vitamin C synthesis.	Mice display symptoms of scurvy such as bone fracture and rachitic rosary, and then die by 135 days after receiving the vitamin C–deficient diet.	[58]
ODS rats	Osteogenic disorder Shionogi (ODS) rats have hereditary deficiency in vitamin C synthesis.	Cannot synthesize their own vitamin C, allowing for appropriate deficiency studies using vitamin C supplementation.	[59]
SVCT2$^{+/-}$ mice	Heterozygous for the sodium-dependent vitamin C transporter. Note that full knockout of SVCT2 is embryonically lethal.	These mice express lower numbers of the transporter and consequently have vitamin C levels in the brain that are 30% lower than WT controls.	[53,60]
SVCT2-Tg mice	Carry extra copies of the SVCT2 gene, increasing expression of the transporter by varying degrees.	Increased expression of SVCT2 elevates vitamin C concentration in tissues and offers further protection against oxidative damage.	[54]
SVCT1$^{-/-}$ mice	Knockout of SVCT1, transporter in mice.	Elimination of one of the transporters disrupts proper maintenance of intracellular vitamin C.	[52]

dementia, with nearly 5.7 million Americans living with Alzheimer dementia in 2018 (Alzheimer's Association [United States] 2018). While the histopathology of AD is well-defined by β-amyloid (Aβ) plaques, neurofibrillary tangles, and neuronal cell death, the etiology remains elusive, as research in the field has been unable to confirm these histopathologic changes as the specific cause or consequence of the disease. Currently, few effective treatments exist for either the pathology or cognitive decline, and therefore, any intervention that can delay disease onset would significantly lessen the health burden of the disease. Chronic, subclinical vitamin C deficiency may impact AD pathogenesis via several key mechanisms and is thus an essential neuroprotector against disease progression.

AD was first characterized by the presence of Aβ plaques and neurofibrillary tangles (NFTs) by Alois Alzheimer in 1906 in the postmortem brains of patients who suffered from cognitive impairments. Since then, these two histopathologic hallmarks have largely dominated the focus of AD research. First, Aβ is an endogenous protein produced when the transmembrane amyloid precursor protein (APP) is cleaved by secretase enzymes [61]. APP is normally cleaved in a nonamyloidogenic pathway by α-secretase followed by γ-secretase. The insoluble amyloid oligomers for which AD is characterized are the result of a shift to the amyloidogenic cleavage by β-secretase followed by γ-secretase [62]. In the brains of AD patients, the burden of toxic protein aggregates combined with elevated oxidative stress lead to a global loss of neurons. The loss of neurons in the hippocampus and entorhinal cortex due to toxic proteins and oxidative stress is a strong correlate for the memory loss associated with AD. Second, hyperphosphorylation of the microtubule-associated tau protein prevents tau from stabilizing microtubule assembly, promoting the formation of NFTs [61,63]. These intracellular protein aggregates dramatically disrupt microtubule cytoskeleton and intracellular transport [64]. Evidence that oxidative stress elevates tau phosphorylation is well founded and highlights the potential for antioxidant therapeutic options such as vitamin C [65,66]. While studies of the relationship between tau hyperphosphorylation and vitamin C in particular are scarce, the available data ranging from cell culture to animal models and clinical samples tie the etiological pathways together and emphasize the importance of oxidative dysfunction in the further generation of pathologic damage.

Vitamin C and Molecular Mechanisms of AD

Oxidative Stress and β-Amyloid

Aβ peptides and oligomers are toxic to cells, activate glial cell response, and consequently increase oxidative stress. Under in vitro conditions, Aβ alone is sufficient to induce apoptosis in neurons [67]. Additionally, Aβ can cause impairment of the electron transport chain and render neurons more vulnerable to oxidative stress. Increases in oxidative stress have been shown to upregulate presenilin-1 (PSEN1) and BACE1, both of which are major protein complexes of β- and γ-secretase enzymes responsible for amyloidogenic cleavage of the amyloid precursor protein [68–71]. This shift induces Aβ accumulation, which subsequently promotes reactive oxygen species (ROS) production [72]. Further, in vitro studies confirm that Aβ-induced cell death can be mediated via oxidative stress [73,74], thus creating a cyclical relationship that drives disease progression. Adequate vitamin C may also inhibit the production of Aβ by attenuating oxidative stress [75–79]. Over 160 genetically modified rodent models have been designed to mimic some aspect of the histopathologic and cognitive deficits characteristic of human patients with AD (AD Research Models|Alzforum. Retrieved February 8, 2019, from https://www.alzforum.org/research-models). Such models have yielded crucial insight into disease pathogenesis and progression and have allowed experimental manipulation to further investigate the role of oxidative stress in

AD pathology and progression. Many, if not all, of the most commonly used models also exhibit elevated oxidative stress [80–83], presumably as a consequence of induced disease state through genetic mutation. Additionally, it is important to consider the unavoidable caveats of AD mouse models, which are modeled after inheritable forms of AD and may not accurately represent disease pathogenesis in sporadic cases of AD in patients. While convenient for laboratory research, faster disease onset due to mutational burden, as observed in the 5xFAD model compared with single- and double-mutation models, lessens the contribution of age-related changes in the mice.

Mitochondria are the main source of cellular ROS generation and are also uniquely susceptible to damage caused by oxidative stress. The mitochondrial cascade hypothesis of AD proposes that baseline mitochondrial dysfunction due to inherited and environmental factors influences APP expression, processing, and Aβ accumulation [84]. Many AD mouse models have illustrated mitochondrial dysfunction as a result of soluble Aβ [85–88]. Mitochondria obtained from 4-month-old SVCT2$^{+/-}$ mice (with approximately 30% lower brain levels of vitamin C than wild-type controls) consumed less oxygen and have lower membrane potential compared to wild type [25]. Interestingly, acute administration of vitamin C increased oxygen consumption in wild-type isolates. In contrast, the same study showed that mitochondria obtained from 4-month-old APP/PSEN1 mice consumed more oxygen and had higher membrane potential but produced lower amounts of ATP compared to WT isolates. Mitochondrial isolates from both APP/PSEN1 and SVCT2$^{+/-}$ models had elevated levels of ROS compared to WT controls. Electron microscopy evidence showed that high vitamin C supplementation (3.3 g/L) protects against abnormal mitochondrial morphologies associated with AD observed in 6-month-old Gulo$^{-/-}$ 5xFAD mice with lower vitamin C supplementation (0.6 g/L) [78]. These studies serve as evidence that insufficient vitamin C can compromise mitochondrial function, leading to elevated oxidative stress and dwindling energy reserves even in the early stages of AD.

A common approach for therapeutic dietary intervention in AD mice is to use antioxidant combination diets to provide greater antioxidant protection with considerations of recycling mechanisms among different antioxidants.

Supplementation studies with food cocktails containing vitamins C and E, among other pharmacologic cocktails, showed decreased soluble Aβ and Aβ plaque formation [89,90]. However, such findings are not universal and may depend on specific experimental parameters such as the age of the animal at intervention [91]. Dietary supplementation with vitamins C, E, and other antioxidants rescued memory impairments in rodents with oxidative stress and learning deficits due to APP/PSEN1 mutations, melamine treatment, and hypoxia [92–95]. These data support the notion that a healthy diet rich in antioxidants such as vitamin C is crucial to delaying the cognitive deficits typically seen in these models. However, these types of studies cannot assess the role of vitamin C alone, particularly in the case of deficiency. Studies involving chronic deficiency, with proper quantification are needed to further confirm these findings.

Acute intraperitoneal administration of vitamin C in 12- and 24-month-old APP/PSEN1 mice improved cognitive deficits in AD mouse models but did not diminish Aβ plaque formation [94,96]. In contrast, chronic, high vitamin C supplementation (a high 3.3 g/L compared to a lower 0.6 g/L in drinking water) beginning at 2 months of age led to a 60% decrease in plaque burden in the hippocampus and 40% decrease in the cortex, respectively, by 6 months of age in Gulo$^{-/-}$ 5xFAD mice [78]. The promise of this study should be treated cautiously, as the 5xFAD model is a highly aggressive AD mouse model that underrepresents the burden of age on disease progression. The attenuation of histopathology in these younger mice by supplemental intervention may simply be due to the dramatic differences in brain plasticity in earlier ages. Additionally, the two supplementation groups were neither true deficiency-rescue, as 0.6 g/L vitamin C supplementation is more than sufficient to maintain healthy vitamin C concentrations in Gulo$^{-/-}$ mice. Nonetheless, this study emphasizes the importance of combining genetic manipulation and diet to properly modify brain vitamin C levels and the potential role for vitamin C in AD progression. Since mice are capable of synthesizing their own vitamin C, preventing upregulation of vitamin C synthesis is critical to modeling a true human AD state, especially given that decreased vitamin C is associated with increased oxidative stress and earlier disease onset [60,97].

Mice that are heterozygous for the globally expressed sodium-dependent vitamin C transporter (SVCT2$^{+/-}$) have decreased expression of the transporter and, consequently, have an approximate 30% decrease in brain vitamin C levels compared to wild-type littermates, because it is the only vitamin C transporter expressed in the brain [53,60] (Table 10.1). Double transgenic SVCT2$^{+/-}$ APP/PSEN1 mice were found to have increased markers of brain oxidative stress (higher malondialdehyde, protein carbonyls, and F$_2$-isoprostanes) and Aβ plaque aggregation (in the hippocampus and cortex) [97]. Memory deficits were assessed with several behavioral tasks including an olfactory memory task, Y-maze alternation, conditioned fear, and Morris water maze, with some early deficits evident from 6 months of age in the low vitamin C APP/PSEN1 mice. These findings suggest that vitamin C deficiency during the early stages of disease development can play a substantial role in accelerating amyloid pathogenesis, which is likely modulated by key oxidative stress pathways. Despite this range of data, there is no consensus on the mechanism of vitamin C in AD progression, and it is likely through multiple functional roles. However, given that the presence of oxidative species has been shown to induce Aβ and NFT assembly, vitamin C may prevent AD pathology by scavenging reactive molecules before they impact Aβ and tau pathologies.

Glutamate–Ascorbic Acid Exchange Mechanism and Seizure Susceptibility

Epileptic activity is associated with neurodegenerative disease, particularly with AD [98,99]. Even mild seizures create excitotoxic damage that can influence disease progression. Following glutamate clearance by astrocytes, the cells swell and trigger the opening of volume-regulated anion channels (VRACs), allowing simultaneous release of vitamin C into the synapse [35,100]. The released vitamin C protects against oxidative damage in the synapse. The oxidized form of vitamin C, dehydroascorbate, is then recycled in the astrocyte via glucose transporters (GLUTs). Disruption of this pathway by altered expression or function of GLT-1 or by decreased vitamin C availability may mediate glutamate toxicity within the synapse, especially in neurodegenerative disorders that are associated with excitotoxicity and seizures, such as

AD. Low vitamin C intake dramatically increased mortality in Gulo$^{-/-}$ APP/PSEN1 mice with mortality attributed to spontaneous seizures [101]. This was tested more directly in SVCT2$^{+/-}$ APP/PSEN1, which also exhibit increased mortality and were more susceptible to pharmacologically induced seizures [102]. The mice showed faster seizure onset latency following treatment with kainic acid (10 mg/kg) and more ictal events following pentylenetetrazol (50 mg/kg) treatment. This study also showed that vitamin C deficiency alone increased the severity of these pharmacologically induced seizures. Vitamin C deficiency alone in Gulo$^{-/-}$ mice led to the same increased sensitivity to kainic acid as measured by seizure onset time, behavioral measures of activity, and electroencephalogram (EEG) recordings [38]. Further, vitamin C deficiency in Gulo$^{-/-}$ mice altered expression of GLT-1 and EAAC1 providing a mechanistic link between vitamin C deficiency and cognitive decline in AD.

Clinical and Epidemiological Evidence for Involvement of Vitamin C in AD

Multiple clinical and cross-sectional studies investigating the relationship between AD and vitamin C status have been reviewed together to determine overall effects [43,103,104]. Randomized clinical trials have not yet shown an association between vitamin C antioxidant therapeutic activity and a delay in AD neurodegeneration [105], possibly owing to large methodological challenges in these types of studies. One study found that despite similar vitamin C intake, plasma vitamin C levels are lower in AD patients, and that this lower level correlates to the patient's degree of cognitive impairment [106]. A meta-analysis proposed that antioxidant defenses are more aggressively consumed in AD patients, shown by elevated oxidative stress biomarkers and decreased antioxidant capability in their plasma [107]. A separate meta-analysis specifically evaluating nutrient status in plasma and cerebrospinal fluid (CSF) found that patients with AD have lower availability of docosahexaenoic acid, choline, vitamin B12, folate, vitamin C, and vitamin E compared to age-matched controls [108]. A study with over 5000 participants indicated that high intake of vitamin C was associated with lower risk of AD after a mean follow-up period of 6 years [109]. Similarly, a more recent investigation

showed the use of vitamins E and C reduced risk of both cognitive decline and AD development in persons over the age of 65 years compared to those not taking vitamin supplements [110].

There is further support for the proposal that nutritional antioxidant supplementation, such as vitamins C and E, may protect against AD-associated cognitive decline [111–114], but equivocal evidence also exists in cohort studies (reviewed in [43,115]). In a study that investigated the role of vitamins E and C and the risk of AD, people over the age of 65 years who were given vitamin E or C supplements did not have AD over the follow-up period averaging 4 years, suggesting that the use of higher-dose vitamin E and C may lower the risk of AD development in aged individuals [116]. However, another study by the same group conducted within a resident community aged 65 years and older found that increased intake of vitamin C was not significantly associated with risk of this disease during the follow-up period averaging 4 years [117]. This work was performed in subjects who were free of AD at baseline, and who were not given additional vitamin supplements. Instead, participants filled out a questionnaire, which was used to evaluate differences in dietary vitamin C consumption. Additionally, combined supplementation with vitamins E and C was reported to reduce prevalence and incidence of AD in a cross-sectional and prospective study with individuals 65 years or older. However, vitamin C without vitamin E supplementation did not reduce the risk of developing AD [114]. While the clinical data can be contradictory, in combination with animal and cell culture data, such studies provide compelling evidence in favor of supplementation as a simple, safe, and cost-effective therapeutic intervention to preserve antioxidant capacity in the elderly and protect against disease-related oxidative damage.

HUNTINGTON DISEASE

Background, Presentation, Etiology

Huntington disease is an autosomal dominant neurodegenerative disease with cognitive, psychiatric, and motor impairments affecting 10.6 per 100,000 individuals worldwide [118]. Symptoms typically manifest mid-life (<50 years of age) with a median survival of 18 years following symptom onset [119]. Cognitive deficits present as executive dysfunction, difficulty multitasking, memory loss, and difficulty learning. Psychiatric disturbances include depression, apathy, suicidal ideation, anxiety, irritability, and agitation. Motor impairments, the hallmark symptom of HD, present differently depending on the stage of the disease. Early on, chorea is typical, while in late-stage HD, rigidity, dystonia, and dyskinesia are prevalent. Cognitive and psychiatric symptoms often precede motor symptoms, although this is often identified in hindsight. Disease onset is defined by the point in time when characteristic motor symptoms emerge [119,120]. There is no cure and current treatment, such as the monoamine-depleting drug tetrabenazine, focuses on alleviating symptoms with no effect on progression of the disease [121,122].

HD is caused by an expanded (≥40) cytosine-adenine-guanine (CAG) trinucleotide repeat in exon 1 of the Huntingtin (HTT) gene [123,124]. This genetic mutation encodes for an expanded polyglutamine (polyQ) region near the N-terminus of the HTT protein. The precise function of wild-type (WT) HTT protein is uncertain. It is essential for development, demonstrated by embryonic lethality observed in mice homozygous null for Htt [125], and broadly necessary for neural maintenance [126,127]. HTT is a large protein (348 kDa), and through interactions with nearly 200 proteins [128–130], it plays a role in vesicle trafficking and axonal transport [131–133], transcriptional regulation [134–137], autophagy [138–140], and cell survival [125,136,141]. Many of these functions and protein-protein interactions, particularly those related to transcription regulation and cell signaling, depend on the nonexpanded polyQ tract (<35 repeats) present in wild-type HTT [119,142,143]. When mutated, the elongated polyQ tract not only interferes with normal HTT function but also leads to the formation of toxic cytosolic and nuclear protein aggregates [133,144–147].

Whether the mutation leads predominately to a loss of wild-type protein function or a toxic gain-of-function remains a contested topic, though both likely contribute to pathology [119,123,146,148–150]. This could, in part, explain why HD manifests as a neurodegenerative disease with mid-life onset despite continuous HTT expression throughout development. Reduced wild-type HTT function results in abnormal neuron development and could subsequently render these cells more vulnerable to mutant HTT (mHTT) aggregate toxicity; neurodevelopment and neurodegeneration are not mutually exclusive [151]. For example, striatal medium spiny neurons (MSNs), the primary cell type that degenerates in HD, rely on brain-derived neurotrophic factor (BDNF) produced by cortical neurons to be delivered via axonal transport, a process impaired by mHTT [132]. Throughout aging, mHTT aggregates accumulate in all cell types, but particularly vulnerable striatal cells atrophy, contributing to signs and symptoms of the disease [151–153]. Prevailing hypotheses of the underlying molecular mechanisms of neuronal dysfunction in HD implicate roles for increased oxidative damage and glutamate excitotoxicity [154,155], offering critical areas in which vitamin C deficiency may contribute to disease progression.

Vitamin C and Molecular Mechanisms of HD

Oxidative Stress

There is substantial evidence documenting increased oxidative stress in HD, although whether redox imbalance itself contributes to disease progression or is the result of other pathologic mechanisms is debated [156,157]. Several studies analyzing postmortem HD patient brain tissue show increased oxidative stress compared to age-matched controls. Positive markers of oxidative damage to protein, DNA, and lipids, as well as induction of antioxidant defense mechanisms, have been identified in the striatum and regions of cortex [158–161]. However, increased oxidative damage in postmortem HD patient brains is not universally reported [162]. Discrepancies between findings may partially be explained by the challenges of postmortem sample analyses. Confounds such as manner of death, time or season of death, intervening medications, length of postmortem interval prior to analysis, and subject-to-subject variation can all substantially influence the outcome of biochemical assays on postmortem tissue [163]. Although these variables are usually addressed in studies, they do not always contribute to the interpretation of results [164,165]. HD patients also tend to exhibit elevated levels of oxidative stress in peripheral tissues, including blood. Increases in lipid peroxidation and malondialdehyde (MDA), 8-oxo-2'-deoxyguanosine (8-OHdG), carbonylated proteins, as well as decreased levels of reduced glutathione (GSH) and diminished efficiency of antioxidant defense mechanisms have been reported in the blood and leukocytes of HD patients [166–169]. Some of these markers, such as 8-OHdG, even correlate with the severity of disease and may be useful biomarkers for disease progression [167].

Similar to findings in patient blood and postmortem tissue, animal and cellular models of HD show elevated markers of oxidative stress. The R6/2 mouse model was the first transgenic mouse model of HD and remains one of the most widely used. R6/2 mice express the N-terminal fragment of the Htt gene with 144–150 CAG repeats, which results in a progressive behavioral and neurological HD-like phenotype with a shortened life span of 12–18 weeks [170]. Studies examining R6/2 mice have reported increases in 8-OHdG in the urine, plasma, and striatum [171], oxidatively modified proteins [172], and lipid peroxidation measured by MDA levels [173]. The YAC128 mouse model of HD expresses full-length human HTT with 128 CAG repeats and develops motor and cognitive deficits characteristic of HD, as well as striatal-specific neuronal loss and a normal life span [170]. Unlike the R6/2 mouse model, increases in protein oxidation and lipid peroxidation were not detected in YAC128 mice until 12 months of age [174]. However, YAC128 mice still exhibit increased oxidative stress as measured by depleted reduced GSH and elevated levels of oxidized glutathione (GSSG) [175]. Increased oxidative stress has also been reported in a *Caenorhabditis elegans* model of HD, which expresses a polyQ fragment (40 CAG repeats), but no part of the Htt gene [176]. Striatal cell lines derived from knock-in HD mice with 111 CAG repeats (STHdhQ111/Q111 cells) also show evidence of altered antioxidant profile [177].

Given that oxidative stress is elevated in both patients and HD models, many studies have investigated the effectiveness of compounds that combat oxidative stress as potential therapeutics. The extracellular fluid of the striatum, the region that selectively degenerates in HD, has a high vitamin C concentration under normal healthy conditions [178]. Sufficient vitamin C is necessary in this region, and destruction of striatal extracellular vitamin C with infusion of ascorbate oxidase in rats suppressed behavioral activation [179]. Much of the work concerning vitamin C biology in mouse models of HD comes from the Rebec et al. [37,178–180]. Voltammetry studies indicate that vitamin C release in the striatum of postanesthesia, awake-behaving mice is significantly disrupted in R6/2 mice compared to wild-type littermates. When the mice were under anesthesia, there was not a significant difference in vitamin C levels between HD and controls, but a 25%–50% deficit emerged upon behavioral activation. In a subsequent study, vitamin C treatment for 4 weeks via intraperitoneal injection in 6-week-old R6/2 mice was sufficient to abolish the defect in striatal vitamin C release in awake-behaving animals, such that there was no difference between vitamin C–treated HD mice compared to both vehicle- and vitamin C–treated wild-type littermates [180]. Additional evidence for the use of antioxidant supplementation as a therapeutic intervention in HD has yielded promising results. For example, twice-daily oral administration of the fumaric acid ester dimethylfumarate (DMF) attenuated motor impairment and neurological pathology in both R6/2 and YAC128 mice, presumably through detoxification pathways activated by the induction of the transcription factor nuclear factor E2-related factor (Nrf2) [181], which is reportedly disrupted in HD STHdhQ111/Q111 cell lines [182]. Additionally, a diet rich in the antioxidant lipoic acid exerted modestly beneficial effects in R6/2 mice by significantly improving survival [183]. Lipoic acid acts as an antioxidant directly by scavenging hydroxyl radicals but can also regenerate other antioxidants such as vitamins C and E [184]. Future work should directly address a role for vitamin C deficiency in HD.

Glutamate Homeostasis and Vitamin C Release

It has been suggested that the lower levels of vitamin C release are likely not due to corticostriatal degeneration but rather a malfunction in glutamate uptake [185]. In line with this hypothesis, R6/2 mice have reduced expression of glutamate transporters such as GLT-1 [186] and glutamate clearance, and recycling is reportedly impaired in R6/2 mice and postmortem patient brains [187,188]. In the knock-in mouse model of HD, which expresses a chimeric exon 1 of the human/mouse Htt gene with 140 CAG repeats inserted into the mouse Htt gene, the same deficit in striatal vitamin C release between HD and WT was not observed. Instead, a sex difference between HD male and HD female mice but not WT male and female mice was noted and attributed to varying levels of behavioral activation [189]. Typically, increased extracellular vitamin C is associated with relocation of neuronal SVCT2 to the plasma membrane, presumably to permit additional uptake into neurons, if needed. In striatal neurons derived from knock-in mice expressing the mutant Huntingtin, the same translocation of SVCT2 was not observed, leaving the cells at risk of excitotoxic damage [36]. In the same STHdhQ111 cell line, vitamin C was also shown to modulate neuronal glucose uptake, suggesting a further role for acutely altered ROS levels in neuronal metabolism [36].

The vitamin C supplementation paradigm adopted by Rebec and colleagues also claimed to attenuate HD behavioral phenotypes as measured by stereotypic grooming behavior and locomotor activity in an open field [180]. Rotarod performance was not assessed in this group of mice, and thus, the potential for vitamin C treatment to provide therapeutic effects on motor coordination is unknown. Mechanistically, the positive effect of vitamin C treatment in HD mice has been linked to the glutamate recycling system. This hypothesis is supported by evidence showing ceftriaxone-induced upregulation of GLT-1 improves motor phenotypes and the glutamate clearance defect in R6/2 mice [190,191]. Changes in striatal vitamin C release in response to upregulation of GLT-1 have not been examined in these mice, nor have oxidative stress measures been evaluated directly. Interestingly, HD behavioral phenotypes were not exacerbated in R6/2 mice that were heterozygous null for GLT-1 [192], and real-time imaging methods of glutamate clearance in R6/2 mice fail to reproduce the defect in glutamate uptake found using synaptosomal preparation methods by others [193]. However, the evidence that vitamin C supplementation or reinstatement of appropriate glutamate clearance can attenuate pathologic phenotypes suggests that both oxidative stress and excitotoxity may be contributing to disease.

Although it is unclear whether the increased oxidative stress primarily contributes to, or is a consequence of, pathology in HD, there is strong evidence supporting a role for elevated oxidative stress and aberrant antioxidant defense mechanisms. Neither neural nor peripheral vitamin C levels have been reported in HD patient populations. However, based on work from Rebec et al., a difference in striatal vitamin C release may be the primary issue rather than total vitamin C concentrations [185], which would not be measurable in postmortem tissue. While changes in vitamin C levels have not been investigated, postmortem studies show that glutamate uptake is impaired in HD [188]. Unfortunately, glutamatergic transmission has not yet been studied in living HD patients via methodologies such as positron emission tomography (PET) [194]. Based on the strong connections between and involvement of vitamin C and glutamatergic signaling in animal models of HD, it would greatly benefit the field to investigate these systems in human populations, particularly HD patients.

Clinical and Epidemiological Evidence for Involvement of Vitamin C in HD

Vitamin C treatment has not yet been examined in the clinical setting as a therapeutic for HD patients. To our knowledge, there are few studies regarding vitamin supplementation or specific dietary status in HD patients. In fact, many HD patients are underweight and require additional calories to even maintain a healthy weight [195]; thus, the focus of clinicians and researchers regarding diet in HD has been geared toward understanding metabolic defects rather than specific nutrient deficits. One exception is the Mediterranean diet, which has been studied in HD patient populations. Self-reports of dietary intake were evaluated, and researchers found that HD patients who adhered more strictly to a Mediterranean-style diet, which included higher vitamin C consumption, had better quality of life and lower motor impairment scores than patients who did not follow this diet [196]. This study was correlational, which puts limitations on the interpretation of results. A Mediterranean diet provides ample vitamin C, as it is high in plant-based foods, particularly fresh fruits, vegetables, and grains, but these foods are also high in additional antioxidants, which provide additional benefits [197,198]. However, given that these data are self-reported and dietary guidelines were not assigned to patients, these data do not strongly support a therapeutic effect of dietary vitamin C supplementation. In a small study of 73 HD patients, high doses of vitamin E were observed to confer some benefits on the expression of neurological symptoms, although this was limited to patients in early disease stages [199]. Additional research is needed in the clinical setting to better understand the role of vitamin C in HD and the positive effect supplementation may have on disease onset and progression.

PARKINSON DISEASE

Background, Presentation, Etiology

Parkinson disease (PD) is characterized by progressive motor and nonmotor deficits, the accumulation of proteinaceous inclusions, called Lewy bodies, and the subsequent depletion of dopamine in the striatum. Although both environmental and genetic factors are implicated in the pathogenesis and progression of PD, the etiology of the disease is still unknown. The disease is most often clinically diagnosed by the appearance of motor symptoms, beginning with bradykinesia, muscular rigidity, resting tremor, and postural instability, but nonmotor symptoms may be present long before motor symptoms emerge [200,201]. The motor symptoms occur after significant loss of dopaminergic cells in the pars compacta region of the substantia nigra (SNpc). Vitamin C is a necessary cofactor in the synthesis of L-dopa by the enzyme tyrosine hydroxylase, which is then converted to dopamine and other catecholamines [26]. Several studies have shown decreased antioxidant capacity in the brains of patients with PD, and these cells may be more vulnerable to degeneration owing to increased free radical production during this enzymatic process, and the auto-oxidation of certain neurotransmitters [202]. In addition to its antioxidant role, vitamin C has been shown to drive dopamine differentiation neuronal stem cells derived from the midbrain through hydroxy- and de-methylase enzymes [30,32,203]. As vitamin C reserves are depleted by disease-related oxidative stress, these critical mechanisms are also compromised and thus also contribute to the progression of disease.

Lewy bodies are primarily composed of α-synuclein protein aggregates and, although no consensus has been reached as to the function of this protein, they are believed to play a critical role in the pathology of the disease because they appear long before the emergence of clinical symptoms [204,205]. The toxicity exerted by oligomerized α-synuclein is similar to that of β-amyloid, though the direct mechanism by which α-synuclein causes dopaminergic cell loss is still unknown. α-Synuclein oligomers can permeabilize lipid membranes, activate endoplasmic reticulum stress response, disrupt protein degradation mechanisms, and disrupt energy production by interacting with mitochondria [204]. Oxidative stress can increase phosphorylation of α-synuclein

in in vitro models, increasing the formation of protein inclusions and perpetuating a detrimental cycle [206,207]. In a meta-analysis of 80 studies designed to quantify systemic oxidative stress markers, Wei et al. showed elevated blood levels in 4 (8-OhdG, MDA, nitrite, and ferritin) of the 22 markers of oxidative damage tested, as well as a marked decrease in the antioxidants glutathione, catalase, and uric acid in PD (7212) and healthy control (6037) subjects [18].

Dopamine itself is considered cytotoxic, and the elevated levels of dopamine found in the SNpc can contribute to oxidative stress vulnerability. It is believed that dopamine oxidation by monoamine oxidase or transition metals exerts the toxic effects on cells. Increased mitochondrial oxidative stress leads to an accumulation of oxidized dopamine in a dose-dependent manner in human-derived dopaminergic-neurons lacking either one or both copies of DJ-1, a chaperone protein believed to mediate cellular oxidant defenses [208]. Selective dopaminergic synaptic terminal loss is observed when dopamine is injected into rat striatum and the amount of loss is in correlation with the amount of oxidized dopamine product [209,210]. Further, PD symptomology and progressive cell loss in the SNpc are observed in mice that cannot package dopamine into vesicles due to deficient expression of the vesicular monoamine transporter type 2 [211–213]. Further, the role of vitamin C in epigenetic modification and, in particular, in the differentiation of dopaminergic neurons through TET-mediated epigenesis [30,214] supports the necessity for vitamin C in the formation and maintenance of the dopaminergic cell population. A clear role for vitamin C in stem cell differentiation and preservation is also a critical consideration if stem cells are ever developed into a safe therapy for PD as has been proposed [216].

Vitamin C and Molecular Mechanisms of PD

Oxidative Stress and Mitochondrial Function

Mitochondrial dysfunction became associated with PD after it was observed that accidental inhibition of mitochondrial complex 1 in a subset of drug users injecting synthetic heroin caused specific cases of parkinsonism [217,218]. This led to the discovery that complex 1 activity is modestly decreased in the SNpc of PD patients [219,220]. In a study using subunit-specific and complex 1 immuno-capture to quantify complex 1 macro-assembly, Keeney and colleagues found increased oxidation of the complex 1 catalytic subunits and associated mis-assembly in PD brains compared with age-matched controls [221]. The α-synuclein protein accumulates in mitochondria of the SN, and there is some in vitro evidence to suggest that α-synuclein can disrupt complex 1 activity [204]. Overexpression of α-synuclein in mouse models shows greater susceptibility to mitochondrial toxins that disrupt the electron transport chain [222,223]; however, mice lacking α-synuclein appear to be protected [224]. Loss of catalytic activity in mitochondria not only decreases available energy within cells but also increases the generation of reactive oxygen species as a by-product of incomplete oxidation [225]. Interestingly, mice overexpressing mitochondria-targeted catalase, an endogenously produced antioxidant enzyme that breaks down hydrogen peroxide, did not show the same increase in reactive oxygen species from treatment with complex 1 inhibitors [226].

Exposure to the toxic herbicide paraquat results in parkinsonian symptoms through inhibition of mitochondrial complex 1 and oxidatively mediated selective degeneration of dopaminergic neurons in the SNpc. PC12 cells preloaded with vitamin C were protected against the toxicity after exposure to paraquat [227]. Increased intracellular vitamin C in mice designed to globally overexpress SVCT2 also protected against paraquat-induced oxidative stress as measured by F_2-isoprostanes in the lung tissue [54]. Inhibition of complex 1 of the electron transport chain increases superoxide production, thus overwhelming antioxidant capacity and causing cell loss in the SN and parkinsonism in humans and animal models [14,228]. Chronic administration of complex 1 inhibitor, rotenone, to rats caused specific degeneration in the SNpc and protein inclusions similar to Lewy bodies observed in human patients, directly linking complex 1 dysfunction in mitochondria to the pathologic hallmarks of PD [229–231]. Intracellular vitamin C deficiency also diminishes the catalytic activity of mitochondria, specifically shown by a decrease in oxygen consumption [25]. Presumably, rapid consumption of antioxidants like vitamin C in response to increased oxidative stress could further compromise an already deficient mitochondrial response during high energy demands related to disease progression. Vitamin C may be particularly important for protecting the ratio of reduced to oxidized glutathione (GSH/GSSG), as shown in mitochondrial function in cultured mesencephalic neurons [232].

Glutamate Homeostasis and Excitotoxicity

As the dopamine system deteriorates, other neurotransmitter systems within the basal ganglia are also affected. Specifically, altered glutamatergic transmission appears to be a secondary event in the propagation of excitotoxic cell death in PD [233]. The accumulation of α-synuclein is associated with the dysregulation of glutamate receptor-mediated calcium homeostasis in cultured neurons [234]. α-Synuclein is believed to be involved in synaptic transmission through maintenance of the synaptic pool, though the exact relationship is not yet understood [235,236]. Additionally, inhibition of mitochondrial complex 1 in primary astrocyte cultures, alone or in combination with depletion of the antioxidant glutathione, caused an increase in the extracellular concentrations of glutamate and hydrogen peroxide [237], thereby increasing the likelihood of excitotoxic and oxidative damage. However, vitamin C preloading protected dopamine-like neurons derived from SH-SY5Y cells against glutamate-induced excitotoxicity in a dose-dependent manner [75].

Clinical and Epidemiological Evidence for Involvement of Vitamin C in PD

Overall, there is little agreement about the role for vitamin C in PD within the existing literature from either population-based or clinical studies. Dietary supplementation in an open-label pilot trial extended the time until levodopa treatment was necessary in a small group of 21 patients with PD for 4 years or fewer [238]. Patients were recommended to take 2000–3200 IU vitamin E and 3000 mg vitamin C in four divided doses per day. Time to further treatment was 65 and 59 months in younger- and older-onset groups, compared to a similar matched group that required additional medications at 40 and 24 months, respectively. Early studies have reported higher incidence of low vitamin C in PD patients compared to controls; however, reports were in very low sample sizes, utilized varying methodologies, and yielded conflicting reports. For example, one study in particular utilized the observation of "corkscrew hairs" representing a potential early sign of severe deficiency (scurvy) rather than clinical determination of vitamin C levels [239]. Another report showed no difference in serum vitamin C between PD patients and their spouses, which were employed to provide a matched control for dietary exposures [240]. However, in another study, serum vitamin C (and vitamin E) was lower in patients with vascular PD ($n = 12$) but not in PD groups ($n = 44$) compared to control ($n = 16$) [241]. As is typically the case, particularly in older populations, all groups included cases below the threshold for clinical depletion ($<40 \ \mu mol/L$). Further, these data were not replicated in a second comparison of vascular PD and PD [242]. A more recent study found that peripheral lymphocyte vitamin C levels were associated with severity of PD in one sample of 62 patients [243]. In the Nurses' Health Study and Health Professionals Follow-Up Study, vitamin C intake was not associated with PD risk, although high dietary vitamin E (e.g., nuts) conferred diminished risk compared to a very low intake group [244]. Further analysis of the same data set refuted this earlier hypothesis that antioxidant supplements had any effect on PD risk [245]. In contrast, vitamin E and β-carotene were associated with lower risk for PD, with only a modest effect of vitamin C in two Swedish population-based cohorts with a mean follow-up of 14.9 years [246].

Nevertheless, the ability of vitamin C to support levodopa therapy has been recognized for over four decades [247]. Vitamin C supplementation is associated with better absorption of pharmaceutical L-dopa absorption in elderly PD patients with low bioavailability levels [248]. Delayed motor symptoms in a drosophila model of PD and relief from levodopa toxicity in in vitro models of PD were also observed with vitamin C supplementation [249,250]. These data suggest that the addition of vitamin C to a course of treatment may prove to be beneficial. However, the nonmotor symptoms of PD, such as cognitive and neurobehavioral abnormalities, may affect the physical health and subsequent resilience of patients with PD due to changes in lifestyle and diet [201,251].

AMYOTROPHIC LATERAL SCLEROSIS AND MULTIPLE SCLEROSIS

Amyotrophic Lateral Sclerosis

Amyotrophic lateral sclerosis (ALS) is a fatal motor neuron disorder in which the neurons of the motor cortex, brainstem, and spinal cord progressively degenerate, leading to a debilitating loss of motor function. It is not yet known why motor neurons are specifically targeted, but genetic mutations appear to contribute significantly to the nearly identical manifestation of both the familial and sporadic forms of the disease. As with other neurodegenerative diseases, protein inclusions visible within the cytoplasm of motor neurons are pathologic hallmarks of ALS. These aggregates are composed of highly ubiquitinated protein products made unstable by gene mutations, such as cytoplasmic Cu/Zn superoxide dismutase 1 (Cu/Zn-SOD1) [252]. In fact, the missense and truncation mutations in SOD1 were the first genetic mutations to be associated with ALS due to the abundance of both mutant and wild-type SOD1 found in the aggregates. The discovery of this gene association with ALS along with other genes involved in mitochondrial function suggest that oxidative stress plays a significant role in the pathogenesis of the disease [252,253]. Cu/Zn-SOD1 specifically reduces cytoplasmic superoxide anions to hydrogen peroxide molecules, which are then neutralized by other antioxidant enzymes. At higher concentrations, vitamin C also effectively scavenges superoxide [254], thus providing a secondary antioxidant defense in maintaining redox homeostasis.

Vitamin C and Molecular Mechanisms of ALS

Mitochondrial morphology and distribution are altered in motor neurons of mice carrying familial ALS-linked SOD1 mutations [255]. High-resolution respirometry indicates substantially reduced baseline respiration rates in mitochondria isolated from mice carrying a familial ALS-linked SOD1 mutation at the emergence of motor symptoms; however, this decrease in mitochondrial respiration was attenuated by treatment with a mitochondria-targeted antioxidant (MitoQ) [256]. SVCT2 is expressed on the inner mitochondrial membranes, indicating a specific and localized role for vitamin C in the organelle [23,24]. As described earlier, mitochondrial isolates from SVCT2$^{+/-}$ mice had altered function and increased ROS production compared to wild-type isolates [25]. Not only would a mutation in the SOD1 enzyme compromise antioxidant defenses, but it also could lead to copper-mediated toxicity [11,257]. High doses of the copper chelator trientine in combination with vitamin C administered to mice carrying a familial ALS-linked mutation before the onset of motor symptoms dramatically improved survival and maintained motor performance far longer than controls; however, treatment after the onset of motor symptoms showed no beneficial effects [257]. As mitochondrial dysfunction and oxidative stress accumulate due to the progression of ALS, thus overwhelming the antioxidant capacity within motor neurons, axons begin to degenerate leading to irreparable loss of motor function.

Clinical Evidence for Involvement of Vitamin C in ALS

To date, very few studies have examined dietary antioxidants intake and ALS risk, and the data are largely contradictory. In a 2009 Japanese study, 153 individuals with ALS (18–81 years) and 306 age- and gender-matched controls self-reported dietary intake of fruits, vegetables, and antioxidants. Dietary intake of fruits and vegetables or fruit alone was associated with decreased risk of developing ALS, but no significant dose-response relationship was observed between specific antioxidants vitamin C, vitamin E, or β-carotene and ALS risk [258]. In a meta-analysis of five nutritional cohort studies in the United States, higher dietary intake of β-carotene and lutein had an inverse association with ALS risk, while lycopene, β-cryptoxanthin, and vitamin C showed no association [259]. Data in this study were collected via self-reported dietary questionnaires and, in conjunction with the Japanese cohort study, suggest that avoiding general nutrient deficiency decreases risk of ALS. Studies in which quantification of vitamin C in cerebrospinal fluid (CSF) was examined in ALS patients have also yielded somewhat conflicting results. Blasco et al. reported higher vitamin C concentrations in the CSF of ALS patients (n = 50) compared to controls (n = 44) using 1H nuclear magnetic resonance spectrometry [260]. In a separate study, lower concentrations of vitamin C were measured in the CSF of ALS patients, but electron paramagnetic resonance spectrometry measured more ascorbyl radicals (the one electron oxidized form of vitamin C), in the CSF of ALS patients compared to control subjects in which the radicals were almost undetectable [261]. These apparently contradictory data may

suggest a greater need for antioxidant capacity due to rapid consumption of antioxidant reserves in patients with ALS.

Multiple Sclerosis

Multiple sclerosis (MS) is a chronic degenerative disease of the central nervous system in which progressive demyelination and subsequent axonal damage result in a loss of motor and sensory function. The cause of MS has yet to be defined. However immunologic, environmental, genetic, or a combination of these factors have been implicated in the development of the disease [262]. As with other neurodegenerative conditions, oxidative stress is believed to play a role in the pathology and progression of the disease, but MS is considered an autoimmune disease because of infiltration at the blood-brain barrier and accumulation in the central nervous system by immune cells, particularly T and B lymphocytes. Autoreactive T cells recognize antigens on the myelin sheath as foreign and release cytokines that recruit B cells and macrophages, which then contribute to the breakdown of myelin and axonal degeneration. The resulting prolonged inflammation leads to the demyelinating lesions that are characteristic of the disease. The specific manifestations of the disease are most commonly optic neuritis, spasticity, muscle weakness/fatigue, and issues with gait and imbalance [263]. The number, size, and location of the lesions, which appear to develop in white matter through lymphocytic targeting of myelin sheaths and oligodendrocytes, determine the symptoms that emerge.

Vitamin C and Molecular Mechanisms of MS

Vitamin C is involved in several cellular functions that are directly associated with immune system support. Vitamin C functions as a cofactor to the 2-oxoglutarate–dependent dioxygenase family of enzymes, which includes TET hydroxylases [30,214,264,265]. Thus, vitamin C plays a considerable role in developmental epigenetic modification and may continue to support epigenetic modification and regulation during the disease-driven changes to methylation patterns. T and B cells maintain high intracellular vitamin C concentrations presumably for antioxidant support, though studies show vitamin C is a critical component of lymphocyte programming [266,267]. Specifically, vitamin C treatment increased demethylation and subsequent expression of the critical programming elements in regulatory T cells (Tregs), which are necessary for "self-antigen tolerance" and are believed to prevent autoimmune disorders like MS [268,269]. Additionally, vitamin C status can modulate both pro- and anti-inflammatory cytokines depending on the type of cell and inflammatory stimulation [266]. For example, a recent study reported that depletion of SVCT2 in primary rat microglia using shRNA depleted intracellular vitamin C concentration and increased the production and release of pro-inflammatory cytokines [270]. In contrast, increasing intracellular vitamin C in microglia through overexpression of SVCT2 or pretreatment with vitamin C attenuated the microglial activation and release of pro-inflammatory cytokines and decreased the generation of ROS, compared to control microglia after inflammatory stimulation [270]. These studies suggest that vitamin C has the potential to restore and maintain immune system homeostasis in the presence of prolonged inflammation.

Oligodendrocytes can repair and restore damaged myelin, but not at the speed with which the myelin is destroyed by demyelinating lesions. It has been suggested that increasing the population of oligodendrocytes could attenuate the axonal degeneration caused by chronic lesions [271]. Cerebral injection of fetal glial progenitor cells increased myelination in several regions of the brain and rescued the phenotype in congenitally hypomyelinated shiverer mice [272]. Both mature and premyelinating oligodendrocytes have been visualized in lesions collected from patient autopsy; however, there was little evidence of effective remyelination, presumably due to the disruption in the microenvironment necessary to maintain the relationship between oligodendrocytes and axons [273]. It is long established that vitamin C is necessary for Schwann cell myelination in the peripheral nervous system due to its role in collagen synthesis [274], but vitamin C also appears to play a crucial role in the differentiation of oligodendrocyte progenitor cells into mature oligodendrocytes [275].

Clinical Evidence for Involvement of Vitamin C in MS

A definitive effect of vitamin C status or supplementation in patients with MS has not yet been demonstrated, but one case control study suggests higher intake of vitamin C is associated with lowered risk of developing MS [276], perhaps due to the regulatory effects on the immune system. However, another study was unable to recapitulate these findings using much larger cohorts [277]. Lower plasma vitamin C concentrations were observed in patients with MS, particularly during episodes of relapse, along with an inverse correlation between serum vitamin C and lipid peroxidation [215], suggesting depletion of antioxidant reserves due to increased demand and oxidative stress.

SUMMARY AND CONCLUSIONS

A major challenge of establishing consistent and meaningful conclusions from nutritional intake studies is normalizing the various methodologies and techniques employed by research teams. Animal and cell culture studies have elucidated multiple pathways involved in the pathogenesis of neurodegenerative diseases and show specific roles for vitamin C. However, these have not been translated into diagnostic markers or clinically supported treatments. It is evident that oxidative stress is contributory in each of the degenerative diseases described earlier, and as such, vitamin C is an invaluable component of antioxidant defense in the brain and critical to maintaining redox homeostasis. Although vitamin C can act as a pro-oxidant in the presence of transition metals such as iron, most evidence supports the notion that vitamin C is an antioxidant in most of its biological functions. The additional functions of vitamin C as an enzymatic cofactor also support the neuroprotective response to disease pathology and progression. As new technology is developed and disease-related biomarkers are identified that provide key insights into the etiology of neurodegenerative pathologies, care must be taken to consider the confounding factors that are always present in human studies. The diversity of patient and control populations with regard to individual lifestyle choices, medical history, environmental factors, and genetic factors introduce myriad variables to be considered when designing a study, as each of these factors affect the utilization of vital nutrients within each subject. However, given the array of evidence described earlier, it can be concluded that maintaining sufficient levels of vitamin C throughout life is therefore likely to contribute to preservation of health and to complement other therapeutic interventions.

ABBREVIATIONS

5hmC: 5-hydroxymethylcytosine
5mC: 5-methylcytosine
8-OHdG: 8-oxo-2′-deoxyguanosine
Aβ: amyloid-beta
AD: Alzheimer disease
ALS: amyotrophic lateral sclerosis
APP: amyloid precursor protein
BDNF: brain-derived neurotrophic factor
CAG: cytosine-adenine-guanine
CSF: cerebrospinal fluid
DMF: dimethylfumarate
FAD: familial Alzheimer disease
GSH: glutathione (reduced form)
GSSG: glutathione (oxidized form)
HD: Huntington disease
HTT: *Huntingtin* (gene)
HTT: Huntingtin protein
IP: intraperitoneal
L-dopa: l-3,4-dyhydroxyphenylalanine, levodopa
MDA: malondialdehyde
mHTT: mutant HTT protein
MS: multiple sclerosis
MSN: medium spiny neurons
NE: norepinephrine
NFT: neurofibrillary tangles
PD: Parkinson disease
polyQ: polyglutamine
PSEN1: presenilin 1
PSEN2: presenilin 2
ROS: reactive oxygen species
SNpc: pars compacta of the substantia nigra
SOD1: superoxide dismutase
SVCT2: sodium-dependent vitamin C transporter type 2
TET: ten-eleven translocation dioxygenases
Tregs: regulatory T cells
WT: wild type

REFERENCES

1. Rice, M. E. and Russo-Menna, I. 1998. Differential compartmentalization of brain ascorbate and glutathione between neurons and glia. *Neuroscience* 82, 1213–1223.

2. Harrison, F. E., Green, R. J., Dawes, S. M. and May, J. M. 2010. Vitamin C distribution and retention in the mouse brain. *Brain Res.*, Elsevier B.V. 1348, 181–186.

3. Rice, M. E. 2000. Ascorbate regulation and its neuroprotective role in the brain. *Trends Neurosci.* 23, 209–216.

4. Bürzle, M., Suzuki, Y., Ackermann, D., Miyazaki, H., Maeda, N., Clémençon, B., Burrier, R. and Hediger, M. A. 2013. The sodium-dependent ascorbic acid transporter family SLC23. *Mol. Aspects Med.* 34, 436–454.

5. Harrison, F. E. and May, J. M. 2009, March 15. Vitamin C function in the brain: Vital role of the ascorbate transporter SVCT2. *Free Radic. Biol. Med.*, 46(6), 719–730.

6. Morris, J. C., Roe, C. M., Xiong, C., Fagan, A. M., Goate, A. M., Holtzman, D. M. and Mintun, M. A. 2010. APOE predicts amyloid-beta but not tau Alzheimer pathology in cognitively normal aging. *Ann. Neurol.* 67, 122–131.

7. Rodrigue, K. M., Kennedy, K. M., Devous, M. D., Rieck, J. R., Hebrank, A. C., Diaz-Arrastia, R., Mathews, D. and Park, D. C. 2012. β-Amyloid burden in healthy aging: Regional distribution and cognitive consequences. *Neurology* 78, 387–395.

8. Duyao, M., Ambrose, C., Myers, R., Novelletto, A., Persichetti, F., Frontali, M., Folstein, S., Ross, C., Franz, M. and Abbott, M. 1993. Trinucleotide repeat length instability and age of onset in Huntington's disease. *Nat. Genet.* 4, 387–392.

9. Snell, R. G., MacMillan, J. C., Cheadle, J. P., Fenton, I., Lazarou, L. P., Davies, P., MacDonald, M. E., Gusella, J. F., Harper, P. S. and Shaw, D. J. 1993. Relationship between trinucleotide repeat expansion and phenotypic variation in Huntington's disease. *Nat. Genet.* 4, 393–397.

10. Kocot, J., Luchowska-Kocot, D., Kiełczykowska, M., Musik, I. and Kurzepa, J. 2017. Does vitamin C influence neurodegenerative diseases and psychiatric disorders? *Nutrients* 9, 659.

11. Moretti, M., Fraga, D. B. and Rodrigues, A. L. S. 2017. Preventive and therapeutic potential of ascorbic acid in neurodegenerative diseases. *CNS Neurosci. Ther.* 23, 921–929.

12. Halliwell, B. 2006. Oxidative stress and neurodegeneration: Where are we now? *J. Neurochem.* 97, 1634–1658.

13. Halliwell, B. 2007. Biochemistry of oxidative stress. *Biochem. Soc. Trans.* 35, 1147–1150.

14. Bose, A. and Beal, M. F. 2016. Mitochondrial dysfunction in Parkinson's disease. *J. Neurochem.* 139(Suppl), 216–231.

15. García-Escudero, V., Martín-Maestro, P., Perry, G. and Avila, J. 2013. Deconstructing mitochondrial dysfunction in alzheimer disease. *Oxid. Med. Cell. Longev.* 2013.

16. Grünewald, A., Kumar, K. R. and Sue, C. M. 2019. New insights into the complex role of

mitochondria in Parkinson's disease. *Prog. Neurobiol.*, 177, 73–93.

17. Milakovic, T. and Johnson, G. V. W. 2005. Mitochondrial respiration and ATP production are significantly impaired in striatal cells expressing mutant Huntingtin. *J. Biol. Chem.* 280, 30773–30782.

18. Wei, Z., Li, X., Li, X., Liu, Q. and Cheng, Y. 2018. Oxidative stress in Parkinson's disease: A systematic review and meta-analysis. *Front. Mol. Neurosci.* 11, 236.

19. Byrne, L. M. and Wild, E. J. 2016. Cerebrospinal fluid biomarkers for Huntington's disease. *J. Huntingtons. Dis.* 5, 1–13.

20. Montine, T. J., Markesbery, W. R., Morrow, J. D. and Roberts, L. J. 1998. Cerebrospinal fluid F2-isoprostane levels are increased in Alzheimer's disease. *Ann. Neurol.* 44, 410–413.

21. Montine, T. J., Beal, M. F., Cudkowicz, M. E., O'Donnell, H., McFarland, L., Bachrach, A. F., Zackert, W. E., Roberts, L. J. and Morrow, J. D. 1999. Increased CSF F2-isoprostane concentration in probable AD. *Neurology* 52, 562–565.

22. Zabel, M., Nackenoff, A., Kirsch, W. M., Harrison, F. E., Perry, G. and Schrag, M. 2018. Markers of oxidative damage to lipids, nucleic acids and proteins and antioxidant enzymes activities in Alzheimer's disease brain: A meta-analysis in human pathological specimens. *Free Radic. Biol. Med.* 115, 351–360.

23. Bánhegyi, G., Benedetti, A., Margittai, É., Marcolongo, P., Fulceri, R., Németh, C. E. and Szarka, A. 2014. Subcellular compartmentation of ascorbate and its variation in disease states. *Biochim. Biophys. Acta—Mol. Cell Res.* 1843, 1909–1916.

24. Muñoz-Montesino, C., Roa, F. J., Peña, E., González, M., Sotomayor, K., Inostroza, E., Muñoz, C. A. et al. 2014. Mitochondrial ascorbic acid transport is mediated by a low-affinity form of the sodium-coupled ascorbic acid transporter-2. *Free Radic. Biol. Med.*, Elsevier 70, 241–254.

25. Dixit, S., Fessel, J. P. and Harrison, F. E. 2017. Mitochondrial dysfunction in the APP/PSEN1 mouse model of Alzheimer's disease and a novel protective role for ascorbate. *Free Radic. Biol. Med.* 112.

26. Stone, K. J. and Townsley, B. H. 1973. The effect of L-ascorbate on catecholamine biosynthesis. *Biochem. J.* 131, 611–613.

27. Diliberto, Jr, E. J. and Allen, P. L. 1981. Mechanism of dopamine-beta-hydroxylation.

28. Semidehydroascorbate as the enzyme oxidation product of ascorbate. *J. Biol. Chem.* 256, 3385–3393.

28. May, J. M., Qu, Z.-C., Nazarewicz, R. and Dikalov, S. 2013. Ascorbic acid efficiently enhances neuronal synthesis of norepinephrine from dopamine. *Brain Res. Bull.*, Elsevier Inc. 90, 35–42.

29. Ward, M. S., Lamb, J., May, J. M. and Harrison, F. E. 2013. Behavioral and monoamine changes following severe vitamin C deficiency. *J. Neurochem.*, NIH Public Access 124, 363–375.

30. He, X.-B., Kim, M., Kim, S.-Y., Yi, S.-H., Rhee, Y.-H., Kim, T., Lee, E.-H. et al. 2015. Vitamin C facilitates dopamine neuron differentiation in fetal midbrain through Tet1- and Jmjd3-dependent epigenetic control manner. *Stem Cells* 33(4), 1320–1332.

31. Agathocleous, M., Meacham, C. E., Burgess, R. J., Piskounova, E., Zhao, Z., Crane, G. M., Cowin, B. L. et al. 2017. Ascorbate regulates haematopoietic stem cell function and leukaemogenesis. *Nature* 549, 476–481.

32. Camarena, V. and Wang, G. 2016. The epigenetic role of vitamin C in health and disease. *Cell. Mol. Life Sci.* 73, 1645–1658.

33. Young, J. I., Züchner, S. and Wang, G. 2015. Regulation of the epigenome by vitamin C. *Annu. Rev. Nutr.* 35, 545–564.

34. Grünewald, R. A. 1993. Ascorbic acid in the brain. *Brain Res. Brain Res. Rev.* 18, 123–133.

35. Wilson, J. X., Peters, C. E., Sitar, S. M., Daoust, P. and Gelb, A. W. 2000. Glutamate stimulates ascorbate transport by astrocytes. *Brain Res.* 858, 61–66.

36. Acuña, A. I., Esparza, M., Kramm, C., Beltrán, F. A., Parra, A. V., Cepeda, C., Toro, C. A. et al. 2013. A failure in energy metabolism and antioxidant uptake precede symptoms of Huntington's disease in mice. *Nat. Commun.* 4, 2917.

37. Dorner, J. L., Miller, B. R., Klein, E. L., Murphy-Nakhnikian, A., Andrews, R. L., Barton, S. J. and Rebec, G. V. 2009. Corticostriatal dysfunction underlies diminished striatal ascorbate release in the R6/2 mouse model of Huntington's disease. *Brain Res.*, Elsevier B.V. 1290, 111–120.

38. Mi, D. J., Dixit, S., Warner, T. A., Kennard, J. A., Scharf, D. A., Kessler, E. S., Moore, L. M. et al. 2018. Altered glutamate clearance in ascorbate deficient mice increases seizure susceptibility and contributes to cognitive impairment in APP/PSEN1 mice. *Neurobiol. Aging*, Elsevier Inc 71, 241–254.

39. Cahill, L. E. and El-Sohemy, A. 2009. Vitamin C transporter gene polymorphisms, dietary vitamin C and serum ascorbic acid. *J. Nutrigenet. Nutrigenomics* 2, 292–301.

40. Timpson, N. J., Forouhi, N. G., Brion, M.-J., Harbord, R. M., Cook, D. G., Johnson, P., McConnachie, A. et al. 2010. Genetic variation at the SLC23A1 locus is associated with circulating levels of L-ascorbic acid (Vitamin C). Evidence from 5 independent studies with over 15000 participants. *Am. J. Clin. Nutr.* 92, 375–382.

41. Ortiz-Andrellucchi, A., Sánchez-Villegas, A., Doreste-Alonso, J., de Vries, J., de Groot, L. and Serra-Majem, L. 2009. Dietary assessment methods for micronutrient intake in elderly people: A systematic review. *Br. J. Nutr.* 102(Suppl), S118–S149.

42. Cross, A. J., George, J., Woodward, M. C., Ames, D., Brodaty, H. and Elliott, R. A. 2017. Dietary supplement use in older people attending memory clinics in Australia. *J. Nutr. Health Aging* 21, 46–50.

43. Harrison, F. E. 2012. A critical review of vitamin C for the prevention of age-related cognitive decline and Alzheimer's disease. *J. Alzheimer's Dis.* 29, 711–726.

44. Levine, M., Wang, Y., Padayatty, S. J. and Morrow, J. 2001. A new recommended dietary allowance of vitamin C for healthy young women. *Proc. Natl. Acad. Sci. USA* 98, 9842–9846.

45. Langlois, K., Cooper, M. and Colapinto, C. K. 2016. Vitamin C status of Canadian adults: Findings from the 2012/2013 Canadian Health Measures Survey. *Heal. Reports* 27, 3–10.

46. Nishikimi, M. and Udenfriend, S. 1976. Immunologic evidence that the gene for L-gulono-gamma-lactone oxidase is not expressed in animals subject to scurvy. *Proc. Natl. Acad. Sci. USA* 73, 2066–2068.

47. Hasselholt, S., Tveden-Nyborg, P. and Lykkesfeldt, J. 2015. Distribution of vitamin C is tissue specific with early saturation of the brain and adrenal glands following differential oral dose regimens in guinea pigs. *Br. J. Nutr.* 113, 1539–1549.

48. Hansen, S., Schou-Pedersen, A., Lykkesfeldt, J. and Tveden-Nyborg, P. 2018. Spatial memory dysfunction induced by vitamin C deficiency is associated with changes in monoaminergic neurotransmitters and aberrant synapse formation. *Antioxidants*, Multidisciplinary Digital Publishing Institute (MDPI) 7, 82.

49. Paidi, M. D., Schjoldager, J. G., Lykkesfeldt, J. and Tveden-Nyborg, P. 2014. Chronic vitamin C deficiency promotes redox imbalance in the brain but does not alter sodium-dependent vitamin C transporter 2 expression. *Nutrients* 6, 1809–1822.

50. Mizuchuma, Y., Harauchi, T., Yoshizaki, T. and Makino, S. 1984. A rat mutant unable to synthesize vitamin C. *Experientia* 40, 359–361.

51. Maeda, N., Hagihara, H., Nakata, Y., Hiller, S., Wilder, J. and Reddick, R. 2000. Aortic wall damage in mice unable to synthesize ascorbic acid. *Proc. Natl. Acad. Sci. USA* 97, 841–846.

52. Corpe, C. P., Tu, H., Eck, P., Wang, J., Faulhaber-Walter, R., Schnermann, J., Margolis, S. et al. 2010. Vitamin C transporter Slc23a1 links renal reabsorption, vitamin C tissue accumulation, and perinatal survival in mice. *J. Clin. Invest.* 120, 1069–1083.

53. Sotiriou, S., Gispert, S., Cheng, J., Wang, Y., Chen, A., Hoogstraten-Miller, S., Miller, G. F. et al. 2002. Ascorbic-acid transporter Slc23a1 is essential for vitamin C transport into the brain and for perinatal survival. *Nat. Med.* 8, 514–517.

54. Harrison, F. E., Best, J. L., Meredith, M. E., Gamlin, C. R., Borza, D.-B., May, J. M. and May, J. M. 2012. Increased expression of SVCT2 in a new mouse model raises ascorbic acid in tissues and protects against paraquat-induced oxidative damage in lung. *PLOS ONE* 7, e35623.

55. Harrison, F. E., Yu, S. S., Van Den Bossche, K. L., Li, L., May, J. M. and McDonald, M. P. 2008. Elevated oxidative stress and sensorimotor deficits but normal cognition in mice that cannot synthesize ascorbic acid. *J. Neurochem.* 106, 1198–1208.

56. Tveden-Nyborg, P., Johansen, L. K., Raida, Z., Villumsen, C. K., Larsen, J. O. and Lykkesfeldt, J. 2009. Vitamin C deficiency in early postnatal life impairs spatial memory and reduces the number of hippocampal neurons in guinea pigs. *Am. J. Clin. Nutr.* 90, 540–546.

57. Chatterjee, I. B., Majumder, A. K., Nandi, B. K. and Subramanian, N. 1975. Synthesis and some major functions of vitamin C in animals. *Ann. N. Y. Acad. Sci.* 258, 24–47.

58. Kondo, Y., Sasaki, T., Sato, Y., Amano, A., Aizawa, S., Iwama, M., Handa, S. et al. 2008. Vitamin C depletion increases superoxide generation in brains of SMP30/GNL knockout mice. *Biochem. Biophys. Res. Commun.*, Elsevier Inc. 377, 291–296.

59. Togari, A., Arai, M., Nakagawa, S., Banno, A., Aoki, M. and Matsumoto, S. 1995. Alteration of bone status with ascorbic acid deficiency in ODS (osteogenic disorder Shionogi) rats. *Jpn. J. Pharmacol.* 68, 255–261.

60. Harrison, F. E., Dawes, S. M., Meredith, M. E., Babaev, V. R., Li, L. and May, J. M. 2010. Low vitamin C and increased oxidative stress and cell death in mice that lack the sodium-dependent vitamin C transporter SVCT2. *Free Radic. Biol. Med.* 49, 821–829.

61. Swerdlow, R. H. 2007. Pathogenesis of Alzheimer's disease. *Clin. Interv. Aging* 2, 347–359.

62. Prasansuklab, A. and Tencomnao, T. 2013. Amyloidosis in Alzheimer's disease: The toxicity of amyloid beta (A β), mechanisms of its accumulation and implications of medicinal plants for therapy. *Evidence-Based Complement. Altern. Med.* 2013, 413808.

63. Zhao, Y. and Zhao, B. 2013. Oxidative stress and the pathogenesis of Alzheimer's disease. *Oxid. Med. Cell. Longev.* 2013, 316523.

64. Stamer, K., Vogel, R., Thies, E., Mandelkow, E.-M. and Mandelkow, E. M. 2002. Tau blocks traffic of organelles, neurofilaments, and APP vesicles in neurons and enhances oxidative stress. *J. Cell Biol.* 156, 1051–1063.

65. Alavi Naini, S. M. and Soussi-Yanicostas, N. 2015. Tau hyperphosphorylation and oxidative stress, a critical vicious circle in neurodegenerative tauopathies? *Oxid. Med. Cell. Longev.*, Hindawi 2015, 151979.

66. Mondragón-Rodríguez, S., Perry, G., Zhu, X., Moreira, P. I., Acevedo-Aquino, M. C. and Williams, S. 2013. Phosphorylation of tau protein as the link between oxidative stress, mitochondrial dysfunction, and connectivity failure: Implications for Alzheimer's disease. *Oxid. Med. Cell. Longev.* 2013, 940603.

67. Loo, D. T., Copani, A., Pike, C. J., Whittemore, E. R., Walencewicz, A. J. and Cotman, C. W. 1993. Apoptosis is induced by beta-amyloid in cultured central nervous system neurons. *Proc. Natl. Acad. Sci.*, National Academy of Sciences 90, 7951–7955.

68. Fleming, J. L., Phiel, C. J. and Toland, A. E. 2012. The role for oxidative stress in aberrant DNA methylation in Alzheimer's disease. *Curr. Alzheimer Res.* 9, 1077–1096.

69. Krstic, D. and Knuesel, I. 2013. Deciphering the mechanism underlying late-onset Alzheimer disease. *Nat. Rev. Neurol.*, Nature Publishing Group 9, 25–34.

70. Tong, Y., Zhou, W., Fung, V., Christensen, M. A., Qing, H., Sun, X. and Song, W. 2005. Oxidative stress potentiates BACE1 gene expression and Abeta generation. *J. Neural Transm.* 112, 455–469.

71. Tamagno, E., Bardini, P., Obbili, A., Vitali, A., Borghi, R., Zaccheo, D., Pronzato, M. A. et al. 2002. Oxidative stress increases expression and activity of BACE in NT2 neurons. *Neurobiol. Dis.* 10, 279–288.

72. Hansen, S. N., Tveden-Nyborg, P. and Lykkesfeldt, J. 2014. Does vitamin C deficiency affect cognitive development and function? *Nutrients, Multidisciplinary Digital Publishing Institute (MDPI)* 6, 3818–3846.

73. Abarikwu, S. O., Pant, A. B. and Farombi, E. O. 2012. 4-Hydroxynonenal induces mitochondrial-mediated apoptosis and oxidative stress in SH-SY5Y human neuronal cells. *Basic Clin. Pharmacol. Toxicol.* 110, 441–448.

74. Liu, W., Kato, M., Akhand, A. A., Hayakawa, A., Suzuki, H., Miyata, T., Kurokawa, K., Hotta, Y., Ishikawa, N. and Nakashima, I. 2000. 4-Hydroxynonenal induces a cellular redox status-related activation of the caspase cascade for apoptotic cell death. *J. Cell Sci.* 113, 635–641.

75. Ballaz, S., Morales, I., Rodríguez, M. and Obeso, J. A. 2013. Ascorbate prevents cell death from prolonged exposure to glutamate in an in vitro model of human dopaminergic neurons. *J. Neurosci. Res.* 91, 1609–1617.

76. Choi, S.-M., Kim, B. C., Cho, Y.-H., Choi, K., Chang, J., Park, M., Kim, M.-K., Cho, K. and Kim, J. 2014. Effects of flavonoid compounds on β-amyloid-peptide-induced neuronal death in cultured mouse cortical neurons. *Chonnam Med. J.* 50, 45–51.

77. Huang, J. and May, J. M. 2006. Ascorbic acid protects SH-SY5Y neuroblastoma cells from apoptosis and death induced by β-amyloid. *Brain Res.* 1097, 52–58.

78. Kook, S. Y., Lee, K. M., Kim, Y., Cha, M. Y., Kang, S., Baik, S. H., Lee, H., Park, R. and Mook-Jung, I. 2014. High-dose of vitamin C supplementation reduces amyloid plaque burden and ameliorates pathological changes in the brain of 5XFAD mice. *Cell Death Dis.*, Nature Publishing Group 5, e1083.

79. Murakami, K., Murata, N., Ozawa, Y., Kinoshita, N., Irie, K., Shirasawa, T. and Shimizu, T. 2011. Vitamin C restores behavioral deficits and amyloid-β oligomerization without affecting plaque formation in a mouse model of Alzheimer's disease. *J. Alzheimer's Dis.* 26, 7–18.

80. Bilkei-Gorzo, A. 2014. Genetic mouse models of brain ageing and Alzheimer's disease. *Pharmacol. Ther.* 142, 244–257.

81. Siedlak, S. L., Casadesus, G., Webber, K. M., Pappolla, M. A., Atwood, C. S., Smith, M. A. and Perry, G. 2009. Chronic antioxidant therapy reduces oxidative stress in a mouse model of Alzheimer's disease. *Free Radic. Res.* 43, 156–164.

82. Chou, J. L., Shenoy, D. V., Thomas, N., Choudhary, P. K., Laferla, F. M., Goodman, S. R. and Breen, G. A. M. 2011. Early dysregulation of the mitochondrial proteome in a mouse model of Alzheimer's disease. *J. Proteomics* 74, 466–479.

83. Devi, L. and Ohno, M. 2012. Mitochondrial dysfunction and accumulation of the β-secretase-cleaved C-terminal fragment of APP in Alzheimer's disease transgenic mice. *Neurobiol. Dis.* 45, 417–424.

84. Swerdlow, R. H., Burns, J. M. and Khan, S. M. 2014, August 1. The Alzheimer's disease mitochondrial cascade hypothesis: Progress and perspectives. *Biochim. Biophys. Acta.* 1842(8), 1219–1231.

85. Manczak, M., Mao, P., Calkins, M. J., Cornea, A., Reddy, A. P., Murphy, M. P., Szeto, H. H., Park, B. and Reddy, P. H. 2010. Mitochondria-targeted antioxidants protect against amyloid-beta toxicity in Alzheimer's disease neurons. *J. Alzheimers. Dis.* 20(Suppl 2), S609–S631.

86. Eckert, A., Hauptmann, S., Scherping, I., Rhein, V., Müller-Spahn, F., Götz, J. and Müller, W. E. 2008. Soluble beta-amyloid leads to mitochondrial defects in amyloid precursor protein and tau transgenic mice. *Neurodegener. Dis.* 5, 157–159.

87. Devi, L., Prabhu, B. M., Galati, D. F., Avadhani, N. G. and Anandatheerthavarada, H. K. 2006. Accumulation of amyloid precursor protein in the mitochondrial import channels of human Alzheimer's disease brain is associated with mitochondrial dysfunction. *J. Neurosci.* 26, 9057–9068.

88. Cha, M.-Y., Han, S.-H., Son, S. M., Hong, H.-S., Choi, Y.-J., Byun, J. and Mook-Jung, I. 2012. Mitochondria-specific accumulation of amyloid β induces mitochondrial dysfunction leading to apoptotic cell death. *PLOS ONE* 7, e34929.

89. Broersen, L. M., Kuipers, A. A. M., Balvers, M., van Wijk, N., Savelkoul, P. J. M., de Wilde, M. C., van der Beek, E. M. et al. 2013. A specific multi-nutrient diet reduces Alzheimer-like pathology in young adult AβPPswe/PS1dE9 mice. *J. Alzheimers. Dis.* 33, 177–190.

90. Parachikova, A., Green, K. N., Hendrix, C. and Laferla, F. M. 2010. Formulation of a medical food cocktail for Alzheimer's disease: Beneficial effects on cognition and neuropathology in a mouse model of the disease. *PLOS ONE* 5, e14015.

91. Harrison, F. E., Allard, J., Bixler, R., Usoh, C., Li, L., May, J. M. and McDonald, M. P. 2009. Antioxidants and cognitive training interact to affect oxidative stress and memory in APP/PSEN1 mice. *Nutr. Neurosci.* 12, 203–218.

92. An, L. and Zhang, T. 2014. Vitamins C and E reverse melamine-induced deficits in spatial cognition and hippocampal synaptic plasticity in rats. *Neurotoxicology* 44, 132–139.

93. Fukui, K., Omoi, N.-O., Hayasaka, T., Shinnkai, T., Suzuki, S., Abe, K. and Urano, S. 2002. Cognitive impairment of rats caused by oxidative stress and aging, and its prevention by vitamin E. *Ann. N. Y. Acad. Sci.* 959, 275–284.

94. Harrison, F. E., Hosseini, A. H., McDonald, M. P. and May, J. M. 2009. Vitamin C reduces spatial learning deficits in middle-aged and very old APP/PSEN1 transgenic and wild-type mice. *Pharmacol. Biochem. Behav.* 93, 443–450.

95. Joseph, J. A., Denisova, N. A., Arendash, G., Gordon, M., Diamond, D., Shukitt-Hale, B. and Morgan, D. 2003. Blueberry supplementation enhances signaling and prevents behavioral deficits in an Alzheimer disease model. *Nutr. Neurosci.* 6, 153–162.

96. Kennard, J. A. and Harrison, F. E. 2014. Intravenous ascorbate improves spatial memory in middle-aged APP/PSEN1 and wild type mice. *Behav. Brain Res.*, Elsevier B.V. 264, 34–42.

97. Dixit, S., Bernardo, A., Walker, J. M., Kennard, J. A., Kim, G. Y., Kessler, E. S. and Harrison, F. E. 2015. Vitamin C deficiency in the brain impairs cognition, increases amyloid accumulation and deposition, and oxidative stress in APP/PSEN1 and normally aging mice. *ACS Chem. Neurosci.* 6, 570–581.

98. Vossel, K. A., Beagle, A. J., Rabinovici, G. D., Shu, H., Lee, S. E., Naasan, G., Hegde, M. et al. 2013. Seizures and epileptiform activity in the early stages of Alzheimer disease. *JAMA Neurol.* 70, 1158–1166.

99. Vossel, K. A., Tartaglia, M. C., Nygaard, H. B., Zeman, A. Z. and Miller, B. L. 2017. Epileptic activity in Alzheimer's disease: Causes and clinical relevance. *Lancet. Neurol.* 16, 311–322.

100. Rebec, G. V. 2013. Dysregulation of corticostriatal ascorbate release and glutamate

uptake in transgenic models of Huntington's disease. *Antioxid. Redox Signal.* 19, 2115–2128.

101. Harrison, F. E., May, J. M. and McDonald, M. P. 2010. Vitamin C deficiency increases basal exploratory activity but decreases scopolamine-induced activity in APP/PSEN1 transgenic mice. *Pharmacol. Biochem. Behav.*, NIH Public Access 94, 543–552.

102. Warner, T. A., Kang, J. Q., Kennard, J. A. and Harrison, F. E. 2015. Low brain ascorbic acid increases susceptibility to seizures in mouse models of decreased brain ascorbic acid transport and Alzheimer's disease. *Epilepsy Res.* 110, 20–25.

103. Zhao, C., Noble, J. M., Marder, K., Hartman, J. S., Gu, Y. and Scarmeas, N. 2018. Dietary patterns, physical activity, sleep, and risk for dementia and cognitive decline. *Curr. Nutr. Rep.* 7, 335–345.

104. Rocaspana-García, M., Blanco-Blanco, J., Arias-Pastor, A., Gea-Sánchez, M. and Piñol-Ripoll, G. 2018. Study of community-living Alzheimer's patients' adherence to the Mediterranean diet and risks of malnutrition at different disease stages. *PeerJ* 6, e5150.

105. Monacelli, F., Acquarone, E., Giannotti, C., Borghi, R. and Nencioni, A. 2017. Vitamin C, aging and Alzheimer's disease. *Nutrients*, Multidisciplinary Digital Publishing Institute (MDPI) 9.

106. Rivière, S., Birlouez-Aragon, I., Nourhashémi, F. and Vellas, B. 1998. Low plasma vitamin C in Alzheimer patients despite an adequate diet. *Int. J. Geriatr. Psychiatry* 13, 749–754.

107. Schrag, M., Mueller, C., Zabel, M., Crofton, A., Kirsch, W. M., Ghribi, O., Squitti, R. and Perry, G. 2013. Oxidative stress in blood in Alzheimer's disease and mild cognitive impairment: A meta-analysis. *Neurobiol. Dis.*, Elsevier B.V. 59, 100–110.

108. de Wilde, M. C., Vellas, B., Girault, E., Yavuz, A. C. and Sijben, J. W. 2017. Lower brain and blood nutrient status in Alzheimer's disease: Results from meta-analyses. *Alzheimer's Dement.* (New York, N. Y.) 3, 416–431.

109. Engelhart, M. J., Geerlings, M. I., Ruitenberg, A., Van Swieten, J. C., Hofman, A., Witteman, J. C. M. and Breteler, M. M. B. 2002. Dietary intake of antioxidants and risk of Alzheimer disease. *J. Am. Med. Assoc.*, American Medical Association 287, 3223–3229.

110. Basambombo, L. L., Carmichael, P.-H., Côté, S. and Laurin, D. 2017. Use of vitamin E and C supplements for the prevention of cognitive decline. *Ann. Pharmacother.* 51, 118–124.

111. Bowman, G. L., Dodge, H., Frei, B., Calabrese, C., Oken, B. S., Kaye, J. A. and Quinn, J. F. 2009. Ascorbic acid and rates of cognitive decline in Alzheimer's disease. *J. Alzheimers. Dis.* 16, 93–98.

112. Gale, C. R., Martyn, C. N. and Cooper, C. 1996. Cognitive impairment and mortality in a cohort of elderly people. *BMJ* 312, 608–611.

113. Goodwin, J. S., Goodwin, J. M. and Garry, P. J. 1983. Association between nutritional status and cognitive functioning in a healthy elderly population. *JAMA* 249, 2917–2921.

114. Zandi, P. P., Anthony, J. C., Khachaturian, A. S., Stone, S. V., Gustafson, D., Tschanz, J. T., Norton, M. C., Welsh-Bohmer, K. A., Breitner, J. C. S. and Cache County Study Group. 2004. Reduced risk of Alzheimer disease in users of antioxidant vitamin supplements: The Cache County Study. *Arch. Neurol.* 61, 82–88.

115. Harrison, F. E., Bowman, G. L. and Polidori, M. C. 2014, April 24. Ascorbic acid and the brain: Rationale for the use against cognitive decline. *Nutrients*, Multidisciplinary Digital Publishing Institute (MDPI).

116. Morris, M. C., Beckett, L. A., Scherr, P. A., Hebert, L. E., Bennett, D. A., Field, T. S. and Evans, D. A. 1998. Vitamin E and vitamin C supplement use and risk of incident Alzheimer disease. *Alzheimer Dis. Assoc. Disord.* 12, 121–126.

117. Morris, M. C., Evans, D. A., Bienias, J. L., Tangney, C. C., Bennett, D. A., Aggarwal, N., Wilson, R. S. and Scherr, P. A. 2002. Dietary intake of antioxidant nutrients and the risk of incident Alzheimer disease in a biracial community study. *J. Am. Med. Assoc.* 287, 3230–3237.

118. Morrison, P., Harding-Lester, S. and Bradley, A. 2011. Uptake of Huntington disease predictive testing in a complete population. *Clin. Genet.* 80, 281–286.

119. Bates, G. P., Dorsey, R., Gusella, J. F., Hayden, M. R., Kay, C., Leavitt, B. R., Nance, M. et al. 2015. Huntington disease. *Nat. Rev. Dis. Prim.* 1, 15005.

120. Novak, M. J. U. and Tabrizi, S. J. 2011. Huntington's disease: Clinical presentation and treatment. *Int. Rev. Neurobiol.* 98, 297–323.

121. Frank, S. 2014. Treatment of Huntington's disease. *Neurotherapeutics* 11, 153–160.

122. Yero, T. and Rey, J. A. 2008. Tetrabenazine (Xenazine), an FDA-Approved treatment option for Huntington's disease—Related chorea. *P T* 33, 690–694.

123. Finkbeiner, S. 2011. Huntington's disease. *Cold Spring Harb. Perspect. Biol.* 3, 1–24.

124. Potter, N. T., Spector, E. B. and Prior, T. W. 2004. Technical standards and guidelines for Huntington disease testing. *Genet. Med.* 6, 61–65.

125. Zeitlin, S., Liu, J. P., Chapman, D. L., Papaioannou, V. E. and Efstratiadis, A. 1995. Increased apoptosis and early embryonic lethality in mice nullizygous for the Huntington's disease gene homologue. *Nat. Genet.* 11, 155–163.

126. Bano, D., Zanetti, F., Mende, Y. and Nicotera, P. 2011. Neurodegenerative processes in Huntington's disease. *Cell Death Dis.* 2, e228.

127. Clabough, E. B. D. 2013. Huntington's disease: The past, present, and future search for disease modifiers. *Yale J. Biol. Med.* 86, 217–233.

128. Guo, Q., Huang, B., Cheng, J., Seefelder, M., Engler, T., Pfeifer, G., Oeckl, P. et al. 2018. The cryo-electron microscopy structure of Huntingtin. *Nature*, Nature Publishing Group 555, 117–120.

129. Saudou, F. and Humbert, S. 2016. The biology of Huntingtin. *Neuron* 89, 910–926.

130. Schulte, J. and Littleton, J. T. 2011. The biological function of the Huntingtin protein and its relevance to Huntington's disease pathology. *Curr. Trends Neurol.* 5, 65–78.

131. Colin, E., Zala, D., Liot, G., Rangone, H., Borrell-Pagès, M., Li, X.-J., Saudou, F. and Humbert, S. 2008. Huntingtin phosphorylation acts as a molecular switch for anterograde/retrograde transport in neurons. *EMBO J.* 27, 2124–2134.

132. Gauthier, L. R., Charrin, B. C., Borrell-Pagès, M., Dompierre, J. P., Rangone, H., Cordelières, F. P., De Mey, J. et al. 2004. Huntingtin controls neurotrophic support and survival of neurons by enhancing BDNF vesicular transport along microtubules. *Cell* 118, 127–138.

133. Trushina, E., Dyer, R. B., Ii, J. D. B., Eide, L., Tran, D. D., Vrieze, B. T., Mcpherson, P. S., et al. 2004. Mutant Huntingtin impairs axonal trafficking in mammalian neurons in vivo and in vitro. *Mol. Cell. Biol.* 24, 8195–8209.

134. Luthi-Carter, R. and Cha, J. H. J. 2003. Transcriptional dysregulation in Huntington's disease. *Clin. Neurosci. Res.* 3, 165–177.

135. Zuccato, C., Belyaev, N., Conforti, P., Ooi, L., Tartari, M., Papadimou, E., MacDonald, M., et al. 2007. Widespread disruption of repressor element-1 silencing transcription factor/neuron-restrictive silencer factor occupancy at its target genes in Huntington's disease. *J. Neurosci.* 27, 6972–6983.

136. Zuccato, C., Ciammola, A., Rigamonti, D., Leavitt, B. R., Goffredo, D., Conti, L., MacDonald, M. E., Timmusk, T., Sipione, S. and Cattaneo, E. 2001. Loss of Huntingtin-mediated BDNF gene transcription in Huntington's disease 293, 493–498.

137. Zuccato, C., Tartari, M., Crotti, A., Goffredo, D., Valenza, M., Conti, L., Cataudella, T. et al. 2003. Huntingtin interacts with REST/NRSF to modulate the transcription of NRSE-controlled neuronal genes. *Nat. Genet.* 35, 76–83.

138. Gelman, A., Rawet-Slobodkin, M. and Elazar, Z. 2015. Huntingtin facilitates selective autophagy. *Nat. Cell Biol.*, Nature Publishing Group 17, 214–215.

139. Wong, Y. C. and Holzbaur, E. L. F. 2014. The regulation of autophagosome dynamics by Huntingtin and HAP1 is disrupted by expression of mutant Huntingtin, leading to defective cargo degradation. *J. Neurosci.* 34, 1293–1305.

140. Zheng, S., Clabough, E. B. D., Sarkar, S., Futter, M., Rubinsztein, D. C. and Zeitlin, S. O. 2010. Deletion of the Huntingtin polyglutamine stretch enhances neuronal autophagy and longevity in mice. *PLoS Genet.* 6.

141. Rigamonti, D., Bauer, J. H., De-Fraja, C., Conti, L., Sipione, S., Sciorati, C., Clementi, E. et al. 2000. Wild-type Huntingtin protects from apoptosis upstream of caspase-3. *J. Neurosci.* 20, 3705–3713.

142. Schaefer, M. H., Wanker, E. E. and Andrade-Navarro, M. A. 2012. Evolution and function of CAG/polyglutamine repeats in protein-protein interaction networks. *Nucleic Acids Res.* 40, 4273–4287.

143. Totzeck, F., Andrade-Navarro, M. A. and Mier, P. 2017. The protein structure context of polyQ regions. *PLOS ONE* 12, 2–11.

144. Bao, J., Sharp, A. H., Wagster, M. V., Becher, M., Schilling, G., Ross, C. A., Dawson, V. L. and Dawson, T. M. 1996. Expansion of polyglutamine repeat in Huntingtin leads to abnormal protein interactions involving calmodulin. *Proc. Natl. Acad. Sci. USA* 93, 5037–5042.

145. Benn, C. L., Sun, T., Sadri-Vakili, G., McFarland, K. N., DiRocco, D. P., Yohrling, G. J., Clark, T. W., Bouzou, B. and Cha, J.-H. J. 2008. Huntingtin modulates transcription, occupies gene promoters in vivo, and binds directly to DNA in a polyglutamine-dependent manner. *J. Neurosci.* 28, 10720–10733.

146. Cattaneo, E., Zuccato, C. and Tartari, M. 2005. Normal Huntingtin function: An alternative

approach to Huntington's disease. *Nat. Rev. Neurosci.* 6, 919–930.

147. Schaffar, G., Breuer, P., Boteva, R., Behrends, C., Tzvetkov, N., Strippel, N., Sakahira, H., Siegers, K., Hayer-Hartl, M. and Hartl, F. U. 2004. Cellular toxicity of polyglutamine expansion proteins: Mechanism of transcription factor deactivation. *Mol. Cell, Cell Press* 15, 95–105.

148. Arrasate, M. and Finkbeiner, S. 2012. Protein aggregates in Huntington's disease. *Exp. Neurol.*, Elsevier Inc. 238, 1–11.

149. Van Raamsdonk, J. M., Pearson, J., Murphy, Z., Hayden, M. R. and Leavitt, B. R. 2006. Wild-type Huntingtin ameliorates striatal neuronal atrophy but does not prevent other abnormalities in the YAC128 mouse model of Huntington disease. *BMC Neurosci.* 7, 1–9.

150. Adegbuyiro, A., Sedighi, F., Pilkington, A. W., Groover, S. and Legleiter, J. 2017. Proteins containing expanded polyglutamine tracts and neurodegenerative disease. *Biochemistry* 56, 1199–1217.

151. Nopoulos, P. C. 2016. Huntington disease: A single-gene degenerative disorder of the striatum. *Dialogues Clin. Neurosci.* 18, 91.

152. Koyuncu, S., Fatima, A., Gutierrez-Garcia, R. and Vilchez, D. 2017. Proteostasis of Huntingtin in health and disease. *Int. J. Mol. Sci.* 18, 11–13.

153. Rikani, A. A., Choudhry, Z., Choudhry, A. M., Rizvi, N., Ikram, H., Mobassarah, N. J. and Tulli, S. 2014. The mechanism of degeneration of striatal neuronal subtypes in Huntington disease. *Ann. Neurosci.* 21, 112–114.

154. Browne, S. E. and Beal, M. F. 2006. Oxidative damage in Huntington's disease pathogenesis. *Antioxid. Redox Signal.* 8, 2061–2073.

155. Liu, Z., Ren, Z., Zhang, J., Chuang, C. C., Kandaswamy, E., Zhou, T. and Zuo, L. 2018. Role of ROS and nutritional antioxidants in human diseases. *Front. Physiol.* 9, 1–14.

156. Browne, S. E., Ferrante, R. J. and Beal, M. F. 1999. Oxidative stress in Huntington's disease. *Brain Pathol.* 9, 147–163.

157. Kumar, A. and Ratan, R. R. 2016. Oxidative stress and Huntington's disease: The good, The bad, and the ugly. *J Huntingtons Dis* 5, 217–237.

158. Browne, S. E. and Beal, M. F. 2004. The energetics of Huntington's disease. *Neurochem. Res.* 29, 531–46.

159. Lee, J., Kozaras, B., Del Signore, S. J., Cormier, K., McKee, A., Ratan, R. R., Kowall, N. W. and Ryu, H. 2011. Modulation of lipid peroxidation and mitochondrial function by nordihydroguaiaretic acid (NDGA) improves neuropathology in Huntington's disease mice. *Acta Neuropathol.* 121, 487–498.

160. Polidori, M. C., Mecocci, P., Browne, S. E., Senin, U. and Beal, M. F. 1999. Oxidative damage to mitochondrial DNA in Huntington's disease parietal cortex. *Neurosci. Lett.* 272, 53–56.

161. Sorolla, M. A., Reverter-Branchat, G., Tamarit, J., Ferrer, I., Ros, J. and Cabiscol, E. 2008. Proteomic and oxidative stress analysis in human brain samples of Huntington disease. *Free Radic. Biol. Med.* 45, 667–678.

162. Alam, Z. I., Halliwell, B. and Jenner, P. 2000. No evidence for increased oxidative damage to lipids, proteins, or DNA in Huntington's disease. *J. Neurochem.* 75, 840–846.

163. Eckman, J., Dixit, S., Nackenoff, A., Schrag, M. and Harrison, F. E. 2018. Oxidative stress levels in the brain are determined by post-mortem interval and ante-mortem vitamin C state but not Alzheimer's disease status. *Nutrients* 10.

164. Lewis, D. A. 2002. The human brain revisited: Opportunities and challenges in postmortem studies of psychiatric disorders. *Neuropsychopharmacology* 26, 143–154.

165. Schut, M. H., Patassini, S., Kim, E. H., Bullock, J., Waldvogel, H. J., Faull, R. L. M., Pepers, B. A., Den Dunnen, J. T., Van Ommen, G. J. B. and Van Roon-Mom, W. M. C. 2017. Effect of post-mortem delay on N-terminal Huntingtin protein fragments in human control and Huntington disease brain lysates. *PLOS ONE* 12, 1–13.

166. Chen, C.-M., Wu, Y.-R., Cheng, M.-L., Liu, J.-L., Lee, Y.-M., Lee, P.-W., Soong, B.-W. and Chiu, D. T.-Y. 2007. Increased oxidative damage and mitochondrial abnormalities in the peripheral blood of Huntington's disease patients. *Biochem. Biophys. Res. Commun.* 359, 335–340.

167. Long, J. D., Matson, W. R., Juhl, A. R., Leavitt, B. R. and Paulsen, J. S. 2012. 8OHdG as a marker for Huntington disease progression. *Neurobiol. Dis.*, Elsevier Inc. 46, 625–634.

168. Sánchez-López, F., Tasset, I., Agüera, E., Feijóo, M., Fernández-Bolaños, R., Sánchez, F. M., Ruiz, M. C., Cruz, A. H., Gascón, F. and Túnez, I. 2012. Oxidative stress and inflammation biomarkers in the blood of patients with Huntington's disease. *Neurol. Res.* 34, 721–724.

169. Túnez, I., Sánchez-López, F., Agüera, E., Fernández-Bolaños, R., Sánchez, F. M. and Tasset-Cuevas, I. 2011. Important role of

oxidative stress biomarkers in Huntington's disease. *J. Med. Chem.* 54, 5602–5606.

170. Ferrante, R. J. 2009. Mouse models of Huntington's disease and methodological considerations for therapeutic trials. *Biochim. Biophys. Acta - Mol. Basis Dis.*, Elsevier B.V. 1792, 506–520.

171. Bogdanov, M. B., Andreassen, O. A., Dedeoglu, A., Ferrante, R. J. and Beal, M. F. 2001. Increased oxidative damage to DNA in a transgenic mouse model of Huntington's disease. *J. Neurochem.* 79, 1246–1249.

172. Perluigi, M., Poon, H. F., Maragos, W., Pierce, W. M., Klein, J. B., Calabrese, V., Cini, C., De Marco, C. and Butterfield, D. A. 2005. Proteomic analysis of protein expression and oxidative modification in R6/2 transgenic mice. *Mol. Cell. Proteomics* 4, 1849–1861.

173. Johri, A., Calingasan, N. Y., Hennessey, T. M., Sharma, A., Yang, L., Wille, E., Chandra, A. and Beal, M. F. 2012. Pharmacologic activation of mitochondrial biogenesis exerts widespread beneficial effects in a transgenic mouse model of Huntington's disease. *Hum. Mol. Genet.* 21, 1124–1137.

174. Brocardo, P. S., McGinnis, E., Christie, B. R. and Gil-Mohapel, J. 2016. Time-course analysis of protein and lipid oxidation in the brains of Yac128 Huntington's disease transgenic mice. *Rejuvenation Res.* 19, 140–148.

175. Hong, C., Seo, H., Kwak, M., Jeon, J., Jang, J., Jeong, E. M., Myeong, J. et al. 2015. Increased TRPC5 glutathionylation contributes to striatal neuron loss in Huntington's disease. *Brain* 138, 3030–3047.

176. Machiela, E., Dues, D. J., Senchuk, M. M. and Van Raamsdonk, J. M. 2016. Oxidative stress is increased in *C. elegans* models of Huntington's disease but does not contribute to polyglutamine toxicity phenotypes. *Neurobiol. Dis.*, Elsevier Inc. 96, 1–11.

177. Ribeiro, M., Rosenstock, T. R., Cunha-Oliveira, T., Ferreira, I. L., Oliveira, C. R. and Rego, A. C. 2012. Glutathione redox cycle dysregulation in Huntington's disease knock-in striatal cells. *Free Radic. Biol. Med.* 53, 1857–1867.

178. Basse-Tomusk, A. and Rebec, G. V. 1991. Regional distribution of ascorbate and 3,4-dihydroxyphenylacetic acid (DOPAC) in rat striatum. *Brain Res.* 538, 29–35.

179. Rebec, G. V. and Wang, Z. 2001. Behavioral activation in rats requires endogenous ascorbate release in striatum. *J Neurosci.* 21, 668–675.

180. Rebec, G. V., Barton, S. J., Marseilles, A. M. and Collins, K. 2003. Ascorbate treatment attenuates the Huntington behavioral phenotype in mice. *Neuroreport* 14, 1263–1265.

181. Ellrichmann, G., Petrasch-Parwez, E., Lee, D. H., Reick, C., Arning, L., Saft, C., Gold, R. and Linker, R. A. 2011. Efficacy of fumaric acid esters in the R6/2 and YAC128 models of huntington's disease. *PLOS ONE* 6, 1–11.

182. Jin, Y. N., Yu, Y. V., Gundemir, S., Jo, C., Cui, M., Tieu, K. and Johnson, G. V. W. 2013. Impaired mitochondrial dynamics and Nrf2 signaling contribute to compromised responses to oxidative stress in striatal cells expressing full-length mutant Huntingtin. *PLOS ONE* 8.

183. Andreassen, O. A., Ferrante, R. J., Dedeoglu, A. and Beal, M. F. 2001. Lipoic acid improves survival in transgenic mouse models of Huntington's disease. *Neuroreport* 12, 3371–3373.

184. Packer, L., Witt, E. H. and Tritschler, H. J. 1995. Alpha-lipoic acid as a biological antioxidant. *Free Radic. Biol. Med.* 19, 227–250.

185. Rebec, G. V., Barton, S. J. and Ennis, M. D. 2002. Dysregulation of ascorbate release in the striatum of behaving mice expressing the Huntington's disease gene. *J. Neurosci.* 22, U39–U43.

186. Estrada-Sánchez, A. M., Montiel, T., Segovia, J. and Massieu, L. 2009. Glutamate toxicity in the striatum of the R6/2 Huntington's disease transgenic mice is age-dependent and correlates with decreased levels of glutamate transporters. *Neurobiol. Dis.*, Elsevier Inc. 34, 78–86.

187. Behrens, P. F., Franz, P., Woodman, B., Lindenberg, K. S. and Landwehrmeyer, G. B. 2002. Impaired glutamate transport and glutamate-glutamine cycling: Downstream effects of the Huntington mutation. *Brain* 125, 1908–1922.

188. Cross, A. J., Slater, P. and Reynolds, G. P. 1986. Reduced high-affinity glutamate uptake sites in the brains of patients with Huntington's disease. *Neurosci. Lett.* 67, 198–202.

189. Dorner, J. L., Miller, B. R., Barton, S. J., Brock, T. J. and Rebec, G. V. 2007. Sex differences in behavior and striatal ascorbate release in the 140 CAG knock-in mouse model of Huntington's disease. *Behav. Brain Res.* 178, 90–97.

190. Miller, B. R., Dorner, J. L., Shou, M., Sari, Y., Barton, S. J., Sengelaub, D. R., Kennedy, R. T. and Rebec, G. V. 2008. Up-regulation of GLT1 expression increases glutamate uptake and attenuates the Huntington's disease phenotype in the R6/2 mouse. *Neuroscience* 153, 329–337.

191. Sari, Y., Prieto, A. L., Barton, S. J., Miller, B. R. and Rebec, G. V. 2010. Ceftriaxone-induced up-regulation of cortical and striatal GLT1 in the R6/2 model of Huntington's disease. *J. Biomed. Sci.* 17, 1–5.

192. Petr, G. T., Schultheis, L. A., Hussey, K. C., Sun, Y., Dubinsky, J. M., Aoki, C. and Rosenberg, P. A. 2013. Decreased expression of GLT-1 in the R6/2 model of Huntington's disease does not worsen disease progression Geraldine. *Eur. J. Neurosci.* 38, 2477–2490.

193. Parsons, M. P., Vanni, M. P., Woodard, C. L., Kang, R., Murphy, T. H. and Raymond, L. A. 2016. Real-time imaging of glutamate clearance reveals normal striatal uptake in Huntington disease mouse models. *Nat. Commun.*, Nature Publishing Group 7, 1–12.

194. Roussakis, A. A. and Piccini, P. 2015. PET imaging in Huntington's disease. *J. Huntingtons. Dis.* 4, 287–296.

195. Brotherton, A., Campos, L., Rowell, A., Zoia, V., Simpson, S. A. and Rae, D. 2012. Nutritional management of individuals with Huntington's disease: Nutritional guidelines. *Neurodegener. Dis. Manag.* 2, 33–43.

196. Rivadeneyra, J., Cubo, E., Gil, C., Calvo, S., Mariscal, N. and Martínez, A. 2016. Factors associated with Mediterranean diet adherence in Huntington's disease. *Clin. Nutr. ESPEN* 12, e7–e13.

197. Hernández-Ruiz, A., García-Villanova, B., Guerra-Hernández, E., Amiano, P., Sánchez, M. J., Dorronsoro, M. and Molina-Montes, E. 2018. Comparison of the dietary antioxidant profiles of 21 a priori defined Mediterranean diet indexes. *J. Acad. Nutr. Diet.* 118, 2254–2268.e8.

198. Simopoulos, A. P. 2001. The mediterranean diets: What is so special about the diet of greece? The scientific evidence. *J. Nutr.* 131, 3065S–3073S.

199. Peyser, C. E., Folstein, M., Chase, G. A., Starkstein, S., Brandt, J., Cockrell, J. R., Bylsma, F., Coyle, J. T., McHugh, P. R. and Folstein, S. E. 1995. Trial of d-alpha-tocopherol in Huntington's Disease. *Am J Psychiatry* 152, 1771–1775.

200. Sveinbjornsdottir, S. 2016. The clinical symptoms of Parkinson's disease. *J. Neurochem.* 139, 318–324.

201. Jankovic, J. 2008. Parkinson's disease: Clinical features and diagnosis. *J. Neurol. Neurosurg. Psychiatry* 79, 368–376.

202. Sutachan, J. J., Casas, Z., Albarracin, S. L., Stab, B. R., Samudio, I., Gonzalez, J., Morales, L. and Barreto, G. E. 2012. Cellular and molecular mechanisms of antioxidants in Parkinson's disease. *Nutr. Neurosci.* 15, 120–126.

203. Minor, E. A., Court, B. L., Young, J. I. and Wang, G. 2013. Ascorbate induces ten-eleven translocation (Tet) methylcytosine dioxygenase-mediated generation of 5-hydroxymethylcytosine. *J. Biol. Chem.* 288, 13669–13674.

204. Villar-Piqué, A., Lopes da Fonseca, T. and Outeiro, T. F. 2016. Structure, function and toxicity of alpha-synuclein: The Bermuda triangle in synucleinopathies. *J. Neurochem.* 139, 240–255.

205. Spillantini, M. G., Schmidt, M. L., Lee, V. M., Trojanowski, J. Q., Jakes, R. and Goedert, M. 1997. Alpha-synuclein in Lewy bodies. *Nature* 388, 839–840.

206. Smith, W. W., Margolis, R. L., Li, X., Troncoso, J. C., Lee, M. K., Dawson, V. L., Dawson, T. M., Iwatsubo, T. and Ross, C. A. 2005. Alpha-synuclein phosphorylation enhances eosinophilic cytoplasmic inclusion formation in SH-SY5Y cells. *J. Neurosci.* 25, 5544–5552.

207. Reeve, A., Simcox, E. and Turnbull, D. 2014. Ageing and Parkinson's disease: Why is advancing age the biggest risk factor? *Ageing Res. Rev.*, Elsevier B.V. 14, 19–30.

208. Burbulla, L. F., Song, P., Mazzulli, J. R., Zampese, E., Wong, Y. C., Jeon, S., Santos, D. P. et al. 2017. Dopamine oxidation mediates mitochondrial and lysosomal dysfunction in Parkinson's disease. *Science* 357, 1255–1261.

209. Hastings, T. G., Lewis, D. A. and Zigmond, M. J. 1996. Role of oxidation in the neurotoxic effects of intrastriatal dopamine injections. *Proc. Natl. Acad. Sci. USA* 93, 1956–1961.

210. Rabinovic, A. D., Lewis, D. A. and Hastings, T. G. 2000. Role of oxidative changes in the degeneration of dopamine terminals after injection of neurotoxic levels of dopamine. *Neuroscience* 101, 67–76.

211. Caudle, W. M., Richardson, J. R., Wang, M. Z., Taylor, T. N., Guillot, T. S., McCormack, A. L., Colebrooke, R. E., Di Monte, D. A., Emson, P. C. and Miller, G. W. 2007. Reduced vesicular storage of dopamine causes progressive nigrostriatal neurodegeneration. *J. Neurosci.* 27, 8138–8148.

212. Taylor, T. N., Caudle, W. M., Shepherd, K. R., Noorian, A., Jackson, C. R., Iuvone, P. M., Weinshenker, D., Greene, J. G. and Miller, G. W. 2009. Nonmotor symptoms of Parkinson's disease revealed in an animal model with

reduced monoamine storage capacity. *J. Neurosci.* 29, 8103–8113.

213. Taylor, T. N., Caudle, W. M. and Miller, G. W. 2011. VMAT2-Deficient mice display nigral and extranigral pathology and motor and nonmotor symptoms of Parkinson's disease. *Parkinsons. Dis.* 2011, 124165.

214. Yin, R., Mao, S. Q., Zhao, B., Chong, Z., Yang, Y., Zhao, C., Zhang, D. et al. 2013. Ascorbic acid enhances TET-mediated 5-methylcytosine oxidation and promotes DNA demethylation in mammals. *J. Am. Chem. Soc.* 135, 10396–10403.

215. Besler, H. T., Comoğlu, S. and Okçu, Z. 2002. Serum levels of antioxidant vitamins and lipid peroxidation in multiple sclerosis. *Nutr. Neurosci.* 5, 215–220.

216. Wulansari, N., Kim, E. H., Sulistio, Y. A., Rhee, Y. H., Song, J. J. and Lee, S. H. 2017. Vitamin C-induced epigenetic modifications in donor NSCs establish midbrain marker expressions critical for cell-based therapy in Parkinson's disease. *Stem Cell Reps.,* Elsevier. 9, 1192–1206.

217. Heikkila, R. E., Hess, A. and Duvoisin, R. C. 1984. Dopaminergic neurotoxicity of 1-methyl-4-phenyl-1,2,5,6-tetrahydropyridine in mice. *Science* 224, 1451–1453.

218. Langston, J. W., Ballard, P., Tetrud, J. W. and Irwin, I. 1983. Chronic parkinsonism in humans due to a product of meperidine-analog synthesis. *Science* 219, 979–980.

219. Schapira, A. H., Cooper, J. M., Dexter, D., Jenner, P., Clark, J. B. and Marsden, C. D. 1989. Mitochondrial complex I deficiency in Parkinson's disease. *Lancet (London, England)* 1, 1269.

220. Schapira, A. H. V. 2007. Mitochondrial dysfunction in Parkinson's disease. *Cell Death Differ.* 14, 1261–1266.

221. Keeney, P. M. 2006. Parkinson's disease brain mitochondrial complex i has oxidatively damaged subunits and is functionally impaired and misassembled. *J. Neurosci.* 26, 5256–5264.

222. Thomas, B. and Beal, M. F. 2010. Mitochondrial therapies for Parkinson's disease. *Mov. Disord.* 25(Suppl 1), S155–S160.

223. Dauer, W. and Przedborski, S. 2003. Parkinson's disease: Mechanisms and models. *Neuron* 39, 889–909.

224. Klivenyi, P., Siwek, D., Gardian, G., Yang, L., Starkov, A., Cleren, C., Ferrante, R. J., Kowall, N. W., Abeliovich, A. and Beal, M. F. 2006. Mice lacking alpha-synuclein are resistant to mitochondrial toxins. *Neurobiol. Dis.* 21, 541–548.

225. Murphy, M. P. 2009. How mitochondria produce reactive oxygen species. *Biochem. J.* 417, 1–13.

226. Perier, C., Bové, J., Dehay, B., Jackson-Lewis, V., Rabinovitch, P. S., Przedborski, S. and Vila, M. 2010. Apoptosis-inducing factor deficiency sensitizes dopaminergic neurons to parkinsonian neurotoxins. *Ann. Neurol.* 68, 184–192.

227. Izumi, Y., Ezumi, M., Takada-Takatori, Y., Akaike, A. and Kume, T. 2014. Endogenous dopamine is involved in the herbicide paraquat-induced dopaminergic cell death. *Toxicol. Sci.* 139, 466–478.

228. Langston, J. W. and Ballard, P. 1984. Parkinsonism induced by 1-methyl-4-phenyl-1,2,3,6-tetrahydropyridine (MPTP): Implications for treatment and the pathogenesis of Parkinson's disease. *Can. J. Neurol. Sci.* 11, 160–165.

229. Betarbet, R., Canet-Aviles, R. M., Sherer, T. B., Mastroberardino, P. G., McLendon, C., Kim, J.-H., Lund, S. et al. 2006. Intersecting pathways to neurodegeneration in Parkinson's disease: Effects of the pesticide rotenone on DJ-1, alpha-synuclein, and the ubiquitin-proteasome system. *Neurobiol. Dis.* 22, 404–420.

230. Drolet, R. E., Cannon, J. R., Montero, L. and Greenamyre, J. T. 2009. Chronic rotenone exposure reproduces Parkinson's disease gastrointestinal neuropathology. *Neurobiol. Dis.* 36, 96–102.

231. Cannon, J. R., Tapias, V., Na, H. M., Honick, A. S., Drolet, R. E. and Greenamyre, J. T. 2009. A highly reproducible rotenone model of Parkinson's disease. *Neurobiol. Dis.* 34, 279–290.

232. Ehrhart, J. and Zeevalk, G. D. 2003. Cooperative interaction between ascorbate and glutathione during mitochondrial impairment in mesencephalic cultures. *J. Neurochem.* 86, 1487–1497.

233. Ambrosi, G., Cerri, S. and Blandini, F. 2014. A further update on the role of excitotoxicity in the pathogenesis of Parkinson's disease. *J. Neural Transm.* 121, 849–859.

234. Hüls, S., Högen, T., Vassallo, N., Danzer, K. M., Hengerer, B., Giese, A. and Herms, J. 2011. AMPA-receptor-mediated excitatory synaptic transmission is enhanced by iron-induced α-synuclein oligomers. *J. Neurochem.* 117, 868–878.

235. Marques, O. and Outeiro, T. F. 2012. Alpha-synuclein: From secretion to dysfunction and death. *Cell Death Dis.* 3, e350.

236. Lundblad, M., Decressac, M., Mattsson, B. and Björklund, A. 2012. Impaired neurotransmission caused by overexpression of α-synuclein in nigral dopamine neurons. *Proc. Natl. Acad. Sci. USA* 109, 3213–3219.

237. McNaught, K. S. and Jenner, P. 2000. Extracellular accumulation of nitric oxide, hydrogen peroxide, and glutamate in astrocytic cultures following glutathione depletion, complex I inhibition, and/or lipopolysaccharide-induced activation. *Biochem. Pharmacol.* 60, 979–988.

238. Fahn, S. 1992. A pilot trial of high-dose alpha-tocopherol and ascorbate in early Parkinson's disease. *Ann. Neurol.* 32 Suppl, S128–S132.

239. Yapa, S. C. 1992. Detection of subclinical ascorbate deficiency in early Parkinson's disease. *Public Health* 106, 393–395.

240. Fernandez-Calle, P., Jimenez-Jimenez, F. J., Molina, J. A., Cabrera-Valdivia, F., Vazquez, A., Garcia Urra, D., Bermejo, F., Cruz Matallana, M. and Codoceo, R. 1993. Serum levels of ascorbic acid (vitamin C) in patients with Parkinson's disease. *J. Neurol. Sci.* 118, 25–28.

241. Paraskevas, G. P., Kapaki, E., Petropoulou, O., Anagnostouli, M., Vagenas, V. and Papageorgiou, C. 2003. Plasma levels of antioxidant vitamins C and E are decreased in vascular parkinsonism. *J. Neurol. Sci.* 215, 51–55.

242. Iwasaki, Y., Igarashi, O., Ichikawa, Y., Kawabe, K. and Ikeda, K. 2004. Vitamins A, C and E in vascular parkinsonism. *J. Neurol. Sci.* 227, 149; author reply 151–2.

243. Ide, K., Yamada, H., Umegaki, K., Mizuno, K., Kawakami, N., Hagiwara, Y., Matsumoto, M. et al. 2015. Lymphocyte vitamin C levels as potential biomarker for progression of Parkinson's disease. *Nutrition,* Elsevier Inc. 31, 406–408.

244. Zhang, S. M., Hernán, M. A., Chen, H., Spiegelman, D., Willett, W. C. and Ascherio, A. 2002. Intakes of vitamins E and C, carotenoids, vitamin supplements, and PD risk. *Neurology* 59, 1161–1169.

245. Hughes, K. C., Gao, X., Kim, I. Y., Rimm, E. B., Wang, M., Weisskopf, M. G., Schwarzschild, M. A. and Ascherio, A. 2016. Intake of antioxidant vitamins and risk of Parkinson's disease. *Mov. Disord.* 31, 1909–1914.

246. Yang, F., Wolk, A., Håkansson, N., Pedersen, N. L. and Wirdefeldt, K. 2017. Dietary antioxidants and risk of Parkinson's disease in two population-based cohorts. *Mov. Disord.* 32, 1631–1636.

247. Sacks, W. and Simpson, G. M. 1975. Letter: Ascorbic acid in levodopa therapy. *Lancet (London, England)* 1, 527.

248. Nagayama, H., Hamamoto, M., Ueda, M., Nito, C., Yamaguchi, H. and Katayama, Y. 2004. The effect of ascorbic acid on the pharmacokinetics of levodopa in elderly patients with Parkinson disease. *Clin. Neuropharmacol.* 27, 270–273.

249. Khan, S., Jyoti, S., Naz, F., Shakya, B., Rahul Afzal, M. and Siddique, Y. H. 2012. Effect of L-Ascorbic acid on the climbing ability and protein levels in the brain of *drosophila* model of Parkinson's disease. *Int. J. Neurosci.* 122, 704–709.

250. Pardo, B., Mena, M. A., Fahn, S. and de Yébenes, J. G. 1993. Ascorbic acid protects against levodopa-induced neurotoxicity on a catecholamine-rich human neuroblastoma cell line. *Mov. Disord.* 8, 278–284.

251. Daff, M., Stevenson, A. and Nance, M. 2017. Vitamin C deficiency in psychiatric illness. *Asian J. Psychiatr.* 28, 97.

252. Peters, O. M., Ghasemi, M. and Brown, R. H. 2015. Emerging mechanisms of molecular pathology in ALS. *J. Clin. Invest.* 125, 1767–1779.

253. Feitosa, C. M., da Silva Oliveira, G. L., do Nascimento Cavalcante, A., Morais Chaves, S. K. and Rai, M. 2018. Determination of parameters of oxidative stress in vitro models of Neurodegenerative Diseases-A review. *Curr. Clin. Pharmacol.* 13, 100–109.

254. Jackson, T. S., Xu, A., Vita, J. A. and Keaney, J. F. 1998. Ascorbate prevents the interaction of superoxide and nitric oxide only at very high physiological concentrations. *Circ. Res.* 83, 916–922.

255. Vande Velde, C., McDonald, K. K., Boukhedimi, Y., McAlonis-Downes, M., Lobsiger, C. S., Bel Hadj, S., Zandona, A., Julien, J.-P., Shah, S. B. and Cleveland, D. W. 2011. Misfolded SOD1 associated with motor neuron mitochondria alters mitochondrial shape and distribution prior to clinical onset. *PLOS ONE* 6, e22031.

256. Miquel, E., Cassina, A., Martínez-Palma, L., Souza, J. M., Bolatto, C., Rodríguez-Bottero, S., Logan, A. et al. 2014. Neuroprotective effects of the mitochondria-targeted antioxidant MitoQ in a model of inherited amyotrophic lateral sclerosis. *Free Radic. Biol. Med.,* Elsevier 70, 204–213.

257. Nagano, S., Fujii, Y., Yamamoto, T., Taniyama, M., Fukada, K., Yanagihara, T. and Sakoda, S. 2003. The efficacy of trientine or ascorbate alone

compared to that of the combined treatment with these two agents in familial amyotrophic lateral sclerosis model mice. *Exp. Neurol.* 179, 176–180.

258. Okamoto, K., Kihira, T., Kobashi, G., Washio, M., Sasaki, S., Yokoyama, T., Miyake, Y., Sakamoto, N., Inaba, Y. and Nagai, M. 2009. Fruit and vegetable intake and risk of amyotrophic lateral sclerosis in Japan. *Neuroepidemiology* 32, 251–256.

259. Fitzgerald, K. C., O'Reilly, É. J., Fondell, E., Falcone, G. J., McCullough, M. L., Park, Y., Kolonel, L. N. and Ascherio, A. 2013. Intakes of vitamin C and carotenoids and risk of amyotrophic lateral sclerosis: Pooled results from 5 cohort studies. *Ann. Neurol.* 73, 236–245.

260. Blasco, H., Corcia, P., Moreau, C., Veau, S., Fournier, C., Vourc'h, P., Emond, P. et al. 2010. 1H-NMR-based metabolomic profiling of CSF in early amyotrophic lateral sclerosis. *PLOS ONE* 5, e13223.

261. Spasojević, I., Stević, Z., Nikolić-Kokić, A., Jones, D. R., Blagojević, D. and Spasić, M. B. 2010. Different roles of radical scavengers—Ascorbate and urate in the cerebrospinal fluid of amyotrophic lateral sclerosis patients. *Redox Rep.* 15, 81–86.

262. Dobson, R. and Giovannoni, G. 2019. Multiple sclerosis—A review. *Eur. J. Neurol.* 26, 27–40.

263. Browning, V., Joseph, M. and Sedrak, M. 2012. Multiple sclerosis: A comprehensive review for the physician assistant. *J. Am. Acad. PAs* 25, 24–29.

264. Blaschke, K., Ebata, K. T., Karimi, M. M., Zepeda-Martínez, J. A., Goyal, P., Mahapatra, S., Tam, A. et al. 2013. Vitamin C induces Tet-dependent DNA demethylation and a blastocyst-like state in ES cells. *Nature* 500, 222–226.

265. Chen, J., Guo, L., Zhang, L., Wu, H., Yang, J., Liu, H., Wang, X. et al. 2013. Vitamin C modulates TET1 function during somatic cell reprogramming. *Nat. Genet.*, Nature Publishing Group 45, 1504–1509.

266. Carr, A. C. and Maggini, S. 2017. Vitamin C and immune function. *Nutrients* 9.

267. Manning, J., Mitchell, B., Appadurai, D. A., Shakya, A., Pierce, L. J., Wang, H., Nganga, V. et al. 2013. Vitamin C promotes maturation of T-cells. *Antioxid. Redox Signal.* 19, 2054–2067.

268. Sasidharan Nair, V., Song, M. H. and Oh, K. I. 2016. Vitamin C facilitates demethylation of the Foxp3 enhancer in a Tet-dependent manner. *J. Immunol.* 196, 2119–2131.

269. Nikolouli, E., Hardtke-Wolenski, M., Hapke, M., Beckstette, M., Geffers, R., Floess, S., Jaeckel, E. and Huehn, J. 2017. Alloantigen-induced regulatory T Cells generated in presence of vitamin C display enhanced stability of Foxp3 expression and promote skin allograft acceptance. *Front. Immunol.* 8, 748.

270. Portugal, C. C., Socodato, R., Canedo, T., Silva, C. M., Martins, T., Coreixas, V. S. M., Loiola, E. C., et al. 2017. Caveolin-1-mediated internalization of the vitamin C transporter SVCT2 in microglia triggers an inflammatory phenotype. *Sci. Signal.* 10.

271. Ontaneda, D., Thompson, A. J., Fox, R. J. and Cohen, J. A. 2017. Progressive multiple sclerosis: Prospects for disease therapy, repair, and restoration of function. *Lancet (London, England)* 389, 1357–1366.

272. Windrem, M. S., Schanz, S. J., Guo, M., Tian, G.-F., Washco, V., Stanwood, N., Rasband, M. et al. 2008. Neonatal chimerization with human glial progenitor cells can both remyelinate and rescue the otherwise lethally hypomyelinated shiverer mouse. *Cell Stem Cell* 2, 553–565.

273. Chang, A., Tourtellotte, W. W., Rudick, R. and Trapp, B. D. 2002. Premyelinating oligodendrocytes in chronic lesions of multiple sclerosis. *N. Engl. J. Med.* 346, 165–173.

274. Eldridge, C. F., Bunge, M. B., Bunge, R. P. and Wood, P. M. 1987. Differentiation of axon-related Schwann cells in vitro. I. Ascorbic acid regulates basal lamina assembly and myelin formation. *J. Cell Biol.* 105, 1023–1034.

275. Guo, Y.-E., Suo, N., Cui, X., Yuan, Q. and Xie, X. 2018. Vitamin C promotes oligodendrocytes generation and remyelination. *Glia* 66, 1302–1316.

276. Ghadirian, P., Jain, M., Ducic, S., Shatenstein, B. and Morisset, R. 1998. Nutritional factors in the aetiology of multiple sclerosis: A case-control study in Montreal, Canada. *Int. J. Epidemiol.* 27, 845–852.

277. Zhang, S. M., Hernán, M. A., Olek, M. J., Spiegelman, D., Willett, W. C. and Ascherio, A. 2001. Intakes of carotenoids, vitamin C, and vitamin E and MS risk among two large cohorts of women. *Neurology* 57, 75–80.

CHAPTER ELEVEN

Vitamin C and the Brain

L. John Hoffer

DOI: 10.1201/9780429442025-11

CONTENTS

INTRODUCTION

Vitamin C plays essential roles in brain metabolism and neurotransmission, and it protects neural tissue from the adverse effects of oxidative stress. This review analyzes the clinical evidence that vitamin C deficiency affects mental function, explains the role of vitamin C deficiency and vitamin C therapy in diseases of the brain, including mental illness, and considers the possibility that, when administered in pharmacologic doses, vitamin C improves mood and cognitive function. Recommendations are offered for rational clinical practice and future clinical research.

VITAMIN C: BRAIN PHYSIOLOGY AND HOMEOSTASIS

Physiology

The brain requires vitamin C to metabolize substrates and synthesize neurotransmitters, regulate their release, and modify their actions [1–12]. Vitamin C also protects the brain from oxidative damage [2–4,6,9,11–20]. In rodent models, severe vitamin C deficiency impairs monoamine synthesis and neurotransmission and impairs memory, cognition, and behavior [21–23], but does not seem to induce anxiety [21]. Moderate vitamin C deficiency diminishes muscular strength and impairs agility without affecting memory or cognition [23].

Homeostasis

Any analysis of the clinical consequences of vitamin C deficiency and vitamin C therapy must start by defining normal and deficient vitamin C status. Hypovitaminosis C is defined as a plasma vitamin C concentration <28.4 μmol/L; marginal vitamin C deficiency as a plasma vitamin C concentration <28.4 μmol/L but >11.4 μmol/L, and definite biochemical deficiency as a concentration <11.4 μmol/L. Hypovitaminosis C occurs in 10% of the general populations of healthy societies, in 30% of cigarette smokers, and in 60% of acutely ill hospitalized patients [24–27]. The disease of terminal vitamin C deficiency, scurvy, is uncommon but not rare in socially equitable and economically stable societies. Scurvy typically develops when the plasma vitamin C concentration falls and remains below 11.4 μmol/L. (The values in these widely used reference ranges were originally intended as convenient benchmarks expressed in milligrams per liter [mg/L]. Thus, the lower limit of the normal reference range, 28.4 μmol/L, is 5 mg/L; the lower limit of the marginal range, 11.4 μmol/L, is 2 mg/L.)

Vitamin C's concentration in cerebrospinal fluid (CSF) is approximately three times higher than in the bloodstream. Intraneuronal and glial concentrations are approximately five times higher yet [8,26,28]. Thus, a normal human plasma vitamin C concentration of 50 μmol/L corresponds to ~200 μmol/L in the CSF and 1 mmol/L in the brain. The brain conserves its vitamin C store more effectively than the other tissues of the body [28–30]. This protective mechanism may allow the brain to function more effectively than other tissues in the presence of vitamin C deficiency.

As plasma vitamin C concentrations increase within the normal range, CSF concentrations also increase, but less than proportionately. As plasma vitamin C concentrations decrease, CSF concentrations remain approximately stable until the plasma concentration falls below ~28.4 μmol/L. As plasma concentrations drop further below normal, CSF concentrations decrease increasingly severely [1,8,31–37].

In summary, a level of vitamin C consumption that is sufficient to maintain normal plasma vitamin C concentrations is associated with normal (approximately threefold higher) vitamin C concentrations in the CSF and much higher concentrations in neurons. Insufficient or deficient plasma vitamin C concentrations are associated with subnormal CSF concentrations. We don't know if subnormal CSF concentrations reduce brain concentrations, or reduce them enough to impair human brain metabolism, neurotransmitter synthesis or function, or weaken the brain's antioxidant defenses.

To make matters more complicated, animal models of vitamin C deficiency do not necessarily apply to the human condition [38]. As Levine and Padayatty have pointed out, the signs and symptoms of scurvy can be traced to impaired activities of specific vitamin C–requiring enzymes [26]. Does human scurvy deplete neurons of vitamin C enough to impair neurotransmitter synthesis, release, and regulation? States of vitamin C deficiency that impair neurotransmitter synthesis, action, and behavior in animal models may not occur in humans, if they succumb to its hemorrhagic complications before their brain becomes severely depleted of the vitamin.

The homeostatic mechanisms that stabilize CSF and brain vitamin C concentrations in vitamin C deficiency also buffer them against large increases in plasma concentration [8,10]. Using a novel nuclear magnetic resonance (NMR) imaging technique [39], Terpstra et al. infused a bolus of 3.4 g sodium ascorbate into the veins of normal volunteers [40]. Plasma vitamin C concentrations increased from 80 to 600 μmol/L, but brain vitamin C content did not change 2, 6, 10, and 24 hours after the vitamin injection. Notwithstanding this important evidence, a case report of a child with untreatable optic glioma who responded to high-dose intravenous vitamin C [41] sounds a note of caution about drawing firm conclusions from this limited evidence.

Vitamin C Deficiency and the Mind–Body Problem

The brain normally maintains an even keel of vitamin C content as it navigates the choppy seas of variable dietary supply. How severe does vitamin C deficiency have to be to disrupt brain function? When vitamin C deficiency disease occurs, does it primarily affect the brain—a sensitive, discerning organ—or primarily disrupt and damage the other tissues of the body, which send distress signals to the brain that disturb cognition and mood? Formal assessment of nutritional influences on human cognition presents many challenges [10,42]. For example, marginal vitamin C deficiency could subtly impair physical performance [43–46] or immunity [47,48] in ways that create anxiety or distress in some, predisposed people but not others. High-dose intravenous vitamin C could reduce fatigue and improve quality of life in people with cancer by improving their overall physical well-being without exerting any direct effect on brain metabolism or neurotransmission [49,50]. The clinical evidence bearing on this question is considered next.

MENTAL EFFECTS OF VITAMIN C DEFICIENCY

The disease of terminal vitamin C deficiency, scurvy, manifests in various ways. The commonest and most specific signs of scurvy are follicular hyperkeratosis, corkscrew hairs, and hemorrhagic phenomena such as perifollicular hemorrhage, petechiae, purpura, and bruising in the skin, gums, and occasionally in joint spaces and elsewhere. The diagnosis of scurvy is confirmed by documenting a history of vitamin C deficiency and observing prompt clinical improvement after appropriate vitamin C provision. It is helpful (but usually unnecessary) to document a plasma vitamin C concentration <11.4 µmol/L.

Historical accounts, case reports, and clinical reviews commonly describe fatigue, lassitude, subjective weakness, apathy or emotional irritability, and anorexia as cardinal symptoms of scurvy. Indeed, apathy, irritability, and psychomotor retardation have been described for centuries as heralding the onset of scurvy [51]. A classic report of 19 cases of scurvy stated that "all patients complained of fatigue, weakness and anorexia for months or years and had noted bruises usually related to slight trauma for a few weeks or several years" [52]. A clinical review prepared for the World Health Organization claims that in adults, full-blown scurvy "is preceded by a period of latent subclinical scurvy the early symptoms of which include lassitude, weakness and irritability; vague, dull aching pains in the muscles or joints of the legs and feet and weight loss" [53]. One reviewer asserts that the psychological symptoms of scurvy stem from impaired brain function [51]. It is therefore plausible that fatigue could be a symptom of latent or subclinical vitamin C deficiency [54] when it is considered that nonspecific mental symptoms are harbingers of many metabolic and endocrine diseases.

But where do mental symptoms like fatigue and mood disturbance fit into the symptom spectrum of human vitamin C deficiency? Do they indicate dysfunction of a vitamin C–deficient brain, or a normal brain's emotional responses to the metabolic dysfunction, damage, and inflammation caused by vitamin C deficiency elsewhere in the body, or a combination of these effects? This question is pertinent for two reasons. First, the evidence that the brain sequesters vitamin C more effectively than other organs increases the likelihood that vitamin C deficiency has little direct effect on it. Second, it bears on the scientific plausibility of the popular claim that vitamin C administration to people with normal or nearly normal vitamin C status relieves fatigue by improving brain function.

Case Reports

One way to evaluate the relationship between vitamin C deficiency and brain function is to determine how frequently fatigue, lassitude, subjective weakness, or mood disturbance occur in people with scurvy, and the relationship of these mental symptoms with its somatic manifestations.

I searched MEDLINE from 1946 to August 2019 for obtainable and interpretable case reports of adult scurvy published in English or French, using the Medical Subject Headings (MeSH) terms *ascorbic acid deficiency* or *scurvy* (MeSH or key word) or by pairing *ascorbic acid* (MeSH) or *vitamin C* (key word) with *fatigue*. Many more articles were identified by scanning the reference citations in these articles. Articles were deemed interpretable when they contained enough information to judge the presence or absence of fatigue, lassitude, subjective weakness or mood disturbance, the typical skin or mouth lesions of scurvy, musculoskeletal abnormalities (leg or joint pain or effusion), anemia, and dyspnea. The search identified 132 case reports that described 267 patients (198 men and 69 women; 74% men) [52,54–183].

Characteristic skin or mouth lesions were present in every case of scurvy, because they were its identifying features. Fatigue or a related symptom was noted in 107 patients (40%), anemia in 200 patients (75%), and signs of joint or musculoskeletal disease in 75 patients (28%).

Of the 200 patients with anemia, 76 had noteworthy fatigue and 124 did not. Of the 67 patients without anemia, 35 had fatigue and 32 did not. Of the 108 patients with fatigue, 73 had anemia and 35 did not.

The predominance of scorbutic men over women is consistent with the greater frequency of hypovitaminosis C in men [184], a phenomenon that has been attributed to sex differences in lifestyle and self-care and to differences in body composition [185].

The main conclusion of this analysis is that while noteworthy fatigue and related subjective symptoms, including anorexia, occur often in scurvy, they are far from universal. Fully 60% of patients with scurvy either did not experience fatigue or failed to complain of it, or the authors of the case reports did not deem it sufficiently striking or disproportionate to the clinical setting and the patient's physical disability to merit mention or comment. Failure to note fatigue does not necessarily mean it is absent, and the case reports varied in the amount of detail they provided; fatigue could have been underreported. But it is important to avoid the logical error of circular reasoning by asserting that because a patient has scurvy the patient must experience fatigue, and because the patient experiences fatigue, it must be a necessary feature of scurvy. Going by the text of the case reports, more than half of people with scurvy do not experience unusual or noteworthy fatigue, even though most of them also suffer from anemia, a well-known cause of fatigue. (The 75% prevalence of anemia in these case reports is in line with previous conclusions based on smaller numbers of cases [52,56,57,60,76,186].) It should be noted that many, if not most, of the patients described in the case reports suffered from starvation disease and multiple micronutrient deficiencies. Deficiencies of thiamine [187], folic acid, and niacin are known to impair mood or cognitive and neurological function and frequently coexist with scurvy [52,188].

In conclusion, this review of case reports fails to confirm the common claim that fatigue or mood disturbance are necessary, consistent, or sensitive symptoms of scurvy. When fatigue does occur, it could be fostered by the primary physical or mental disease that led to scurvy, and be exacerbated by social isolation, poverty and drug abuse, anorexia, and the consumption of a diet deficient both in vitamin C and other nutrients. Fatigue and mood disturbance may accompany but not represent primary symptoms of scurvy.

Historical accounts and summaries of case reports repeatedly highlight the intensity of the fatigue and mood disturbance of scurvy. Perhaps it is not the presence of fatigue and mood disturbance, but rather its remarkable intensity, and because it improves so dramatically when scurvy is treated, that leads some authors to regard it as a primary symptom of the disease, and because these symptoms improve so dramatically when scurvy is treated. As noted by Walker [65], "a striking feature in all our patients was their severe depressive state at the time of admission. This cleared within a few days of starting vitamin C therapy. Initially we attributed this change in mood to the relief of pain, but in Case 1 the depression was cured long before the pain in her ulcerated legs settled." Several early authors stress that mental depression is part of the clinical picture of scurvy and is cured by treating it [57,59].

These observations suggest that psychological symptoms are inconstant in scurvy. When psychological symptoms develop, they may be caused by peripheral somatic lesions rather than primary brain dysfunction. Perhaps when mental symptoms do arise they are experienced more intensely by a vitamin C–deficient brain. It must be conceded that case reports and clinical impressions are unreliable vehicles for documenting the mental symptoms of scurvy, for they are confounded by the primary disease or disorder that led to the patient's scurvy and the many other nutritional and micronutrient deficiencies that accompany it.

Experimental Human Scurvy

The systematic review of case reports in the previous section indicates that fatigue and lassitude are common in scurvy but not an obligatory feature of it. When fatigue does occur, it could be a symptom of the primary disease that led to the patient's vitamin C deficiency or the consequence of other nutritional deficiencies that accompany it. More reliable information is provided by formal clinical trials of vitamin C deficiency induced in healthy people.

In a classic self-experiment, Crandon, a young surgical resident in Boston, Massachusetts, continued full-time work while consuming a vitamin C–deficient diet that reduced his plasma vitamin C concentration to zero after 6 weeks. In the first of three publications, he described weight loss (with a reduction in resting metabolic rate) and mild anemia (corrected using an iron supplement) but "no increased fatigue on exertion as measured by tests of work output, while conceding that "subjectively there was a mild lassitude" [189]. In a second article, published the same year, he contradicted his earlier one by asserting that his experience confirmed other accounts of scurvy in that it was characterized by languor and incapacity for work. He first noticed easy fatigability and lassitude near the end of the third month of vitamin C deprivation; these symptoms became increasingly severe as time went on, and his exercise capacity became impaired [190]. He subsequently described lack of energy as one of his initial symptoms. He found the symptom vague and difficult to describe; it included lassitude, a desire for sleep, and a marked disinclination to exertion [191].

Farmer described a 7-month clinical trial of vitamin C and B depletion in healthy men, ages 20–30 years, carried out in the United States [192]. Two of the participants consumed a control diet, five consumed a diet deficient in vitamins C and B, and five consumed the same deficient diet supplemented with vitamin B. Plasma vitamin C concentrations fell to zero by 70 days. After 3–5 months of deficiency, error rates on a choice reaction time task increased; this observation was interpreted as indicating a decrease in interest or motivation. Emotions became more labile. Work output on a bicycle ergometer decreased. Severe fatigue developed after 5 months. Except for follicular hyperkeratosis, the typical skin and mouth lesions of scurvy were absent, even though wound healing was impaired. The observations in this trial are not well reported.

The largest and only double-blind clinical trial of experimental vitamin C deficiency was the British Medical Research Council study carried out in Sheffield in 1944 [193–195]. The trial enrolled 19 healthy men and one healthy woman, 10 of whom were made vitamin C deficient. The first sign of ill health was follicular hyperkeratosis, which developed after 21 weeks of vitamin C deprivation. Muscle coordination was unimpaired, but the volunteers required slightly more time to complete their task, an observation interpreted as evidence of increased fatigue. There were no indications of serious psychiatric disturbances. An attention test, carried out as an indicator of apathy, was unaffected by vitamin C deficiency. All symptoms of vitamin C deficiency were prevented or cured by 10 mg vitamin C per day [193]. In a summary of this trial, one of its authors, the noted biochemist H.A. Krebs, remarked on the absence of complaints of general pains or weakness among the vitamin C–deficient volunteers [196].

In the famous Iowa Study, prisoner volunteers from the Iowa State Penitentiary participated in two clinical trials of vitamin C deficiency lasting several months [184]. Among many other measurements, observations were recorded about their behavior, cognition, and mood. In a report of the first of two studies, the authors pointed out that, as in any prolonged metabolic study, it was difficult to differentiate between subjective complaints and actual symptoms of deficiency, and particularly problematic for the prisoner volunteers, whose special social and emotional problems inclined them to complain of trivial conditions and exaggerate any discomfort. The investigators nonetheless judged, on intuitive grounds, that muscular fatigability, aching, and mild general malaise developed insidiously around the same time that objective manifestations of scurvy became evident. No emotional changes attributable to vitamin C deficiency were observed [197].

In a second trial in which five prisoner volunteers participated, it was necessary to deliver the vitamin C–deficient diet three times daily by gastric tube because of its extreme unpalatability. Fatigue, muscular fatigue and pain, and emotional disturbance became apparent after approximately 90 days of vitamin C deprivation; these symptoms coincided with the appearance of the physical signs of scurvy [198]. Detailed formal psychological tests were administered after days 23, 72, and 107 of vitamin C deprivation. Individual items of the Minnesota Multiphasic Personality Inventory indicated increased fatigue, lassitude, and depression after approximately 72 days of deficiency, coincident with the appearance of follicular hyperkeratosis, gingival edema, and hemorrhage. By day 72, test scores for hypochondriasis increased moderately, whereas depression and hysteria scores increased slightly. At some time between days 72 and 107, all three test scores (and the social inversion score) increased yet further. These personality changes preceded decrements in psychomotor performance, arousal, and motivation [199]. When the psychological test scores were analyzed in relation to plasma vitamin C concentrations, there were no differences in cognition or coordination test scores for plasma vitamin C concentrations of 11.3 versus 119 μmol/L.

When considered from a modern ethical and psychological perspective, the ethical integrity and scientific reliability of the psychological test scores reported in this clinical experiment are defective. The prisoners were treated in a coercive, inhumane way, and the investigators demonstrated cultural and emotional bias against them. The trial was unblinded. Nevertheless, objective evidence of fatigue and mood disturbance emerged only when physical signs of scurvy developed. Despite claims sometimes made about it to the contrary, this study does not support the assertion that emotional fatigue or mood disturbance are prodromal or sentinel symptoms of impending scurvy.

Twenty-three healthy men participated in a 10-week double-blind study of combined restriction of vitamin C and three of the B vitamins. By week 6 of vitamin C deprivation, plasma vitamin C concentration had fallen to 9 μmol/L. The 12 vitamin-deficient men experienced no adverse effects on health, ordinary physical activity, or mental performance, although there was a significant decrease in aerobic power and the time of onset of blood lactate accumulation during exercise [44]. (Either thiamine or vitamin C deficiency could account for this decrement in physical performance [43,45,46].)

In a pharmacokinetic study of vitamin C depletion and repletion, normal male volunteers consumed a vitamin C–deficient but otherwise adequate diet for several weeks.

Consistent with other reports, plasma vitamin C levels fell below 20 μmol/L after 3 weeks of vitamin C deficiency [200–202], at which point mild but consistent feelings of fatigue and irritability were reported by six of the seven volunteers; physical signs of scurvy were absent. The mental symptoms of three volunteers disappeared within 1 week after vitamin C therapy began. There were no differences in psychometric test scores at the lowest and highest vitamin C doses [26].

In conclusion, and contrary to common opinion, there is little or no good evidence that fatigue and mood disturbance herald or precede the physical manifestations of scurvy in experimental volunteers, who either fail to experience fatigue or develop it at the same time the classic skin lesions and other hemorrhagic manifestations of scurvy appear. This analysis leaves open the possibility that vitamin C deficiency may not induce abnormal fatigue or mood change in nonstressed individuals, but these symptoms may be experienced more intensely when triggered by physiologic or even emotional stresses.

For example, in a combined historical review and analysis of experimental vitamin C deficiency trials, Norris concluded that physical exertion and physical stress increase vitamin C requirements and promote the development of scurvy [195]; other reviewers conclude the opposite [203,204]. Current evidence suggests that even mild vitamin C deficiency can reduce peak physical performance [43,45,46]. Perhaps some people, but not others, perceive this adverse somatic effect and are emotionally disturbed by it.

Movement Disorder

The dopamine-containing neurons of the basal motor nuclei are particularly susceptible to oxidative destruction in a process that is accelerated by vitamin C deficiency [2,11,17,51,205]. Destruction of these neurons causes a movement disorder typical of Parkinson disease. The hypothesis that clinically encountered hypovitaminosis C directly impairs brain function is supported by case reports of patients with scurvy or near scurvy who developed parkinsonian movement disorders that disappeared when their vitamin C deficiency was corrected [51,152,206,207]. Patients admitted to a residential treatment center and found to have the symptoms of early Parkinson disease were also highly likely to have hypovitaminosis C and corkscrew hairs [208].

A pilot clinical trial suggested that the combination of high-dose vitamin E and C delayed the progression of Parkinson disease [209,210]. Subsequent large clinical trials of antioxidants to delay the progression of Parkinson disease tested only high-dose vitamin E and were negative. Scholarly reviews of this topic ignore the plausible hypothesis that the combination vitamin E and C, as used in the early pilot trial, would be more effective than high-dose vitamin E alone [17].

Mental Effects of Treating Hypovitaminosis C

The fatigue associated with scurvy is frequently reported as dramatically remitting when vitamin C is provided. Does providing vitamin C to people with marginal vitamin C status reduce fatigue or improve mood?

Schorah et al. carried out two double-blind randomized controlled trials (RCTs) in which 1000 mg/d vitamin C or placebo were provided to long-term inpatients with a high prevalence of hypovitaminosis C. In the first trial, 118 patients with an average plasma vitamin C concentration of 11 μmol/L received 1000 mg vitamin C or placebo daily for 1 month. Vitamin C status improved in the active treatment group and was accompanied by borderline statistically significant reductions of apathy and improved well-being [211]. In the second trial, 94 elderly long-term care patients with average plasma vitamin C concentration 10 μmol/L received 1000 mg vitamin C or placebo for 2 months. The treated patients experienced slight improvements in body weight and reductions in purpura and petechial hemorrhages but no improvement in mood or mobility [212].

Gosney et al. [213] carried out an 8-week placebo-controlled RCT of micronutrient supplementation in 73 elderly nursing home residents among whom depression and anxiety were highly prevalent; two-thirds of the participants had hypovitaminosis C (average concentration 20 μmol/L). After 8 weeks of micronutrient therapy that included 240 mg vitamin C per day, the average plasma vitamin C concentration had increased to 64 μmol/L. There was no overall improvement in psychological symptoms. A post hoc analysis indicated that patients with an initially high depression score experienced an important reduction in the score if in the active but not if in the placebo group.

Clausen et al. [214] carried out a year-long double-blind placebo-controlled trial that examined mental performance and psychological scores of 94 elderly nursing home residents before and after consuming an antioxidant vitamin cocktail that included 270 mg vitamin C per day. Antioxidant supplementation had little effect on mental performance, and there were few associations between plasma vitamin concentrations and mental functioning (plasma vitamin C was not measured). There were slight but statistically significant improvements in psychological scores in the vitamin-treated patients.

We carried out one small open clinical trial [215] and two small double-blind RCTs that measured the effect of vitamin C therapy (500 mg twice daily for approximately 1 week) on mood in acutely ill surgical and medical inpatients with a high prevalence of hypovitaminosis C. These clinical trials compared the effects of vitamin C with those of a safe and clinically plausible (but subsequently determined to be inadequate) dose of vitamin D [216,217]. All three trials indicated a prompt 50%–70% reduction in mood disturbance or psychological distress in patients treated with vitamin C but little or no change in patients treated with vitamin D.

These preliminary indications of rapid and dramatic improvements in mood and psychological distress in vitamin C–treated acutely ill hospitalized patients with hypovitaminosis C stand in contrast to the negative or mostly negative findings reported in other, larger, and longer-term clinical trials of vitamin C therapy in clinically stable, long-term care patients with hypovitaminosis C.

Conclusions

Fatigue, mood disturbance, and related symptoms are striking features of scurvy, but only some patients experience or report them. As many as 60% of case reports of scurvy fail to mention or report fatigue. It is possible that patients in these case reports did experience fatigue, but they or the authors of the case reports did not consider it disproportionate enough to their general clinical condition to merit comment. When fatigue and mood disturbance occur in people with scurvy, they may be at least partly caused by coexistent anemia, systemic inflammation, disability, drug intoxication, protein-energy malnutrition, and other micronutrient deficiencies. Some but not all studies of experimental vitamin C deficiency describe fatigue as a symptom. Fatigue or mood disturbance were specifically absent in the largest and most rigorous, double-blind trial of experimental human scurvy. Except for one small study, in which fatigue was reported very early in vitamin C deficiency, the clinical evidence indicates that when fatigue develops, it does so at the same time as the physical manifestations of scurvy and hence could reflect the response of a normally functioning brain to peripheral tissue damage.

This analysis raises the possibility that when tissue damage, systemic inflammation, or even emotional distress cause fatigue, concurrent vitamin C deficiency increases its intensity. Thus, in roughly the same way that starving, severely thiamine-deficient people may develop Wernicke encephalopathy when glucose is infused, or people with adrenal insufficiency experience a clinical crisis when exposed to trauma or surgery, or people with pellagra develop a phototoxic skin rash when exposed to sunlight, or people with scurvy develop hemorrhagic gingivitis only when they have teeth to chew with, the exaggerated fatigue and mood disturbance of scurvy occur when people are physically or emotionally stressed in ways that trigger "normal" fatigue.

Very few clinical trials have tested the mental effects of correcting hypovitaminosis C. None of them are conclusive, and their results are discordant. Our own three small clinical trials yielded consistent and reproducible evidence that correcting hypovitaminosis C in acutely ill, hospitalized patients rapidly reduces emotional distress, whereas three larger and longer-term trials indicated little or no benefit from vitamin C provision to clinically stable long-term nursing home patients with a high prevalence of hypovitaminosis C.

It is impossible to draw conclusions from such scanty evidence. Active-care, acutely ill patients are different from chronic nursing home patients. Does correcting systemic inflammation-associated hypovitaminosis C in acutely ill patients improve their mood by ameliorating distress signals sent from the peripheral tissues to the brain? Must one be experiencing at least moderately severe mental distress for it to improve when hypovitaminosis C is corrected? Are the forces driving the low mood of chronic nursing home patients so overwhelmingly strong that normalizing their somatic or brain vitamin C stores is futile?

These same questions arise when one considers the mental effects of vitamin C deficiency and its treatment in diseases of the brain, including mental illness.

BRAIN DISEASES

This section provides information about the mental effects of vitamin C deficiency and therapy in acute brain injury, ischemic brain infarction, delirium, dementia, and mental illness.

Acute Brain Injury

Plasma vitamin C concentrations are reduced in patients with brain trauma, and reflect the severity of the injury [218]. More importantly, the vitamin C content of CSF is severely depleted in head-injured adults (77 versus 203 µmol/L in control samples) [219] and infants (54 versus 164 µmol/L) [220]. There is promising, but extremely limited, clinical evidence that antioxidant therapy reduces neurological symptoms and improves recovery in patients with traumatic brain injury [221]. Razmkon et al. compared the effects of low-dose intravenous ascorbic acid (500 daily for 7 days), high-dose ascorbic acid (10 g on days 1 and 4 followed by 4 g on days 5, 6, and 7), vitamin E (400 IU daily), or placebo on clinical outcomes and brain edema in young men with severe brain trauma; the results were inconclusive [222]. This topic was recently reviewed [223]. The most biologically plausible therapy would employ a combination of micronutrients, including vitamin C [224,225].

The plasma vitamin C concentrations of 15 patients with acute bacterial meningitis were extremely low and similar to those in 14 comparison patients with other neurological diseases (headache, seizure, transient ischemic attack, or facial palsy without meningitis; 10.3 versus 9.3 µmol/L); but remarkably, their CSF vitamin C concentrations were fantastically reduced (11.9 versus 144 µmol/L) [226]. Vitamin C concentrations in plasma (16 versus 76 µmol/L) and CSF (66 versus 218 µmol/L) were dramatically reduced in 11 adults with septic encephalopathy as compared with 14 healthy individuals [227]. Serum and CSF vitamin C concentrations of patients with tick-borne encephalitis were normal and similar to those in normal individuals [228].

Ischemic Brain Infarction

Plasma vitamin C concentrations decrease immediately after an acute ischemic stroke [229] and reflect the severity of the injury [218]. In view of basic evidence that vitamin C adequacy or administration could prevent or mitigate the effects of acute ischemic brain injury [20,230], Rabadi and Kristal [231] evaluated the effects of vitamin C supplementation on functional recovery after an ischemic stroke in a retrospective, case-control study of 23 patients with ischemic stroke treated with vitamin C matched with 23 other patients with ischemic stroke who were not vitamin C supplemented. No significant differences in outcome were observed. Lagowska-Lenard et al. administered 500 mg vitamin C or placebo intravenously for 10 days to patients immediately after an ischemic stroke; no acute or long-term clinical benefit was observed [232].

Delirium

Unlike with certain B vitamin deficiencies [233,234], the possibility that hypovitaminosis C could contribute to delirium appears never to have been investigated. A MEDLINE search from 1948 to 2019 revealed no publications dealing with the potential role of hypovitaminosis C in precipitating or worsening delirium (or alcohol-withdrawal delirium), despite the biological plausibility it could do so and the high prevalence of hypovitaminosis C in delirium-prone people [235].

Dementia

Despite continuing interest in antioxidant therapy to delay or slow the progression of dementia [11,18–20,236–238], there is neither strong nor consistent observational and clinical evidence that supplements of vitamin C or other antioxidants prevent cognitive decline or slow the progression of dementia [239]. Nor is it obvious that vitamin C deficiency accelerates neuronal death. A vitamin C–deficient diet reduced plasma ascorbate and dramatically reduced CSF and brain ascorbate in aging guinea pigs but did not accelerate the progression of old age–related brain pathology [240]. Associations between hypovitaminosis C and cognitive dysfunction [241] and lowered plasma vitamin C concentrations in dementia [10] could be explained by inadequate vitamin C consumption or concurrent deficiencies of other micronutrients. However, a recent cross-sectional observational trial indicated that plasma concentrations of vitamin C (and other antioxidants) were substantially lower than normal in both people with Alzheimer disease and those with mild cognitive impairment, a condition that precedes dementia and would not be predicted to be associated with poor nutritional

status. This interesting observation argues against dietary deficiency as the sole explanation for hypovitaminosis C in dementia [242]. A large, recent prospective observational trial indicated that vitamin C and E supplements were associated with a slower rate of cognitive decline in people with dementia [243].

This field of nutritional investigation is confounded by heterogeneity in study design, short observation periods, and varying definitions and evaluation methods [18,244]. It is important, but not always appreciated or acted on, to document the vitamin C status of the patients enrolled in observational and clinical trials. For example, in one study, plasma and CSF vitamin C concentrations of patients with Alzheimer disease were similar to those of control patients [36], whereas in another, long-term observational study of elderly people, poor vitamin C status was strongly associated with cognitive dysfunction, stroke, and death. Only one-third of the patients in the latter study had a normal vitamin C concentration [241].

Having earlier found that one month of supplementation with vitamin C and E increased vitamin C and E concentrations in the CSF and reduced CSF lipid peroxidation [245], Arlt et al. carried out an RCT of vitamin C (1000 mg/d) and E (400 IU/d) administered to 12 patients with Alzheimer disease (11 patients served in the control group). Vitamin supplementation increased CSF antioxidant vitamin concentrations after 1 month and 1 year of therapy but nevertheless failed to slow the progression of Alzheimer disease [246].

Mental Illness

There is increasing recognition of the role of appropriate nutrition in the prevention and amelioration of mental illnesses [247–251]. Does hypovitaminosis C worsen mental illness? Does mental illness increase vitamin C metabolism and its nutritional requirement? Does vitamin C therapy reduce the symptoms of mental illness?

Although unrecognized by most psychiatrists and other physicians, hypovitaminosis C and even scurvy are common in severe mental illness [13,94,108,130,137,155,252–263]. Obvious causes are inadequate diet [264,265], cigarette smoking, and possibly the effects of pharmacotherapy [130,266–268]. It has been suggested that schizophrenia increases vitamin C catabolism, predisposing schizophrenic people to hypovitaminosis C.

This hypothesis is based on the observation that patients with schizophrenia excrete subnormal amounts of vitamin C in their urine following the administration of a test dose [255,269], but the biological and clinical evidence supporting it are weak and inconsistent [270,271]. Mentally ill patients could excrete subnormal amounts of vitamin C for many reasons [258,271,272].

The oxidative stress theory of schizophrenia, initially proposed more than 50 years ago [13], continues to attract interest [2,11,12,273–275]. A holistic nutritional and lifestyle therapy, orthomolecular psychiatry, is motivated by the concept that abnormal brain metabolism in severe mental illness can be improved administering large doses of certain micronutrients, including vitamin C [276–279].

Case reports and small clinical trials indicate, unsurprisingly, that when psychotic people who are deficient in vitamin C [254] (are likely to be [280,281]) receive vitamin C, they improve greatly. The adverse effects of hypovitaminosis C in these patients are likely exacerbated by other micronutrient deficiencies [81,261,282,283].

A small, open clinical trial indicated that high-dose vitamin C therapy potentiated the clinical benefit of the antipsychotic drug, haloperidol [284], but was followed shortly after by a negative report [285]. Vitamin C dramatically improved the symptoms of mental illness of a vitamin C–deficient child suffering from depression and liver disease [286]. One small clinical trial indicated that vitamin C increases the effectiveness of fluoxetine therapy in major depression [287]. A second one indicated no benefit from adding vitamin C ("up to 1000 mg" per day) to citalopram [288]. Neither study determined baseline vitamin C status of the participating patients. Elderly patients suffering from major depression were reported to have higher than normal concentrations of vitamin C in their CSF (304 versus 240 μmol/L). The interpretation of this counterintuitive observation is hampered by the failure to measure plasma vitamin C concentrations [289].

As reviewed (but not always completely) by other authors [11,12,247–249,290], a small number of clinical trials have been published describing the use of vitamin C alone [254,280] or in combination with other micronutrients [291–294] as adjunctive therapy for chronic schizophrenia; the results are inconsistent. As is common in pharmacologic trials involving vitamin C, the interpretation of most of these trials is hindered by the failure to determine the baseline vitamin C status of the participants [24]. In a small, placebo-controlled RCT hindered by a high dropout rate, depressed patients with subnormal plasma vitamin C concentrations benefited from the addition of a daily dose of 500 mg vitamin C to their treatment regimen [295]. There is evidence that the provision of multiple micronutrients (not specifically vitamin C) improves mental function in people with attention deficit disorders [247,290].

PHARMACOLOGIC VITAMIN C AND MENTAL FUNCTION

The mental effects of pharmacologic doses of vitamin C, provided alone or together with other nutrients, have been tested in RCTs involving people with normal vitamin C status.

In a double-blind crossover RCT, the consumption of 160 mg/d of vitamin C or placebo for 4-week periods had no effect on psychomotor or other cognitive function in healthy young men with normal vitamin C status [296]. In two different articles that described the same clinical trial [297,298], healthy ambulatory elderly men and women participated in an approximately year-long clinical trial that tested the effects of daily consumption of an antioxidant vitamin supplement containing 500 mg vitamin C on mood, cognition, and intelligence. One of the articles asserted that despite very few significant differences between the placebo and vitamin groups, increases in plasma vitamin C concentration at 12 months were associated with more positive mood, greater improvements in intellectual functioning, and a reduction in everyday errors of memory, attention, and action. These effects were greatest for those people with more severely depressed mood and lower levels of cognitive function at baseline [297]. The other article concluded that provision of the antioxidant supplement had little or no effect on mental performance [298].

In a 6-month clinical trial, daily consumption of a multiple vitamin supplement containing 600 mg vitamin C had no effect on mood or cognition in elderly people with normal baseline vitamin C status [299]. By contrast, and in disagreement with this trial, three other clinical trials indicated that daily consumption of 500 mg vitamin C as part of a multiple vitamin–mineral supplement reduced fatigue and improved cognitive function in normal adults [300–302].

Participants in the large Age-Related Eye Disease Study (AREDS) received a daily vitamin and mineral supplement containing 500 mg vitamin C, or placebo. After a median of 6.9 years of treatment, there were no differences in any of six cognitive test scores [303].

Healthy young adults recorded their sexual activity and completed the Beck Depression Inventory before and after consuming 3000 mg/d of sustained-release vitamin C or placebo for 14 days [304]. Sexual activity increased in the active treatment group, and their Beck Depression score improved. The latter observation is of doubtful importance, however, since none of the participants were depressed, and the change in the test score was clinically trivial. In a second, similarly designed trial, volunteers randomized to 3000 mg/d vitamin C exhibited less subjective distress and less intense blood pressure changes in response to psychological stress [305]. The authors suggested that vitamin C acts directly on the brain by activating or disinhibiting neurotransmission.

Healthy office workers received a single injection of 10 g vitamin C with normal saline or normal saline alone. A fatigue score was tabulated 2 hours and 1 day after the injection. Fatigue scores decreased in the vitamin C group after 2 hours and remained lower for 1 day, especially in people with a lower baseline plasma vitamin C concentration [306].

Healthy high school students participated in a 14-day clinical trial that tested the effects of 500 mg vitamin C per day or placebo on blood pressure and anxiety. The Beck Anxiety Inventory score decreased significantly from 22 (mild anxiety) to 17 in the vitamin C group but was unchanged in the placebo group [307].

Despite the frequent assertion that pharmacologic doses of vitamin C modify brain function in a way that reduces anxiety and improves mood when administered to people with normal vitamin C status, there is very little convincing evidence supporting it. The scanty clinical trial evidence summarized here is unconvincing that pharmacologic doses of vitamin C have important brain effects in people who are not vitamin C deficient, and the phenomenon of brain homeostasis makes the hypothesis biologically implausible.

It is challenging to design and interpret clinical trials that depend on subjective (or psychologically modifiable) endpoints like fatigue and mood [308]. Placebo effects are complicated, powerful, and subtle. Imperfect blinding of study participants and investigators as to treatment assignment, and imperfectly crafted placebos can confound the results of clinical trials with soft subjective endpoints that are prone to expectation effects. Placebos are frequently imperfect [309]. Study participants could consciously or subconsciously identify physical characteristics of vitamin C (e.g., increased plasma osmolarity) and register effects that are mediated more by expectation than any fundamental physiologic action of the vitamin in the body or brain.

CONCLUSIONS

How Does Vitamin C Deficiency Affect the Brain?

Because the central nervous system has an absolute requirement for vitamin C, it maintains CSF vitamin C concentrations approximately three times higher than in the peripheral circulation and in much higher concentrations yet in neurons and glial cells. Because the brain conserves vitamin C more effectively than other tissues, it may be less adversely affected by dietary vitamin C deficiency than other tissues. Case reports and observations in experimentally induced scurvy do not provide good evidence of primary brain dysfunction. The clinical evidence is most consistent with the hypothesis that the dominant cause of the fatigue, lassitude, and mood disturbance that develop in some, but not all, people with scurvy is somatic tissue damage and the physiologic and emotional response to it. The possibility that scurvy directly causes mental symptoms is, nevertheless, suggested by the common observation that when fatigue does occur it is unusually severe, and it remits almost immediately after vitamin C is provided, a response that seems too rapid to be solely attributable to peripheral tissue repair. Another indication of primary brain dysfunction in vitamin C deficiency emerges from rare case reports of parkinsonian movements in some people with severe hypovitaminosis C and their disappearance after vitamin C provision.

Perhaps vitamin C deficiency does not primarily cause fatigue and mood disturbance but rather intensifies them when they are triggered by the physical effects of scurvy, other nutritional disorders, or the diseases that led the patient to become vitamin C deficient. This notion could explain why clinical trials of vitamin C provision indicated no effect on fatigue and mood disturbance in debilitated, vitamin C–deficient chronic nursing care patients but a major improvement in acute-care, medical and surgical patients who entertain hope of clinical improvement and discharge from hospital. The paucity of clinical trial evidence makes any conclusion unreliable.

Brain Diseases

Severe brain injury drastically depletes the CSF of vitamin C, possibly severely enough to reduce brain vitamin C concentrations. Brain injury from trauma, infection, and inflammation is dangerous and commonly leads to serious permanent disability. There is a strong argument for carrying out clinical trials of high-dose intravenous vitamin C in severe brain injury with the goal of fostering clinical improvement and reducing its complications. Nevertheless, despite the biological and clinical plausibility of this hypothesis, almost no clinical research has been carried out in this area.

Observational and physiologic evidence suggests that normal vitamin C status (possibly in combination with overall good nutritional status) protects the brain against ischemic injury and could limit the extent of an ischemic infarction, but the extremely limited clinical trial evidence currently available does not demonstrate that vitamin C supplementation improves clinical outcomes when commenced immediately after an acute ischemic stroke. These are, in fact, two different hypotheses. Does chronic, lifelong hypovitaminosis C increase the risk of dementia? Does lifelong supplementation with vitamin C (and other antioxidants) in the absence of deficiency reduce the risk of dementia? Despite their biological plausibility, these hypotheses are difficult, if not impossible, to test definitively. The existing observational and clinical trial evidence is inconsistent and unconvincing in any direction.

Hypovitaminosis C is common among people suffering from severe mental illness. Mentally ill people will be affected at least as badly by vitamin C deficiency as mentally normal people, perhaps even more so. Clinical trials are notoriously difficult and unreliable in patients with psychotic mental illness. A few clinical trials have been carried out using low-pharmacologic doses of vitamin C (either alone or with other nutrients) as adjunctive therapy in patients with chronic stable psychotic mental illness or depression but without determining their baseline vitamin C status; the results are inconsistent. They do not support any general conclusion other than the commonplace one that nutritional deficiencies of every kind should be strictly avoided and promptly corrected in everyone, and especially in people already burdened with severe mental illness.

Pharmacologic Vitamin C and Mental Function

The body tightly regulates its plasma, CSF, and brain vitamin C concentrations—at least in health—and this physiologic fact challenges the plausibility of claims that large oral (even intravenous) doses of vitamin C relieve anxiety and improve mental function in people whose baseline vitamin C status was already normal. There is more plausible (but still inconsistent and inclusive) evidence that continuous supplementation with a combination of several micronutrients, including vitamin C, may have cognitive benefits in some people despite their lack of diagnosed deficiencies. The power, complexity, and subtlety of the placebo effect are increasingly apparent. Imperfect blinding of study participants and investigators as to treatment assignment, and imperfectly crafted placebos can confound the results of clinical trials with soft subjective endpoints, like anxiety, well-being, and cognitive symptoms, all of which are highly prone to expectation.

Mind–Body Problem

Patients and their caregivers are more interested in mental and physical well-being than whether the mechanism for it originates in their brain or their body. I have tackled this question in this review because it is pertinent to the plausibility, design, and interpretation of clinical trials investigating the mental effects of vitamin C. The available evidence indicates that, except in certain brain diseases, an intake of vitamin C that is adequate for the body is also adequate for the brain. Vitamin C deficiency should be avoided because it adversely affects the body and, possibly secondarily, the brain. Pharmacologic doses of vitamin C may (or may not) improve a specific somatic disease and secondarily relieve distress, anxiety, and fatigue.

What Is the Best Dose of Vitamin C for the Brain?

What is best for the body is best for the brain. Public health authorities in different countries recommend very different levels of vitamin C consumption by healthy people, from as low as 40 mg to as high as 110 mg/d [27]. Cigarette smokers (and presumably people experiencing equivalent or more severe oxidative stress from disease and systemic inflammation) require more vitamin C. The higher dose recommended in Canada and the United States is based on determinations of the vitamin C intake required to nearly maximize tissue saturation for most people [27]. Several authorities offer plausible physiologic arguments that daily intakes of 200–1000 mg will better guarantee tissue saturation for some individuals [310–313].

Hypovitaminosis C increases the vitamin C requirement until tissue stores are replenished [314,315]. Moreover, people with hypovitaminosis C are at risk of other nutritional deficiencies that must be diagnosed and treated.

Severe tissue injury and systemic inflammation greatly increase the vitamin C dose necessary to normalize plasma and CSF vitamin C concentrations. Normalizing vitamin C concentrations under these conditions may improve clinical outcomes and mental function. People with brain injury or delirium could especially benefit from correction of vitamin C deficiency in their plasma, CSF, and brain.

REFERENCES

1. Spector, R. 1977. Vitamin homeostasis in the central nervous system. N. Engl. J. Med. 296, 1393–1398.
2. Smythies, J. R. 1996. The role of ascorbate in brain: Therapeutic implications. J. Royal Soc. Med. 89, 241.
3. Rice, M. E. 1999. Ascorbate compartmentalization in the CNS. Neurotox. Res. 1, 81–90.
4. Rice, M. E. 2000. Ascorbate regulation and its neuroprotective role in the brain. Trends Neurosci. 23, 209–216.
5. Sandstrom, M. I. and Rebec, G. V. 2007. Extracellular ascorbate modulates glutamate dynamics: Role of behavioral activation. BMC Neurosci. 8, 32.
6. Rebec, G. V. 2007. From interferant anion to neuromodulator: Ascorbate oxidizes its way to respectability. In Electrochemical Methods for Neuroscience (Michael, A. C. and Borland, L. M., eds.). pp. 1–17, CRC Press/Taylor & Francis, Boca Raton (FL).
7. Harrison, F. E. and May, J. M. 2009. Vitamin C function in the brain: Vital role of the ascorbate transporter SVCT2. Free Radic. Biol. Med. 46, 719–730.
8. Spector, R. 2009. Nutrient transport systems in brain: 40 years of progress. J. Neurochem. 111, 315–320.
9. May, J. M. 2012. Vitamin C transport and its role in the central nervous system. Subcell. Biochem. 56, 85–103.
10. Travica, N., Ried, K., Sali, A., Scholey, A., Hudson, I. and Pipingas, A. 2017. Vitamin C status and cognitive function: A systematic review. Nutrients 9, 30.
11. Kocot, J., Luchowska-Kocot, D., Kielczykowska, M., Musik, I. and Kurzepa, J. 2017. Does vitamin C influence neurodegenerative diseases and psychiatric disorders? Nutrients 9, 27.
12. Moretti, M., Fraga, D. B. and Rodrigues, A. L. S. 2017. Ascorbic acid to manage psychiatric disorders. CNS Drugs 09, 09.
13. Hoffer, A. and Osmond, H. 1963. Scurvy and schizophrenia. Dis. Nerv. Syst. 24, 273–285.
14. Majewska, M. D. and Bell, J. A. 1990. Ascorbic acid protects neurons from injury induced by glutamate and NMDA. Neuroreport 1, 194–196.
15. Huang, J., Agus, D. B., Winfree, C. J., Kiss, S., Mack, W. J., McTaggart, R. A., Choudhri, T. F. et al. 2001. Dehydroascorbic acid, a blood-brain barrier transportable form of vitamin C, mediates potent cerebroprotection in experimental stroke. Proc. Natl. Acad. Sci. USA 98, 11720–11724.
16. Castro, M. A., Beltran, F. A., Brauchi, S. and Concha, I. I. 2009. A metabolic switch in brain: Glucose and lactate metabolism modulation by ascorbic acid. J. Neurochem. 110, 423–440.
17. Kincses, Z. T. and Vecsei, L. 2011. Pharmacological therapy in Parkinson's disease: Focus on neuroprotection. CNS Neurosci. Ther. 17, 345–367.
18. Bowman, G. L. 2012. Ascorbic acid, cognitive function, and Alzheimer's disease: A current review and future direction. Biofactors 38, 114–122.
19. Harrison, F. E., Bowman, G. L. and Polidori, M. C. 2014. Ascorbic acid and the brain: Rationale for the use against cognitive decline. Nutrients 6(4), 1752–1781.
20. Hansen, S. N., Tveden-Nyborg, P. and Lykkesfeldt, J. 2014. Does vitamin C deficiency affect cognitive development and function? Nutrients 6, 3818–3846.

21. Dixit, S., Bernardo, A., Walker, J. M., Kennard, J. A., Kim, G. Y., Kessler, E. S. and Harrison, F. E. 2015. Vitamin C deficiency in the brain impairs cognition, increases amyloid accumulation and deposition, and oxidative stress in APP/PSEN1 and normally aging mice. *ACS Chem. Neurosci.* 6, 570–581.

22. Ward, M. S., Lamb, J., May, J. M. and Harrison, F. E. 2013. Behavioral and monoamine changes following severe vitamin C deficiency. *J. Neurochem.* 124, 363–375.

23. Hansen, S. N., Schou-Pedersen, A. M. V., Lykkesfeldt, J. and Tveden-Nyborg, P. 2018. Spatial memory dysfunction induced by vitamin C deficiency Is associated with changes in monoaminergic neurotransmitters and aberrant synapse formation. *Antioxidants* 7, 29.

24. Lykkesfeldt, J. and Poulsen, H. E. 2010. Is vitamin C supplementation beneficial? Lessons learned from randomised controlled trials. *Br. J. Nutr.* 103, 1251–1259.

25. Robitaille, L. and Hoffer, L. J. 2016. A simple method for plasma total vitamin C analysis suitable for routine clinical laboratory use. *Nutr. J.* 15, 40.

26. Padayatty, S. J. and Levine, M. 2016. Vitamin C: The known and the unknown and Goldilocks. *Oral Diseases* 22, 463–493.

27. Granger, M. and Eck, P. 2018. Dietary vitamin C in human health. *Adv. Food Nutr. Res.* 83, 281–310.

28. Lindblad, M., Tveden-Nyborg, P. and Lykkesfeldt, J. 2013. Regulation of vitamin C homeostasis during deficiency. *Nutrients* 5, 2860–2879.

29. Vissers, M. C., Bozonet, S. M., Pearson, J. F. and Braithwaite, L. J. 2011. Dietary ascorbate intake affects steady state tissue concentrations in vitamin C-deficient mice: Tissue deficiency after suboptimal intake and superior bioavailability from a food source (kiwifruit). *Am. J. Clin. Nutr.* 93, 292–301.

30. Hasselholt, S., Tveden-Nyborg, P. and Lykkesfeldt, J. 2015. Distribution of vitamin C is tissue specific with early saturation of the brain and adrenal glands following differential oral dose regimens in guinea pigs. *Br. J. Nutr.* 113, 1539–1549.

31. Ridge, B. D., Fairhurst, E., Chadwick, D. and Reynolds, E. H. 1976. Ascorbic acid concentrations in human plasma and cerebrospinal fluid. *Proc. Nutr. Soc.* 35, 57A–58A.

32. Tallaksen, C. M., Bohmer, T. and Bell, H. 1992. Concentrations of the water-soluble vitamins thiamin, ascorbic acid, and folic acid in serum and cerebrospinal fluid of healthy individuals. *Am. J. Clin. Nutr.* 56, 559–564.

33. Reiber, H., Ruff, M. and Uhr, M. 1993. Ascorbate concentration in human cerebrospinal fluid (CSF) and serum. Intrathecal accumulation and CSF flow rate. *Clin. Chim. Acta* 217, 163–173.

34. Barabas, J., Nagy, E. and Degrell, I. 1995. Ascorbic acid in cerebrospinal fluid—A possible protection against free radicals in the brain. *Arch. Gerontol. Geriatr.* 21, 43–48.

35. Paraskevas, G. P., Kapaki, E., Libitaki, G., Zournas, C., Segditsa, I. and Papageorgiou, C. 1997. Ascorbate in healthy subjects, amyotrophic lateral sclerosis and Alzheimer's disease. *Acta Neurol. Scand.* 96, 88–90.

36. Quinn, J., Suh, J., Moore, M. M., Kaye, J. and Frei, B. 2003. Antioxidants in Alzheimer's disease—Vitamin C delivery to a demanding brain. *J. Alzheimers Dis.* 5, 309–313.

37. Bowman, G. L., Dodge, H., Frei, B., Calabrese, C., Oken, B. S., Kaye, J. A. and Quinn, J. F. 2009. Ascorbic acid and rates of cognitive decline in Alzheimer's disease. *J. Alzheimers Dis.* 16, 93–98.

38. Michels, A. J. and Frei, B. 2013. Myths, artifacts, and fatal flaws: Identifying limitations and opportunities in vitamin C research. *Nutrients* 5, 5161–5192.

39. Emir, U. E., Raatz, S., McPherson, S., Hodges, J. S., Torkelson, C., Tawfik, P., White, T. and Terpstra, M. 2011. Noninvasive quantification of ascorbate and glutathione concentration in the elderly human brain. *NMR Biomed.* 24, 888–894.

40. Terpstra, M., Torkelson, C., Emir, U., Hodges, J. S. and Raatz, S. 2011. Noninvasive quantification of human brain antioxidant concentrations after an intravenous bolus of vitamin C. *NMR Biomed.* 24, 521–528.

41. Mikirova, N., Hunninghake, R., Scimeca, R. C., Chinshaw, C., Ali, F., Brannon, C. and Riordan, N. 2016. High-dose intravenous vitamin C treatment of a child with neurofibromatosis type 1 and optic pathway glioma: A case report. *Am. J. Case Rep.* 17, 774–781.

42. Schmitt, J. A., Benton, D. and Kallus, K. W. 2005. General methodological considerations for the assessment of nutritional influences on human cognitive functions. *Eur. J. Nutr.* 44, 459–464.

43. Johnston, C. S., Swan, P. D. and Corte, C. 1999. Substrate utilization and work efficiency during submaximal exercise in vitamin C depleted-repleted adults. *Int. J. Vitam. Nutr. Res.* 69, 41–44.

44. van der Beek, E. J., van Dokkum, W., Schrijver, J., Wedel, M., Gaillard, A. W., Wesstra, A., van de Weerd, H. and Hermus, R. J. 1988. Thiamin, riboflavin, and vitamins B-6 and C: Impact of combined restricted intake on functional performance in man. *Am. J. Clin. Nutr.* 48, 1451–1462.

45. Johnston, C. S., Barkyoumb, G. M. and Schumacher, S. S. 2014. Vitamin C supplementation slightly improves physical activity levels and reduces cold incidence in men with marginal vitamin C status: A randomized controlled trial. *Nutrients* 6, 2572–2583.

46. Paschalis, V., Theodorou, A. A., Kyparos, A., Dipla, K., Zafeiridis, A., Panayiotou, G., Vrabas, I. S. and Nikolaidis, M. G. 2016. Low vitamin C values are linked with decreased physical performance and increased oxidative stress: Reversal by vitamin C supplementation. *Eur. J. Nutr.* 55, 45–53.

47. Bozonet, S. M., Carr, A. C., Pullar, J. M. and Vissers, M. C. 2015. Enhanced human neutrophil vitamin C status, chemotaxis and oxidant generation following dietary supplementation with vitamin C-rich SunGold kiwifruit. *Nutrients* 7, 2574–2588.

48. Carr, A. C. and Maggini, S. 2017. Vitamin C and immune function. *Nutrients* 9.

49. Carr, A. C., Vissers, M. C. and Cook, J. 2014. Parenteral vitamin C for palliative care of terminal cancer patients. *N. Z. Med. J.* 127, 84–86.

50. Carr, A. C., Vissers, M. C. and Cook, J. S. 2014. The effect of intravenous vitamin C on cancer- and chemotherapy-related fatigue and quality of life. *Front. Oncol.* 4, 283.

51. Brown, T. M. 2015. Neuropsychiatric scurvy. *Psychosomatics* 56, 12–20.

52. Vilter, R. W., Woolford, R. M. and Spies, T. D. 1946. Severe scurvy; a clinical and hematologic study. *J. Lab. Clin. Med.* 31, 609–630.

53. Prinzo, Z. W. 1999. *Scurvy and its Prevention and Control in Major Emergencies*. World Health Organization, Geneva, Switzerland.

54. Gonzalez-Sabin, M., Rodriguez-Diaz, E., Mallo-Garcia, S. and Astola-Hidalgo, I. 2017. Scurvy: An "almost" forgotten disease. *Eur. J. Dermatol.* 27, 539–540.

55. Cameron, D. G. and Mills, E. S. 1942. Scurvy in Montreal. *CMAJ* 46, 548–550.

56. McMillan, R. B. and Inglis, J. C. 1944. Scurvy: A survey of 53 cases. *Br. Med. J.* 2, 233–236.

57. Cutforth, R. H. 1958. Adult scurvy. *Lancet* 1, 454–456.

58. Weary, P. E., Wheeler, C. E. and Cawley, E. P. 1961. Adult scurvy. Report of a case. *Arch. Dermatol.* 83, 657–659.

59. Chazan, J. A. and Mistilis, S. P. 1963. The pathophysiology of scurvy. A report of seven cases. *Am. J. Med.* 34, 350–358.

60. Goldberg, A. 1963. The anaemia of scurvy. *Q. J. Med.* 32, 51–64.

61. Hyams, D. E. and Ross, E. J. 1963. Scurvy, megaloblastic anaemia and osteoporosis. *Br. J. Clin. Pract.* 17, 332–340.

62. Sherlock, P. and Rothschild, E. O. 1967. Scurvy produced by a Zen macrobiotic diet. *JAMA* 199, 794–798.

63. Shafar, J. 1967. Rapid reversion of electrocardiographic abnormalities after treatment in two cases of scurvy. *Lancet* 2, 176–178.

64. Asquith, P., Oelbaum, M. H. and Dawson, D. W. 1967. Scorbutic megaloblastic anaemia responding to ascorbic acid alone. *Br. Med. J.* 4, 402.

65. Walker, A. 1968. Chronic scurvy. *Br. J. Dermatol.* 80, 625–630.

66. Booth, J. B. and Todd, G. B. 1972. Subclinical scurvy—Hypovitaminosis C. *Geriatrics* 27, 130–131.

67. Bevelaqua, F. A., Hasselbacher, P. and Schumacher, H. R. 1976. Scurvy and hemarthrosis. *JAMA* 235, 1874–1876.

68. Walter, J. F. 1979. Scurvy resulting from a self-imposed diet. *West. J. Med.* 130, 177–179.

69. Linaker, B. D. 1979. Scurvy and vitamin C deficiency in Crohn's disease. *Postgrad. Med. J.* 55, 26–29.

70. Price, N. M. 1980. Vitamin C deficiency. *Cutis* 26, 375–377.

71. Leung, F. W. and Guze, P. A. 1981. Adult scurvy. *Ann. Emerg. Med.* 10, 652–655.

72. Dockery, G. L. 1981. Adult vitamin C deficiency. Scurvy—A case report. *J. Am. Podiatry Assoc.* 71, 628–631.

73. Connelly, T. J., Becker, A. and McDonald, J. W. 1982. Bachelor scurvy. *Int. J. Dermatol.* 21, 209–211.

74. Allen, J. I., Naas, P. L. and Perri, R. T. 1982. Scurvy: Bilateral lower extremity ecchymoses and paraparesis. *Ann. Emerg. Med.* 11, 446–448.

75. Ihle, B. U. and Gillies, M. 1983. Scurvy and thrombocytopathy in a chronic hemodialysis patient. *Aust. N. Z. J. Med.* 13, 523.

76. Anonymous. 1983. Scurvy in an old man. *Nutr. Rev.* 41, 152–154.

77. Berger, M. L., Siegel, D. M. and Lee, E. L. 1984. Scurvy as an initial manifestation of Whipple's disease. *Ann. Intern. Med.* 101, 58–59.

78. Warshauer, D. M., Hayes, M. E. and Shumer, S. M. 1984. Scurvy: A clinical mimic of vasculitis. *Cutis* 34, 539–541.

79. Reuler, J. B., Broudy, V. C. and Cooney, T. G. 1985. Adult scurvy. *JAMA* 253, 805–807.

80. Dawes, P. T. and Haslock, I. 1985. Haemarthrosis due to ascorbic acid deficiency. *Br. J. Clin. Pract.* 39, 290–291.

81. Anonymous. 1986. Case records of the Massachusetts General Hospital. Weekly clinicopathological exercises. Case 33-1986. A 62-year-old alcoholic man with confusion, ataxia, and a rash. *N. Engl. J. Med.* 315, 503–508.

82. Morgan, D. R. 1987. A case of bachelor scurvy. *Practitioner* 231, 450, 454–455.

83. Shetty, A. K., Buckingham, R. B., Killian, P. J., Girdany, D. and Meyerowitz, R. 1988. Hemarthrosis and femoral head destruction in an adult diet faddist with scurvy. *J. Rheumatol.* 15, 1878–1880.

84. Sudbury, S. and Ford, P. 1990. Femoral head destruction in scurvy. *J. Rheumatol.* 17, 1108–1110.

85. Grob, J. J., Collet-Villette, A. M., Aillaud, M. F., Capo, C., Farnarier, M. F., Kaplanski, S., Monges, G. et al. 1990. Spontaneous adult scurvy in a developed country: New insight in an ancient disease. *Arch. Dermatol.* 126, 249–251.

86. Wirth, P. B. and Kalb, R. E. 1990. Follicular purpuric macules of the extremities. Scurvy. *Arch. Dermatol.* 126, 385–386, 388–389.

87. Onorato, J. and Lynfield, Y. 1992. Scurvy. *Cutis* 49, 321–322.

88. Gabay, C., Voskuyl, A. E., Cadiot, G., Mignon, M. and Kahn, M. F. 1993. A case of scurvy presenting with cutaneous and articular signs. *Clin. Rheumatol.* 12, 278–280.

89. Oeffinger, K. C. 1993. Scurvy: More than historical relevance. *Am. Fam. Physician.* 48, 609–613.

90. Ghorbani, A. J. and Eichler, C. 1994. Scurvy. *J. Am. Acad. Dermatol.* 30, 881–883.

91. Adelman, H. M., Wallach, P. M., Gutierrez, F., Kreitzer, S. M., Seleznick, M. J., Espinoza, C. G. and Espinoza, L. R. 1994. Scurvy resembling cutaneous vasculitis. *Cutis.* 54, 111–114.

92. Fred, H. L. 1994. Case in point. Scurvy. *Hosp. Pract.* (*Off. Ed.*). 29, 98.

93. Woodruff, P. W., Morton, J. and Russell, G. F. 1994. Neuromyopathic complications in a patient with anorexia nervosa and vitamin C deficiency. *Int. J. Eat. Disorders* 16, 205–209.

94. Shtasel, D. L. and Krell, H. 1995. Scurvy in schizophrenia. *Psychiatr. Serv.* 46, 293.

95. Scully, R. E., Mark, E. J., McNeely, W. F. and McNeely, B. U. 1995. Case records of the Massachusetts General Hospital. Weekly clinicopathological exercises. Case 39-1995. A 72-year-old man with exertional dyspnea, fatigue, and extensive ecchymoses and purpuric lesions. *N. Engl. J. Med.* 333, 1695–1702.

96. Ordoukhanian, E., Bulbul, R., Elenitsas, R. and Von Feldt, J. M. 1995. Scurvy and panniculitis: A case report. *Cutis.* 56, 337–341.

97. Barratt, J. A. and Summers, G. D. 1996. Letter to the editor. Scurvy, osteoporosis and megaloblastic anaemia due to alleged food intolerance. *Br. J. Rheumatol.* 35, 701–702.

98. Mehta, C. L., Cripps, D. and Bridges, A. J. 1996. Systemic pseudovasculitis from scurvy in anorexia nervosa. *Arthritis Rheum.* 39, 532–533.

99. Yalcin, A., Ural, A. U., Beyan, C., Tastan, B., Demiriz, M. and Cetin, T. 1996. Scurvy presenting with cutaneous and articular signs and decrease in red and white blood cells. *Int. J. Dermatol.* 35, 879–881.

100. Ural, A. U. 1997. Anemia related to ascorbic acid deficiency. *Am. J. Hematol.* 56, 69.

101. Leone, J., Delhinger, V., Maes, D., Scheer, C., Pennaforte, J. L., Eschard, J. P. and Etienne, J. C. 1997. Rheumatic manifestations of scurvy. A report of two cases. *Rev. Rhum. Engl. Ed.* 64, 428–431.

102. Fain, O., Mathieu, E. and Thomas, M. 1998. Scurvy in patients with cancer. *BMJ.* 316, 1661–1662.

103. Raynaud, E., Panse, I. and Petit, A. 1998. Case for diagnosis. Scurvy. *Ann. Dermatol. Venereol.* 125, 920–922.

104. Gonzalez-Gay, M. A., Garcia-Porrua, C., Lueiro, M., Fernandez, M. L., Afonso, E., Basanta, D. and Moreno-Lugris, C. 1999. Scurvy can mimick cutaneous vasculitis. Three case reports. *Rev. Rhum. Engl. Ed.* 66, 360–361.

105. Dearaujomartins-Romeo, D., Garcia-Porrua, C. and Gonzalez-Gay, M. A. 1999. Cutaneous vasculitis is not always benign. *Rev. Rhum. Engl. Ed.* 66, 240.

106. Levin, N. A. and Greer, K. E. 2000. Scurvy in an unrepentant carnivore. *Cutis* 66, 39–44.

107. Chaine, B., Cocheton, J. J. and Aractingi, S. 2001. An odd case of abdominal purpura. *Dermatology* 202, 83.

108. Cohen, S. A. and Paeglow, R. J. 2001. Scurvy: An unusual cause of anemia. *J. Am. Board Fam. Pract.* 14, 314–316.

109. Reddy, A. V., Chan, K., Jones, J. I., Vassallo, M. and Auger, M. 1998. Spontaneous bruising in an elderly woman. *Postgrad. Med. J.* 74, 273–275.

110. Stephen, R. and Utecht, T. 2001. Scurvy identified in the emergency department: A case report. *J. Emerg. Med.* 21, 235–237.

111. Pangan, A. L. and Robinson, D. 2001. Hemarthrosis as initial presentation of scurvy. *J. Rheumatol.* 28, 1923–1925.

112. Kieffer, P., Thannberger, P., Wilhelm, J. M., Kieffer, C. and Schneider, F. 2001. Multiple organ dysfunction dramatically improving with the infusion of vitamin C: More support for the persistence of scurvy in our "welfare" society. *Intensive Care Med.* 27, 448.

113. Helms, A. E. and Brodell, R. T. 2002. Scurvy in a patient with presumptive oral lichen planus. *Nutr. Clin. Pract.* 17, 237–239.

114. Vairo, G., Salustri, A., Trambaiolo, P. and D'Amore, F. 2002. Scurvy mimicking systemic vasculitis. *Minerva Med.* 93, 145–150.

115. Christopher, K., Tammaro, D. and Wing, E. J. 2002. Early scurvy complicating anorexia nervosa. *South. Med. J.* 95, 1065–1066.

116. Keenan, S., Mitts, K. G. and Kurtz, C. A. 2002. Scurvy presenting as a medial head tear of the gastrocnemius. *Orthopedics* 25, 689–691.

117. De Luna, R. H., Colley, B. J., III, Smith, K., Divers, S. G., Rinehart, J. and Marques, M. B. 2003. Scurvy: An often forgotten cause of bleeding. *Am. J. Hematol.* 74, 85–87.

118. Pimentel, L. 2003. Scurvy: Historical review and current diagnostic approach. *Am. J. Emerg. Med.* 21, 328–332.

119. Kocak, M., Akbay, G., Eksioglu, M. and Astarci, M. 2003. Case 2: Sudden ecchymosis of the legs with feelings of pain and weakness. Diagnosis: Adult scurvy. *Clin. Exp. Dermatol.* 28, 337–338.

120. Chartier, T. K., Johnson, R. A., Kaminer, M. and Tahan, S. 2003. Palpable purpura in an elderly man. *Arch. Dermatol.* 139, 1363–1368.

121. Nguyen, R. T., Cowley, D. M. and Muir, J. B. 2003. Scurvy: A cutaneous clinical diagnosis. *Australas. J. Dermatol.* 44, 48–51.

122. Chaudhry, S. I., Newell, E. L., Lewis, R. R. and Black, M. M. 2005. Scurvy: A forgotten disease. *Clin. Exp. Dermatol.* 30, 735–736.

123. Francescone, M. A. and Levitt, J. 2005. Scurvy masquerading as leukocytoclastic vasculitis: A case report and review of the literature. *Cutis.* 76, 261–266.

124. Halligan, T. J., Russell, N. G., Dunn, W. J., Caldroney, S. J. and Skelton, T. B. 2005. Identification and treatment of scurvy: A case report. *Oral Surg. Oral Med. Oral Pathol. Oral Radiol. Endod.* 100, 688–692.

125. Des Roches, A., Paradis, L., Paradis, J. and Singer, S. 2006. Food allergy as a new risk factor for scurvy. *Allergy.* 61, 1487–1488.

126. Olmedo, J. M., Yiannias, J. A., Windgassen, E. B. and Gornet, M. K. 2006. Scurvy: A disease almost forgotten. *Int. J. Dermatol.* 45, 909–913.

127. Hatuel, H., Buffet, M., Mateus, C., Calmus, Y., Carlotti, A. and Dupin, N. 2006. Scurvy in liver transplant patients. *J. Am. Acad. Dermatol.* 55, 154–156.

128. Mapp, S. J. and Coughlin, P. B. 2006. Scurvy in an otherwise well young man. *Med. J. Aust.* 185, 331–332.

129. Wang, A. H. and Still, C. 2007. Old world meets modern: A case report of scurvy. *Nutr. Clin. Pract.* 22, 445–448.

130. Arron, S. T., Liao, W. and Maurer, T. 2007. Scurvy: A presenting sign of psychosis. *J. Am. Acad. Dermatol.* 57, S8–S10.

131. Walters, R. W., Vinson, E. N., Soler, A. P. and Burton, C. S. 2007. Scurvy with manifestations limited to a previously injured extremity. *J. Am. Acad. Dermatol.* 57, S48–S49.

132. Chang, C. W., Chen, M. J., Wang, T. E., Chang, W. H., Lin, C. C. and Liu, C. Y. 2007. Scurvy in a patient with depression. *Dig. Dis. Sci.* 52, 1259–1261.

133. Leger, D. 2008. Scurvy: Reemergence of nutritional deficiencies. *Can. Fam. Physician.* 54, 1403–1406.

134. Li, R., Byers, K. and Walvekar, R. R. 2008. Gingival hypertrophy: A solitary manifestation of scurvy. *Am. J. Otolaryngol.* 29, 426–428.

135. Vieira, A. A., Minicucci, M. F., Gaiolla, R. D., Okoshi, M. P., Duarte, D. R., Matsubara, L. S., Inoue, R. M. et al. 2009. Scurvy induced by obsessive-compulsive disorder. *BMJ Case Rep.* 2009.

136. Choh, C. T., Rai, S., Abdelhamid, M., Lester, W. and Vohra, R. K. 2009. Unrecognised scurvy. *BMJ* 339, b3580.

137. Masferrer, E., Canal, L., Alvarez, A. and Jucgla, A. 2009. Gingival hypertrophy and anemia. *Arch. Dermatol.* 145, 195–200.

138. Dolberg, O. J., Elis, A. and Lishner, M. 2010. Scurvy in the 21st century. *Israel Med. Assoc. J.* 12, 183–184.

139. Kocaturk, E., Aktas, S., Kavala, M., Kocak, F., Surucu, M. and Oguz, A. 2010. Scurvy in a housewife manifesting as anemia and ecchymoses. *Eur. J. Dermatol.* 20, 849–850.

140. Swanson, A. M. and Hughey, L. C. 2010. Acute inpatient presentation of scurvy. *Cutis* 86, 205–207.

141. Deligny, C., Dehlinger, V., Goeb, V., Baptiste, G. J. and Arfi, S. 2011. Paradoxical appearance of adult scurvy in Martinique, French West Indies. *Eur. J. Intern. Med.* 22, e7–e8.

142. Dube, M. 2011. Scurvy in a man with schizophrenia. *CMAJ* 183, E760.

143. Mertens, M. T. and Gertner, E. 2011. Rheumatic manifestations of scurvy: A report of three recent cases in a major urban center and a review. *Semin. Arthritis Rheum.* 41, 286–290.

144. Smith, A., Di Primio, G. and Humphrey-Murto, S. 2011. Scurvy in the developed world. *CMAJ* 183, E752–E755.

145. Velandia, B., Centor, R. M., McConnell, V. and Shah, M. 2011. Scurvy is still present in developed countries. *J. Gen. Intern. Med.* 23, 1281–1284.

146. Chapman, J. M. and Marley, J. J. 2011. Scurvy and the ageing population. *Br. Dent. J.* 211, 583–584.

147. Allgaier, R. L., Vallabh, K. and Lahri, S. 2012. Scurvy: A difficult diagnosis with a simple cure. *Afr. J. Emerg. Med.* 2, 20–23.

148. Bernardino, V. R., Mendes-Bastos, P., Noronha, C. and Henriques, C. C. 2012. 2011: The scurvy Odyssey. *BMJ Case Rep.* 17, 17.

149. Kupari, M. and Rapola, J. 2012. Reversible pulmonary hypertension associated with vitamin C deficiency. *Chest* 142, 225–227.

150. Kurtzman, D., Dupont, J., Lian, F. and Curiel-Lewandrowski, C. 2012. Fatigue and lower-extremity ecchymosis in a 36-year-old woman. *Arch. Dermatol.* 148, 1073–1078.

151. Maltos, A. L., Portari, G. V., Saldanha, J. C., Bernardes Junior, A. G., Pardi, G. R. and da Cunha, D. F. 2012. Scurvy in an alcoholic malnourished cirrhotic man with spontaneous bacterial peritonitis. *Clinics (Sao Paulo, Brazil)* 67, 405–407.

152. Noble, M., Healey, C. S., McDougal-Chukwumah, L. D. and Brown, T. M. 2013. Old disease, new look? A first report of parkinsonism due to scurvy, and of refeeding-induced worsening of scurvy. *Psychosomatics* 54, 277–283.

153. Pazzola, G., Possemato, N., Germano, G. and Salvarani, C. 2013. Scurvy mimicking spondyloarthritis in a young man. *Clin. Exp. Rheumatol.* 31, 795.

154. Ciccocioppo, R., Gallia, A., Carugno, A., Gamba, G. and Corazza, G. R. 2013. An unconventional case of scurvy. *Eur. J. Clin. Nutr.* 67, 1336–1337.

155. Fleming, J. D., Martin, B., Card, D. J. and Mellerio, J. E. 2013. Pain, purpura and curly hairs. *Clin. Exp. Dermatol.* 38, 940–942.

156. Ohta, A., Yoshida, S., Imaeda, H., Ohgo, H., Sujino, T., Yamaoka, M., Kanno, R. et al. 2013. Scurvy with gastrointestinal bleeding. *Endoscopy* 45(Suppl 2 UCTN), E147–E148.

157. Yousef, G. M. and Goebel, L. J. 2013. Vitamin C deficiency in an anticoagulated patient. *J. Gen. Intern. Med.* 28, 852–854.

158. Zammit, P. 2013. Vitamin C deficiency in an elderly adult. *J. Am. Geriatr. Soc.* 61, 657–658.

159. Robinson, S., Roth, J. and Blanchard, S. 2013. Light-headedness and a petechial rash. *J. Fam. Pract.* 62, 203–205.

160. Ho, E. Y. and Mathy, C. 2014. Functional abdominal pain causing scurvy, pellagra, and hypovitaminosis A. *F1000Res.* 3, 35.

161. Ong, J. and Randhawa, R. 2014. Scurvy in an alcoholic patient treated with intravenous vitamins. *BMJ Case Rep.* 11, 11.

162. Zipursky, J. S., Alhashemi, A. and Juurlink, D. 2014. A rare presentation of an ancient disease: Scurvy presenting as orthostatic hypotension. *BMJ Case Rep.* 2014, 1–3.

163. Palmer, W. C. and Vazquez-Roche, M. 2014. Scurvy presenting as hematochezia. *Endoscopy* 46(Suppl 1 UCTN), E292.

164. Al-Breiki, S. H. and Al-Zoabi N. M. 2014. Scurvy as the tip of the iceberg. *J. Dermatol. Dermatol. Surg.* 18, 46–48.

165. Meisel, K., Daggubatis, S. and Josephson, S. A. 2015. Scurvy in the 21st century? Vitamin C deficiency presenting to the neurologist. *Neurol. Clin. Pract.* 5, 491–493.

166. Singh, S. M., Richards, S. J. M., Lykins, M. M., Pfister, G. M. and McClain, C. J. M. 2015. An underdiagnosed ailment: Scurvy in a tertiary care academic center. *Am. J. Med. Sci.* 349, 372–373.

167. Levavasseur, M., Becquart, C., Pape, E., Pigeyre, M., Rousseaux, J., Staumont-Salle, D. and Delaporte, E. 2015. Severe scurvy: An underestimated disease. *Eur. J. Clin. Nutr.* 69, 1076–1077.

168. Abbas, F., Ha, L. D., Sterns, R. and von Doenhoff, L. 2016. Reversible right heart failure in scurvy: Rediscovery of an old observation. *Circ. Heart Fail.* 9.

169. Wijkmans, R. A. and Talsma, K. 2016. Modern scurvy. *J. Surg. Case Rep.* 1, 10.

170. Mintsoulis, D., Milman, N. and Fahim, S. 2016. A case of scurvy—Uncommon disease—presenting as panniculitis, purpura, and oligoarthritis. *J. Cutan. Med. Surg.* 20, 592–595.

171. Anderson, M. E., DeNiro, K. L., Harte, B. J. and Sedighi Manesh, R. 2016. The missing element. *J. Hosp. Med.* 11, 879–882.

172. Bonsall, A. 2017. Never surprise a patient with scurvy. *Int. J. Dermatol.* 56, 1488–1489.

173. Lux-Battistelli, C. and Battistelli, D. 2017. Latent scurvy with tiredness and leg pain in alcoholics: An underestimated disease three case reports. *Medicine* 96, e8861.

174. Rotar, Z., Ferkolj, I., Tomsic, M. and Hocevar, A. 2017. Scurvy—Surprisingly not yet extinct. *Vasc. Med.* 22, 351–352.

175. Antonelli, M., Burzo, M. L., Pecorini, G., Massi, G., Landolfi, R. and Flex, A. 2018. Scurvy as cause of purpura in the XXI century: A review on this "ancient" disease. *Eur. Rev. Med. Pharmacol. Sci.* 22, 4355–4358.

176. Baradhi, K. M., Vallabhaneni, S. and Koya, S. 2018. Scurvy in 2017 in the USA. *Baylor Univ. Med. Cent. Proc.* 31, 227–228.

177. Bien, J. Y., Hegarty, R. and Chan, B. 2018. Anemia in scurvy. *J. Gen. Intern. Med.* 33, 2008–2009.

178. Jiang, A. W., Vijayaraghavan, M., Mills, E. G., Prisco, A. R. and Thurn, J. R. 2018. Scurvy, a not-so-ancient disease. *Am. J. Med.* 131, e185–e186.

179. Panchal, S., Schneider, C. and Malhotra, K. 2018. Scurvy in a hemodialysis patient. Rare or ignored? *Hemodial. Int.* 08, 08.

180. Lipner, S. 2018. A classic case of scurvy. *Lancet* 392, 431.

181. Ghulam Ali, S. and Pepi, M. 2018. A very uncommon case of pulmonary hypertension. *Case* 2, 279–281.

182. Uruena-Palacio, S., Ferreyro, B. L., Fernandez-Otero, L. G. and Calo, P. D. 2018. Adult scurvy associated with psychiatric disorders and breast feeding. *BMJ Case Rep.* 30, 30.

183. Bennett, S. E., Schmitt, W. P., Stanford, F. C. and Baron, J. M. 2018. Case 22-2018: A 64-year-old man with progressive leg weakness, recurrent falls, and anemia. *N. Engl. J. Med.* 379, 282–289.

184. Jacob, R. A. 1993. Classic human vitamin C depletion experiments: Homeostasis and requirement for vitamin C. *Nutrition* 9, 74–76.

185. Jungert, A. and Neuhauser-Berthold, M. 2015. The lower vitamin C plasma concentrations in elderly men compared with elderly women can partly be attributed to a volumetric dilution effect due to differences in fat-free mass. *Br. J. Nutr.* 113, 859–864.

186. Cox, E. V. 1968. The anemia of scurvy. *Vitam. Horm.* 26, 635–652.

187. Villacieros-Alvarez, J., Chicharro, P., Trillo, S. and Barbosa, A. 2017. Scurvy and Wernicke's encephalopathy: An underdiagnosed association? *Neurologia* 15, 15.

188. Mitra, M. L. 1971. Confusional states in relation to vitamin deficiencies in the elderly. *J. Am. Geriatr. Soc.* 19, 536–545.

189. Crandon, J. H. and Lund, C. C. 1940. Vitamin C deficiency in an otherwise normal adult. *N. Engl. J. Med.* 222, 748–752.

190. Crandon, J. H., Lund, C. C. and Dill, D. B. 1940. Experimental human scurvy. *N. Engl. J. Med.* 223, 353–369.

191. Crandon, J. H., Mikal, S. and Landeau, B. R. 1953. Ascorbic–acid deficiency in experimental and surgical subjects. *Proc. Nutr. Soc.* 12, 273–279.

192. Farmer, J. C. 1944. Some aspects of vitamin C metabolism. *Fed. Proc.* 3, 179–188.

193. Vitamin C Subcommittee of the Accessory Food Factors Committee, M. R. C. 1948. Vitamin-C requirement of human adults. *Lancet* 1, 853–858.

194. Pemberton, J. 2006. Medical experiments carried out in Sheffield on conscientious objectors to military service during the 1939-45 war. *Int. J. Epidemiol.* 35, 556–558.

195. Norris, J. 1983. The "scurvy disposition": Heavy exertion as an exacerbating influence on scurvy in modern times. *Bull. Hist. Med.* 57, 325–338.

196. Krebs, H. A. 1953. The Sheffield experiment on the vitamin C requirement of human adults. *Proc. Nutr. Soc. India* 12, 237–246.

197. Hodges, R. E., Baker, E. M., Hood, J., Sauberlich, H. E. and March, S. C. 1969. Experimental scurvy in man. *Am. J. Clin. Nutr.* 22, 535–548.

198. Hodges, R. E., Hood, J., Canham, J. E., Sauberlich, H. E. and Baker, E. M. 1971. Clinical manifestations of ascorbic acid deficiency in man. *Am. J. Clin. Nutr.* 24, 432–443.

199. Kinsman, R. A. and Hood, J. 1971. Some behavioral effects of ascorbic acid deficiency. *Am. J. Clin. Nutr.* 24, 455–464.

200. Jacob, R. A., Kelley, D. S., Pianalto, F. S., Swendseid, M. E., Henning, S. M., Zhang, J. Z., Ames, B. N., Fraga, C. G. and Peters, J. H. 1991. Immunocompetence and oxidant defense during ascorbate depletion of healthy men. *Am. J. Clin. Nutr.* 54, 1302S–1309S.

201. Levine, M., Conry-Cantilena, C., Wang, Y., Welch, R. W., Washko, P. W., Dhariwal, K. R., Park, J. B. et al. 1996. Vitamin C pharmacokinetics in healthy volunteers: Evidence for a recommended dietary allowance. *Proc. Natl. Acad. Sci. USA* 93, 3704–3709.

202. Levine, M., Wang, Y., Padayatty, S. J. and Morrow, J. 2001. A new recommended dietary allowance of vitamin C for healthy young women. *Proc. Natl. Acad. Sci. USA* 98, 9842–9846.

203. Kark, R. M. 1953. Ascorbic acid in relation to cold, scurvy, ACTH and surgery. *Proc. Nutr. Soc.* 12, 279–293.

204. Baker, E. M. 1967. Vitamin C requirements in stress. *Am. J. Clin. Nutr.* 20, 583–593.

205. Pardo, B., Mena, M. A., Fahn, S. and Garcia de Yebenes, J. 1993. Ascorbic acid protects against levodopa-induced neurotoxicity on a catecholamine-rich human neuroblastoma cell line. *Mov. Disord.* 8, 278–284.

206. Quiroga, M. J., Carroll, D. W. and Brown, T. M. 2014. Ascorbate- and zinc-responsive parkinsonism. *Ann. Pharmacother.* 48, 1515–1520.

207. Wright, A. D., Stevens, E., Ali, M., Carroll, D. W. and Brown, T. M. 2014. The neuropsychiatry of scurvy. *Psychosomatics* 55, 179–185.

208. Yapa, S. C. 1992. Detection of subclinical ascorbate deficiency in early Parkinson's disease. *Public Health* 106, 393–395.

209. Fahn, S. 1991. An open trial of high-dosage antioxidants in early Parkinson's disease. *Am. J. Clin. Nutr.* 53, 380S–382S.

210. Fahn, S. 1992. A pilot trial of high-dose alpha-tocopherol and ascorbate in early Parkinson's disease. *Ann. Neurol.* 32 Suppl, S128–S132.

211. Schorah, C. J., Newill, A., Scott, D. L. and Morgan, D. B. 1979. Clinical effects of vitamin C in elderly inpatients with low blood-vitamin-C levels. *Lancet* 1, 403–405.

212. Schorah, C. J., Tormey, W. P., Brooks, G. H., Robertshaw, A. M., Young, G. A., Talukder, R. and Kelly, J. F. 1981. The effect of vitamin C supplements on body weight, serum proteins, and general health of an elderly population. *Am. J. Clin. Nutr.* 34, 871–876.

213. Gosney, M. A., Hammond, M. F., Shenkin, A. and Allsup, S. 2008. Effect of micronutrient supplementation on mood in nursing home residents. *Gerontology* 54, 292–299.

214. Clausen, J., Nielsen, S. A. and Kristensen, M. 1989. Biochemical and clinical effects of an antioxidative supplementation of geriatric patients. A double blind study. *Biol. Trace Elem. Res.* 20, 135–151.

215. Evans-Olders, R., Eintracht, S. and Hoffer, L. J. 2009. Metabolic origin of hypovitaminosis C in acutely hospitalized patients. *Nutrition* 26, 1070–1074.

216. Zhang, M., Robitaille, L., Eintracht, S. and Hoffer, L. J. 2011. Vitamin C provision improves mood in acutely hospitalized patients. *Nutrition* 27, 530–533.

217. Wang, Y., Liu, X. J., Robitaille, L., Eintracht, S., MacNamara, E. and Hoffer, L. J. 2013. Effects of vitamin C and vitamin D administration on mood and distress in acutely hospitalized patients. *Am. J. Clin. Nutr.* 98, 705–711.

218. Polidori, M. C., Mecocci, P. and Frei, B. 2001. Plasma vitamin C levels are decreased and correlated with brain damage in patients with intracranial hemorrhage or head trauma. *Stroke* 32, 898–902.

219. Brau, R. H., Garcia-Castineiras, S. and Rifkinson, N. 1984. Cerebrospinal fluid ascorbic acid levels in neurological disorders. *Neurosurgery* 14, 142–146.

220. Bayir, H., Kagan, V. E., Tyurina, Y. Y., Tyurin, V., Ruppel, R. A., Adelson, P. D., Graham, S. H., Janesko, K., Clark, R. S. and Kochanek, P. M. 2002. Assessment of antioxidant reserves and oxidative stress in cerebrospinal fluid after severe traumatic brain injury in infants and children. *Pediatr. Res.* 51, 571–578.

221. Shen, Q., Hiebert, J. B., Hartwell, J., Thimmesch, A. R. and Pierce, J. D. 2016. Systematic review of traumatic brain injury and the impact of antioxidant therapy on clinical outcomes. *Worldviews Evid. Based Nurs.* 13, 380–389.

222. Razmkon, A., Sadidi, A., Sherafat-Kazemzadeh, E., Mehrafshan, A., Jamali, M., Malekpour, B. and Saghafinia, M. 2011. Administration of vitamin C and vitamin E in severe head injury: A randomized double-blind controlled trial. *Clin. Neurosurg.* 58, 133–137.

223. Leichtle, S. W., Sarma, A. K., Strein, M., Yajnik, V., Rivet, D., Sima, A. and Brophy, G. M. 2019. High-dose intravenous ascorbic acid: Ready for prime time in traumatic brain injury? *Neurocrit. Care* 22, 22.

224. Institute of Medicine (US) Committee on Nutrition, Trauma, and the Brain; Erdman, J., Oria, M., Pillsbury, L., editors. 2011. *Nutrition and Traumatic Brain Injury: Improving Acute and Subacute Health Outcomes in Military Personnel.* National Academies Press, Washington, DC.

225. Vonder Haar, C., Peterson, T. C., Martens, K. M. and Hoane, M. R. 2016. Vitamins and nutrients as primary treatments in experimental brain injury: Clinical implications for nutraceutical therapies. *Brain Res.* 1640, 114–129.

226. Kastenbauer, S., Koedel, U., Becker, B. F. and Pfister, H. W. 2002. Oxidative stress in bacterial meningitis in humans. *Neurology* 58, 186–191.

227. Voigt, K., Kontush, A., Stuerenburg, H. J., Muench-Harrach, D., Hansen, H. C. and Kunze, K. 2002. Decreased plasma and cerebrospinal fluid ascorbate levels in patients with septic encephalopathy. *Free Radic. Res.* 36, 735–739.

228. Pancewicz, S. A., Skrzydlewska, E., Kondrusik, M., Zajkowska, J., Grygorczuk, S., Swierzbinska, R. and Moniuszko, A. 2008. Serum and cerebrospinal fluid concentration of vitamins A, E and C in patients with tick-borne encephalitis. *Przegl. Epidemiol.* 62(Suppl 1), 93–98.

229. Sanchez-Moreno, C., Dashe, J. F., Scott, T., Thaler, D., Folstein, M. F. and Martin, A. 2004. Decreased levels of plasma vitamin C and increased concentrations of inflammatory and oxidative stress markers after stroke. *Stroke* 35, 163–168.

230. Sanchez-Moreno, C., Jimenez-Escrig, A. and Martin, A. 2009. Stroke: Roles of B vitamins, homocysteine and antioxidants. *Nutr. Res. Rev.* 22, 49–67.

231. Rabadi, M. H. and Kristal, B. S. 2007. Effect of vitamin C supplementation on stroke recovery: A case-control study. *Clin. Interv. Aging* 2, 147–151.

232. Lagowska-Lenard, M., Stelmasiak, Z. and Bartosik-Psujek, H. 2010. Influence of vitamin C on markers of oxidative stress in the earliest period of ischemic stroke. *Pharmacol. Rep.* 62, 751–756.

233. Sanford, A. M. and Flaherty, J. H. 2014. Do nutrients play a role in delirium? *Curr. Opin. Clin. Nutr. Metab. Care* 17, 45–50.

234. Attaluri, P., Castillo, A., Edriss, H. and Nugent, K. 2018. Thiamine deficiency: An important consideration in critically ill patients. *Am. J. Med. Sci.* 356, 382–390.

235. Torbergsen, A. C., Watne, L. D., Frihagen, F., Wyller, T. B., Brugaard, A. and Mowe, M. 2015. Vitamin deficiency as a risk factor for delirium. *Eur. Geriatr. Med.* 6, 314–318.

236. Heo, J. H., Hyon, L. and Lee, K. M. 2013. The possible role of antioxidant vitamin C in Alzheimer's disease treatment and prevention. *Am. J. Alzheimers Dis. Other Demen.* 28, 120–125.

237. Monacelli, F., Acquarone, E., Giannotti, C., Borghi, R. and Nencioni, A. 2017. Vitamin C, aging and Alzheimer's disease. *Nutrients* 9, 27.

238. Moretti, M., Fraga, D. B. and Rodrigues, A. L. S. 2017. Preventive and therapeutic potential of ascorbic acid in neurodegenerative diseases. *CNS Neurosci. Ther.* 23, 921–929.

239. Grodstein, F., Chen, J. and Willett, W. C. 2003. High-dose antioxidant supplements and cognitive function in community-dwelling elderly women. *Am. J. Clin. Nutr.* 77, 975–984.

240. Tveden-Nyborg, P., Hasselholt, S., Miyashita, N., Moos, T., Poulsen, H. E. and Lykkesfeldt, J. 2012. Chronic vitamin C deficiency does not accelerate oxidative stress in ageing brains of guinea pigs. *Basic Clin. Pharmacol. Toxicol.* 110, 524–529.

241. Gale, C. R., Martyn, C. N. and Cooper, C. 1996. Cognitive impairment and mortality in a cohort of elderly people. *BMJ* 312, 608–611.

242. Rinaldi, P., Polidori, M. C., Metastasio, A., Mariani, E., Mattioli, P., Cherubini, A., Catani, M., Cecchetti, R., Senin, U. and Mecocci, P. 2003. Plasma antioxidants are similarly depleted in mild cognitive impairment and in Alzheimer's disease. *Neurobiol. Aging* 24, 915–919.

243. Basambombo, L. L., Carmichael, P. H., Cote, S. and Laurin, D. 2017. Use of vitamin E and C supplements for the prevention of cognitive decline. *Ann. Pharmacother.* 51, 118–124.

244. Krause, D. and Roupas, P. 2015. Effect of vitamin intake on cognitive decline in older adults: Evaluation of the evidence. *J. Nutr. Health Aging* 19, 745–753.

245. Kontush, A., Mann, U., Arlt, S., Ujeyl, A., Luhrs, C., Muller-Thomsen, T. and Beisiegel, U. 2001. Influence of vitamin E and C supplementation on lipoprotein oxidation in patients with Alzheimer's disease. *Free Radic. Biol. Med.* 31, 345–354.

246. Arlt, S., Muller-Thomsen, T., Beisiegel, U. and Kontush, A. 2012. Effect of one-year vitamin C- and E-supplementation on cerebrospinal fluid oxidation parameters and clinical course in Alzheimer's disease. *Neurochem. Res.* 37, 2706–2714.

247. Rucklidge, J. J. and Kaplan, B. J. 2013. Broad-spectrum micronutrient formulas for the treatment of psychiatric symptoms: A systematic review. *Expert Rev. Neurother.* 13, 49–73.

248. Arroll, M. A., Wilder, L. and Neil, J. 2014. Nutritional interventions for the adjunctive treatment of schizophrenia: A brief review. *Nutr. J.* 13, 91.

249. Firth, J., Stubbs, B., Sarris, J., Rosenbaum, S., Teasdale, S., Berk, M. and Yung, A. R. 2017. The effects of vitamin and mineral supplementation on symptoms of schizophrenia: A systematic review and meta-analysis. *Psychol. Med.* 1.

250. Jonsson, B. H., Winzer, R. and Gornitzki, C. 2018. Letter to the editor: Are older studies lost in database searches for systematic reviews? *Psychol. Med.* 48, 1218–1219.

251. Sarris, J., Logan, A. C., Akbaraly, T. N., Amminger, G. P., Balanza-Martinez, V., Freeman, M. P., Hibbeln, J. et al. and International Society for Nutritional Psychiatry, R. 2015. Nutritional medicine as mainstream in psychiatry. *Lancet Psychiatry* 2, 271–274.

252. Leitner, Z. A. and Church, I. C. 1956. Nutritional studies in a mental hospital. *Lancet* 270, 565–567.

253. Briggs, M. H. 1962. Possible relations of ascorbic acid, ceruloplasmin and toxic aromatic metabolites in schizophrenia. *N. Z. Med. J.* 61, 229–236.

254. Milner, G. 1963. Ascorbic acid in chronic psychiatric patients: A controlled trial. *Br. J. Psychiatry* 109, 294–299.

255. VanderKamp, H. 1966. A biochemical abnormality in schizophrenia involving ascorbic acid. *Int. J. Neuropsychiatry* 2, 204–206.

256. Schorah, C. J., Morgan, D. B. and Hullin, R. P. 1983. Plasma vitamin C concentrations in patients in a psychiatric hospital. *Human Nutr. Clin. Nutr.* 37, 447–452.

257. Carney, M. W. 1986. A case of scurvy. *Br. Med. J. (Clin. Res. Ed).* 293, 883–884.

258. Suboticanec, K., Folnegovic-Smalc, V., Korbar, M., Mestrovic, B. and Buzina, R. 1990. Vitamin C status in chronic schizophrenia. *Biol. Psychiatry* 28, 959–966.

259. Reddy, R., Keshavan, M. and Yao, J. K. 2003. Reduced plasma antioxidants in first-episode patients with schizophrenia. *Schizophr. Res.* 62, 205–212.

260. Dadheech, G., Mishra, S., Gautam, S. and Sharma, P. 2006. Oxidative stress, alpha-tocopherol, ascorbic acid and reduced glutathione status in schizophrenics. *Indian J. Clin. Biochem.* 21, 34–38.

261. Fossitt, D. D. and Kowalski, T. J. 2014. Classic skin findings of scurvy. *Mayo Clin. Proc.* 89, e61.

262. Gabb, G. and Gabb, B. 2015. Scurvy not rare. *Aust. Fam. Physician* 44, 438–440.

263. Brandy-Garcia, A. M., Cabezas-Rodriguez, I., Caravia-Duran, D. and Caminal-Montero, L. 2017. Hemarthrosis and scurvy. *Reumatol. Clin.* 13, 364–365.

264. Davison, K. M. and Kaplan, B. J. 2011. Vitamin and mineral intakes in adults with mood disorders: Comparisons to nutrition standards and associations with sociodemographic and clinical variables. *J. Am. Coll. Nutr.* 30, 547–558.

265. McCreadie, R., Macdonald, E., Blacklock, C., Tilak-Singh, D., Wiles, D., Halliday, J. and Paterson, J. 1998. Dietary intake of schizophrenic patients in Nithsdale, Scotland: Case-control study. *BMJ* 317, 784–785.

266. Gurpegui, M., Aguilar, M. C., Martinez-Ortega, J. M., Diaz, F. J. and de Leon, J. 2004. Caffeine intake in outpatients with schizophrenia. *Schizophr. Bull.* 30, 935–945.

267. Martinez-Ortega, J. M., Jurado, D., Martinez-Gonzalez, M. A. and Gurpegui, M. 2006. Nicotine dependence, use of illegal drugs and psychiatric morbidity. *Addict. Behav.* 31, 1722–1729.

268. Lykkesfeldt, J., Christen, S., Wallock, L. M., Chang, H. H., Jacob, R. A. and Ames, B. N. 2000. Ascorbate is depleted by smoking and repleted by moderate supplementation: A study in male smokers and nonsmokers with matched dietary antioxidant intakes. *Am. J. Clin. Nutr.* 71, 530–536.

269. Pauling, L., Robinson, A. B., Oxley, S. S., Bergeson, M., Harris, A., Cary, P., Blethen, J. and Keaveny, I. T. 1973. Results of a loading test of ascorbic acid, niacinamide, and pyridoxine in schizophrenic subjects and controls. In *Orthomolecular Psychiatry* (Hawkins, D. and Pauling, L., eds). pp. 18–34, W.H. Freeman and Company, San Francisco.

270. Clement, P., Ban, T. A. and Lehmann, H. E. 1975. Ascorbic acid levels in chronic psychotic patients. *Psychopharmacol. Commun.* 1, 415–420.

271. Grant, F. W., Cowen, M. A., Ozerengin, M. F. and Bigelow, N. 1972. Nutritional requirements in mental illness. I. Ascorbic acid retention in schizophrenia. A reexamination. *Biol. Psychiatry* 5, 289–294.

272. Herjanic, M. 1973. Ascorbic acid and schizophrenia. In *Orthomolecular Psychiatry* (Hawkins, D. and Pauling, L., eds). pp. 303–315, W.H. Freeman and Company, San Francisco.

273. Koga, M., Serritella, A. V., Sawa, A. and Sedlak, T. W. 2016. Implications for reactive oxygen species in schizophrenia pathogenesis. *Schizophr. Res.* 176, 52–71.

274. Mahadik, S. P., Pillai, A., Joshi, S. and Foster, A. 2006. Prevention of oxidative stress-mediated neuropathology and improved clinical outcome

by adjunctive use of a combination of antioxidants and omega-3 fatty acids in schizophrenia. *Int. Rev. Psychiatry* 18, 119–131.

275. Boskovic, M., Vovk, T., Kores Plesnicar, B. and Grabnar, I. 2011. Oxidative stress in schizophrenia. *Curr. Neuropharmacol.* 9, 301–312.

276. Pauling, L. 1968. Orthomolecular psychiatry. *Science* 160, 265–271.

277. Anonymous. 1973. Megavitamin therapy— Some personal accounts. *Schizophr. Bull.* 7, 7–9.

278. Hoffer, L. J. 2010. Orthomolecular psychiatry: Past, present and future. *J. Orthomolecular Med.* 25, 56–66.

279. Hoffer, L. J. 2014. Orthomolecular psychiatry: What would Abram Hoffer do? *J. Orthomolecular Med.* 29, 54–66.

280. Beauclair, L., Vinogradov, S., Riney, S. J., Csernansky, J. G. and Hollister, L. E. 1987. An adjunctive role for ascorbic acid in the treatment of schizophrenia? *J. Clin. Psychopharmacol.* 7, 282–283.

281. Sandyk, R. and Kanofsky, J. D. 1993. Vitamin C in the treatment of schizophrenia. *Int. J. Neurosci.* 68, 67–71.

282. DeSantis, J. 1993. Scurvy and psychiatric symptoms. *Perspect. Psychiatr. Care* 29, 18–22.

283. Harrison, R. A., Vu, T. and Hunter, A. J. 2006. Wernicke's encephalopathy in a patient with schizophrenia. *J. Gen. Intern. Med.* 21, C8–C11.

284. Giannini, A. J. 1987. Augmentation of haloperidol by ascorbic acid. *Am. J. Psychiatry* 144, 1207–1209.

285. Straw, G. M., Bigelow, L. B. and Kirch, D. G. 1989. Haloperidol and reduced haloperidol concentrations and psychiatric ratings in schizophrenic patients treated with ascorbic acid. *J. Clin. Psychopharmacol.* 9, 130–132.

286. Cocchi, P., Silenzi, M., Calabri, G. and Salvi, G. 1980. Antidepressant effect of vitamin C. *Pediatrics* 65, 862–863.

287. Amr, M., El-Mogy, A., Shams, T., Vieira, K. and Lakhan, S. E. 2013. Efficacy of vitamin C as an adjunct to fluoxetine therapy in pediatric major depressive disorder: A randomized, double-blind, placebo-controlled pilot study. *Nutr. J.* 12, 31.

288. Sahraian, A., Ghanizadeh, A. and Kazemeini, F. 2015. Vitamin C as an adjuvant for treating major depressive disorder and suicidal behavior, a randomized placebo-controlled clinical trial. *Trials [Electron. Resour.]* 16, 94.

289. Hashimoto, K., Ishima, T., Sato, Y., Bruno, D., Nierenberg, J., Marmar, C. R., Zetterberg, H., Blennow, K. and Pomara, N. 2017. Increased levels of ascorbic acid in the cerebrospinal fluid of cognitively intact elderly patients with major depression: A preliminary study. *Sci. Rep.* 7, 3485.

290. Rucklidge, J. J., Eggleston, M. J. F., Johnstone, J. M., Darling, K. and Frampton, C. M. 2018. Vitamin-mineral treatment improves aggression and emotional regulation in children with ADHD: A fully blinded, randomized, placebo-controlled trial. *J. Child Psychol. Psychiatry* 59, 232–246.

291. Arvindakshan, M., Ghate, M., Ranjekar, P. K., Evans, D. R. and Mahadik, S. P. 2003. Supplementation with a combination of omega-3 fatty acids and antioxidants (vitamins E and C) improves the outcome of schizophrenia. *Schizophr. Res.* 62, 195–204.

292. Dakhale, G. N., Khanzode, S. D., Khanzode, S. S. and Saoji, A. 2005. Supplementation of vitamin C with atypical antipsychotics reduces oxidative stress and improves the outcome of schizophrenia. *Psychopharmacology* 182, 494–498.

293. Sivrioglu, E. Y., Kirli, S., Sipahioglu, D., Gursoy, B. and Sarandol, E. 2007. The impact of omega-3 fatty acids, vitamins E and C supplementation on treatment outcome and side effects in schizophrenia patients treated with haloperidol: An open-label pilot study. *Prog. Neuropsychopharmacol. Biol. Psychiatry* 31, 1493–1499.

294. Bentsen, H., Osnes, K., Refsum, H., Solberg, D. K. and Bohmer, T. 2013. A randomized placebo-controlled trial of an omega-3 fatty acid and vitamins E+C in schizophrenia. *Transl. Psychiatry* 3, e335.

295. Aburawi, S. M., Ghambirlou, F. A., Attumi, A. A., Altubuly, R. A. and Kara, A. A. 2014. Effect of ascorbic acid on mental depression drug therapy: Clinical study. *J. Psychol. Psychother.* 4, 1.

296. Adam, K. 1981. Lack of effect on mental efficiency of extra vitamin C. *Am. J. Clin. Nutr.* 34, 1712–1716.

297. Smith, A. P., Clark, R. E., Nutt, D. J., Haller, J., Hayward, S. G. and Perry, K. 1999. Vitamin C, mood and cognitive functioning in the elderly. *Nutr. Neurosci.* 2, 249–256.

298. Smith, A., Clark, R. E., Nutt, D., Haller, J., Hayward, S. and Perry, K. 1999. Anti-oxidant vitamins and mental performance of the elderly. *Hum. Psychopharmacol.* 14, 459–471.

299. Cockle, S. M., Haller, J., Kimber, S., Dawe, R. A. and Hindmarch, I. 2000. The influence of multivitamins on cognitive function and mood in the elderly. *Aging Ment. Health* 4, 339–353.

300. Haskell, C. F., Robertson, B., Jones, E., Forster, J., Jones, R., Wilde, A., Maggini, S. and Kennedy, D. O. 2010. Effects of a multi-vitamin/mineral supplement on cognitive function and fatigue during extended multi-tasking. *Hum. Psychopharmacol.* 25, 448–461.

301. Kennedy, D. O., Veasey, R., Watson, A., Dodd, F., Jones, E., Maggini, S. and Haskell, C. F. 2010. Effects of high-dose B vitamin complex with vitamin C and minerals on subjective mood and performance in healthy males. *Psychopharmacology* 211, 55–68.

302. Kennedy, D. O., Veasey, R. C., Watson, A. W., Dodd, F. L., Jones, E. K., Tiplady, B. and Haskell, C. F. 2011. Vitamins and psychological functioning: A mobile phone assessment of the effects of a B vitamin complex, vitamin C and minerals on cognitive performance and subjective mood and energy. *Hum. Psychopharmacol.* 26, 338–347.

303. Yaffe, K., Clemons, T. E., McBee, W. L., Lindblad, A. S. and Age-Related Eye Disease Study Research, G. 2004. Impact of antioxidants, zinc, and copper on cognition in the elderly: A randomized, controlled trial. *Neurology* 63, 1705–1707.

304. Brody, S. 2002. High-dose ascorbic acid increases intercourse frequency and improves mood: A randomized controlled clinical trial. *Biol. Psychiatry* 52, 371–374.

305. Brody, S., Preut, R., Schommer, K. and Schurmeyer, T. H. 2002. A randomized controlled trial of high dose ascorbic acid for reduction of blood pressure, cortisol, and subjective responses to psychological stress. *Psychopharmacology* 159, 319–324.

306. Suh, S. Y., Bae, W. K., Ahn, H. Y., Choi, S. E., Jung, G. C. and Yeom, C. H. 2012. Intravenous vitamin C administration reduces fatigue in office workers: A double-blind randomized controlled trial. *Nutr. J.* 11, 7.

307. de Oliveira, I. J., de Souza, V. V., Motta, V. and Da-Silva, S. L. 2015. Effects of oral vitamin C supplementation on anxiety in students: A double-blind, randomized, placebo-controlled trial. *Pak. J. Biol. Sci.* 18, 11–18.

308. Kaptchuk, T. J. and Miller, F. G. 2015. Placebo effects in medicine. *N. Engl. J. Med.* 373, 8–9.

309. Machado, L. A., Kamper, S. J., Herbert, R. D., Maher, C. G. and McAuley, J. H. 2008. Imperfect placebos are common in low back pain trials: A systematic review of the literature. *Eur. Spine J.* 17, 889–904.

310. Carr, A. C. and Frei, B. 1999. Toward a new recommended dietary allowance for vitamin C based on antioxidant and health effects in humans. *Am. J. Clin. Nutr.* 69, 1086–1107.

311. Johnston, C. S. and Cox, S. K. 2001. Plasma-saturating intakes of vitamin C confer maximal antioxidant protection to plasma. *J. Am. Coll. Nutr.* 20, 623–627.

312. Levine, M., Padayatty, S. J. and Espey, M. G. 2011. Vitamin C: A concentration-function approach yields pharmacology and therapeutic discoveries. *Adv. Nutr.* 2, 78–88.

313. Frei, B., Birlouez-Aragon, I. and Lykkesfeldt, J. 2012. Authors' perspective: What is the optimum intake of vitamin C in humans? *Crit. Rev. Food Sci. Nutr.* 52, 815–829.

314. Block, G., Mangels, A. R., Patterson, B. H., Levander, O. A., Norkus, E. P. and Taylor, P. R. 1999. Body weight and prior depletion affect plasma ascorbate levels attained on identical vitamin C intake: A controlled-diet study. *J. Am. Coll. Nutr.* 18, 628–637.

315. Carr, A. C., Pullar, J. M., Bozonet, S. M. and Vissers, M. C. 2016. Marginal ascorbate status (hypovitaminosis C) results in an attenuated response to vitamin C supplementation. *Nutrients* 8, 03.

CHAPTER TWELVE

The Epigenetic Role of Vitamin C in Neurological Development and Disease

Tyler C. Huff and Gaofeng Wang

DOI: 10.1201/9780429442025-12

CONTENTS

INTRODUCTION

Vitamin C is a vital dietary nutrient renowned for its critical role in many physiologic processes. Its functions are broad yet crucial, serving as both a potent antioxidant and a cofactor for many enzymes essential to biological systems. Unlike most animal species, humans are unable to produce vitamin C in the liver and must acquire it through the diet, leaving the fate of many physiologic processes contingent on dietary availability. Dietary vitamin C deficiency most notably causes scurvy due to its necessity in collagen cross-linking. However, vitamin C takes part in many physiologic roles, including a newly discovered function in DNA and histone demethylation. Methylation dynamics have critical implications for health and disease,

particularly in neurological development and the etiology of neuropathologies. This chapter focuses on how vitamin C influences neuronal function and how its newfound role in epigenetic regulation may have profound implications for neurological development and disease.

VITAMIN C IN DNA AND HISTONE METHYLATION

DNA Demethylation

Cellular systems must translate a barrage of extracellular stimuli into functional changes in gene expression to survive and respond appropriately to environmental challenges. The epigenetic landscape serves to interpret these complex extracellular cues and coordinate a pertinent response by regulating relevant expression of the genome. Methylation at the C^5 position of cytosine (5-methylcytosine [5mC]) is the most prevalent epigenetic modification in vertebrates and plays essential roles in development, in transcriptional regulation, and in governing cell identity [1]. These processes are facilitated by methyl-CpG-binding proteins that recognize 5mC and regulate the transcription of methylated genomic loci. The 5mC is considered a stable epigenetic hallmark of DNA and is maintained upon DNA replication by DNA methyltransferase 1 (DNMT1). Therefore, it was long believed that due to proper DNMT1 maintenance, 5mC was a permanent modification in the genome [2]. However, this poses a problem to the cell: How can the cell respond to dynamic environmental stimuli if its most prevalent epigenetic modification is irreversible? It was thus reasoned that there must be a process by which to remove this mark.

Recent work uncovered the answer to this question after discovering that the ten-eleven translocation 1 (TET1) enzyme converted 5-methylcytosine into 5-hydroxymethylcytosine (5hmC) in cultured cells [3]. Subsequent work demonstrated that TET enzymes further convert 5hmC to 5-formylcytosine (5fC) and 5-carboxylcytosine (5caC), two transient modifications that are quickly excised by the DNA base excision repair pathway and ultimately replaced by unmodified cytosine [4,5]. This cycle of conversion from cytosine→5mC→5hmC→5fC →5caC→cytosine comprises active TET-mediated DNA demethylation. TET enzymes belong to the Fe(II)- and 2-oxoglutarate (2OG, also known as α-ketoglutarate)-dependent dioxygenase superfamily, a diverse class of enzymes found in practically all evolutionary taxa [6]. Although their functions greatly vary, each member of this family requires both 2OG and molecular oxygen as cosubstrates for the hydroxylation reaction and utilize Fe(II) as a cofactor. TET enzymes initiate active DNA demethylation by binding molecular oxygen to Fe(II) to form a ferryl iron intermediate Fe(IV), which then hydroxylates 5mC to 5hmC [7,8]. This reaction, however, results in ferric Fe(III) that is unusable by TETs. Without the ability to convert Fe(III) back to catalytically active Fe(II), TET enzymes are stalled and are unable to continue active DNA demethylation. To continue this process, TET enzymes require an additional cofactor capable of reducing Fe(III) back to its catalytically active form, and thus emerges the role of vitamin C.

Vitamin C (L-ascorbic acid) was recently reported to induce active DNA demethylation in cultured cells [9,10]. This finding was later validated in a host of other cell types and confirmed in subsequent animal models [11–13]. Vitamin C, which predominantly exists physiologically as the ascorbate anion, likely induces demethylation by acting as an additional cofactor for TETs. Ascorbate is used by TETs to reduce Fe(III) to catalytically active Fe(II). This conversion promotes continued TET enzyme activity that then allows complete demethylation of modified cytosine. The synergistic effect of ascorbate on TET activity appears to be exclusive, as treatment with other potent antioxidants such as glutathione does not induce 5hmC generation [9,10]. Altogether, this suggests that vitamin C is likely an additional cofactor for TET enzymes. However, with the considerable conservation between TETs and other members of the Fe(II) and 2OG-dependent dioxygenase superfamily, the necessity of ascorbate in other physiologic processes is easily recognized. The classic role of vitamin C in collagen cross-linking is due to its importance for sustaining the activity of prolyl 4-hydroxylase (P4H), another member of the Fe(II) and 2OG-dependent dioxygenase superfamily [14]. With widely distributed functions from oxygen sensing to antibiotic biosynthesis, the role of vitamin C in sustaining processes mediated by this enzyme family is both broad and vital.

Histone Demethylation

Vitamin C may also be critical for maintaining histone methylation dynamics. JmjC domain-containing histone lysine demethylases (KDMs) are the largest enzyme family responsible for histone demethylation and are also members of the Fe(II) and 2OG-dependent dioxygenase superfamily [15]. KDMs are grouped into seven subfamilies based on sequence conservation and the methylated lysine residues that they antagonize. Containing nearly 20 members in total, KDMs are able to recognize and alter the methylation states of all histone lysine residues in order to drive downstream effects that are relevant to both health and disease [16]. KDMs are known to be critical for processes involved in development and mammalian disease states: For instance, JARID2 is known to associate with the Polycomb complex in embryonic stem cells (ESCs) and is essential for ESC differentiation, while KDM3A is critical for mouse spermatogenesis and is implicated in both fertility and obesity [17–19]. The importance of KDMs is wide reaching, and the role of vitamin C in sustaining their activity in critical physiologic processes is becoming more apparent. Vitamin C has been demonstrated to optimize the activity of KDM2, induce KDM-mediated somatic reprogramming, trigger embryonic stem cell demethylation, and alter global DNA and histone methylation to ensure oocyte maturation and developmental competence [20–23]. The role of vitamin C in propagating both DNA and histone demethylation may have a potentially widespread impact on numerous developmental processes and disease states, especially those regarding the nervous system, since neuronal cells contain one of the highest concentrations of intracellular vitamin C [24]. Histone and DNA methylation dynamics underlie nearly every neurological function and disease; thus, the epigenetic role of vitamin C in neural development and pathology deserves elucidation and speculation.

VITAMIN C IN NEUROLOGICAL DEVELOPMENT AND FUNCTION

Vitamin C and Neurons

Vitamin C is tremendously concentrated in nervous system and neuroendocrine tissues: The total ascorbate concentration of mammalian brains ranges between 2 and 10 mM [25–28]. Human adrenal glands, the site of adrenaline synthesis, reach concentrations up to 10 mM, while human cerebrospinal fluid (CSF) steadies around 160 μM [29–31]. The ascorbate concentrations of these tissues are significantly higher than that of nonneural tissue and circulating plasma, which in humans ranges between 40 and 60 μM [32,33]. Furthermore, vitamin C accumulates in these tissues even against a concentration gradient [24]. Vitamin C generally enters cells through sodium-dependent vitamin C transporters (SVCTs) or can enter in its oxidized form (dehydroascorbic acid [DHA]) through glucose transporters (GLUTs). SVCTs seem to be the causal transporters that accumulate vitamin C in neurons, since a lack

of SVCT expression significantly correlates with intracellular vitamin C levels regardless of GLUT expression, and knockout of SVCT2 results in nearly undetectable levels of vitamin C in the embryonic brain and pituitary [34,35]. These findings implicate SVCTs, especially SVCT2, as the principle neuronal vitamin C transporter. Furthermore, astrocytes and other brain glia do not express SVCT2 and consequently contain only 10%–20% the vitamin C concentration of neurons [36–38]. Altogether, these findings suggest that neurons may have a specific need for vitamin C throughout their development and during physiologic activity to perform vital neurological functions.

Work in recent years has described vitamin C as a potentially critical determinant in neuronal development and function. Neurons exhibit a high oxidative metabolism during development and disease and thus require potent antioxidants such as ascorbate to scavenge reactive oxygen species (ROS) and protect cells from oxidative damage [39]. Besides its neuroprotective role, vitamin C may also regulate neuronal maturation and function through epigenetic means. Vitamin C induces differentiation of mouse ESCs and cortical precursor cells into neurons, a transition that is accompanied by upregulation of genes involved in maturation, neurogenesis, and neurotransmission [40–42]. Primary hippocampal neurons lacking SVCT2 exhibit a variety of functional and morphological deficits compared to wild-type cells. Hippocampal cells lacking SVCT2 display stunted neurite outgrowth, diminished glutamate receptor (GluR) clustering, and reduced spontaneous firing, and they are more susceptible to oxidative damage and excitotoxicity [43]. Patch clamp recordings from SVCT knockout mice also show diminished neuronal activity and substantive decreases in firing amplitude and frequency of miniature excitatory postsynaptic currents [44]. Although these data have conventionally been attributed to the antioxidant effect of vitamin C, recent findings regarding its role in DNA and histone demethylation have sparked investigation into how vitamin C controls neuronal maturation and function through epigenetic means. This work is summarized later in this chapter. Collectively, these findings suggest that vitamin C, in part through epigenetic means, may be necessary for the maturation and proper functioning of neurons and their precursors.

NEUROTRANSMITTERS

Catecholamines

Vitamin C has also been implicated in the synthesis, regulation, and transmission of a host of neurotransmitters used in both the central and peripheral nervous systems. The classic and best-described relationship between vitamin C and neurotransmission is that of catecholamine synthesis: Vitamin C is known to serve as an essential cofactor for dopamine β-hydroxylase, the enzyme responsible for converting dopamine into norepinephrine [45,46]. Deficiency in guinea pigs (who, like humans, cannot produce vitamin C endogenously) results in accumulation of dopamine and a subsequent decrease of norepinephrine, presumably due to the diminished activity of dopamine β-hydroxylase [47]. Prolonged vitamin C deficiency has been reported to ultimately deplete dopamine levels and promote the oxidation of catecholamines [48]. This may be linked to a recent finding that embryos from SVCT2 knockout animals exhibit impaired development of midbrain dopaminergic neurons, which coincides with alterations in midbrain levels of 5hmC and H3K27me3, suggesting that vitamin C may be necessary for the development of dopaminergic neurons [49]. Similarly, while adrenal catecholamine levels are decreased as expected, adrenal chromaffin cells from these animals also contain fewer catecholamine storage vesicles and show signs of apoptosis [50]. Moreover, brain regions like the forebrain that are replete with catecholamine innervation coincidentally contain the highest levels of vitamin C [51]. More work is needed to further elucidate the potential role of vitamin C in promoting the development and survival of catecholaminergic neurons.

Glutamate and Acetylcholine

Vitamin C may also modulate the impact and regulation of other major neurotransmitters including glutamate and acetylcholine. Brain acetylcholinesterase levels were found to be decreased in guinea pigs deprived of vitamin C [48]. Rats treated with scopolamine, a drug that inhibits acetylcholinesterase activity, showed restored acetylcholine levels upon eating a vitamin C–rich diet [52]. Moreover, some evidence suggests that vitamin C can induce the release of both catecholamines and acetylcholine

directly from synaptic vesicles [53]. Similarly, it has been considered a neuromodulator of glutamate dynamics and has been shown to thwart neurodegeneration following glutamate excitotoxicity in several cell and animal models [54–56]. Although the exact molecular mechanism driving this phenomenon is unclear, it is thought to be mediated by redox modulation and transcriptional regulation of neuronal glutamate receptors. Furthermore, TET-mediated DNA demethylation and methylation dynamics have been shown to regulate glutamatergic synaptic homeostasis by mediating synaptic scaling and enhancing whole-cell neuronal response to glutamate stimulation [57]. TET activity, and therefore vitamin C, are thus implicated in glutamate signaling and consequently may influence predominantly glutamatergic-driven behaviors including learning and memory. Indeed, TET1 has been previously shown to promote adult hippocampal neurogenesis and underpin learning and memory in mouse models [58]. Although the importance of TET enzymes and 5hmC is discussed later in this chapter, the role of vitamin C in neuronal function is multifaceted, as is its effect on glutamate and acetylcholine metabolism.

Serotonin

Despite the absence of extensive investigation, vitamin C may also play a role in serotonin biosynthesis and the downstream behavior associated with serotonergic signaling. Serotonin is a pervasive neurotransmitter common to many physiologic processes but is most notably associated with mood [59]. The link between serotonin and mood is most salient in mood disorders such as depression. Selective serotonin reuptake inhibitors (SSRIs), which prolong serotonin exposure in the synaptic cleft, remain the most prescribed form of treatment [60]. Vitamin C has also been implicated in mood disorders, as psychological abnormalities have been found to co-occur with vitamin C deficiency [61–63]. Vitamin C administration has been found to improve symptoms of major depressive disorder in both children and adults as well as bolster mood in healthy individuals [64–68]. Additionally, cotreatment of vitamin C and the SSRI fluoxetine has been shown to significantly decrease depressive symptoms in pediatric patients compared to treatment with fluoxetine and placebo [69]. These studies collectively suggest that vitamin C may curb the symptoms of mood disorders such as depression, putatively through regulation of serotonin though the actual mechanism is unknown.

Serotonin is synthesized from the amino acid tryptophan by the enzymes tryptophan hydroxylase (TPH) and aromatic amino acid decarboxylase (DDC), of which the TPH reaction is the rate-limiting factor [70]. Of interest, TPH is a hydroxylase that, like TET and KDM enzymes, requires Fe(II) for full catalytic activity. This would seemingly provide a potential role for ascorbate to sustain the ferrous state of iron and thus prolong the activity of TPH, just as it does for TETs and KDMs. However, to date there is no evidence to suggest that Fe(III) is produced as a result of the hydroxylation reaction, but that instead Fe(II) is maintained in complex with TPH in the resting state [70–71]. Nevertheless, iron predominantly exists in its ferric form physiologically and must be reduced before being utilized by iron-dependent enzymes. Tetrahydropterin, a cofactor for the aromatic amino acid hydroxylases, has been shown to reduce Fe(III) to Fe(II) in tyrosine hydroxylase and has been proposed as the possible physiologic reductant for this enzyme family including TPH [71]. However, ascorbate has also been shown to promote the activity of tyrosine hydroxylase in vitro when in the presence of Fe(III); thus, the potential for ascorbate to promote TPH activity to produce serotonin is not far-fetched [72]. Moreover, scopolamine-treated mice supplemented with dietary ascorbate were found to have significantly higher levels of brain serotonin compared to those without ascorbate [52]. Further investigation is warranted to examine the relationship between ascorbate and serotonin and to determine whether this relationship underlies the therapeutic effects of ascorbate in mood disorders.

5hmC and Neurodevelopment

As noted previously, vitamin C plays a part in many physiologic processes and is now most notably being recognized as an epigenetic modulator of DNA and histone demethylation. Considering the tremendously high intracellular concentration of vitamin C in neurons and neuroendocrine tissues, it is easy to imagine an important role for demethylation in neuronal development and function. Indeed, 5hmC has been shown to accumulate in the hippocampus and cerebellum,

while portions of the human cortex exhibit a 50%–200% increase in 5hmC content throughout aging [73,74]. Similarly, TET enzymes have been implicated in neural development in a host of neuronal subtypes: Differentiation of adult neural stem cells correlates with increased 5hmC content and is thwarted when cells are depleted of TET2 [75]. Overexpression of TET2/3 in neural progenitor cells induces early differentiation while disrupting TET2/3 function abrogates proper development [76]. TET3 has been shown to be crucial in regulating the expression of key developmental genes such as Pax6, Ngn2, and Sox9 in vivo to drive Xenopus development of eye and neural tissues [77]. Additionally, SVCT2 knockout animals display impaired development of midbrain dopaminergic neurons and changes in 5hmC content, further corroborating the impact of vitamin C, TETs, and DNA demethylation on critical neurodevelopmental processes [49].

Considering its involvement in neuronal development, the roles of vitamin C and 5hmC in cognitive faculties and the synaptic underpinnings that drive them have been of great interest for many years. Vitamin C deprivation has been observed to decrease both hippocampal volume and spatial cognition in guinea pigs, thus sparking a recent surge of investigation into the epigenetic mechanisms that underlie these phenomena [78]. TET enzymes appear to be particularly involved in the neurophysiologic underpinnings that drive synaptic plasticity and memory. TET1 knockout mice exhibit decreased levels of cortical and hippocampal 5hmC along with abnormal hippocampal long-term depression (LTD) [79]. These changes coincide with downregulation of genes associated with neuronal activity, impaired memory extinction, and hypermethylation of genes such as NPAS4 that mediate synaptic plasticity and cognition. TET1 overexpression in murine brains drives the expression of neuronal memory-related genes such as Arc and Fos, though mice present with impaired long-term memory in contextual fear conditioning, possibly resulting from inappropriate signaling driven by TET1 overexpression and its downstream effects on these memory genes [80]. Similarly, TET3 has been shown to be necessary for fear extinction in rodents and for accumulating 5hmC in the prefrontal cortex during behavioral adaptation [81]. Depletion of 5hmC readers also seems to negatively affect memory, as murine knockout of UHRF2 depletes cortical and hippocampal 5hmC and results in subsequent impairment of spatial memory acquisition and retention [82]. Considering the consequences of 5hmC depletion on cognitive faculties, one could speculate that 5hmC and TET enzymes may play a more fundamental role in regulating the synaptic machinery that drives these functions. Indeed, 5hmC is found to be enriched in both rodent and human genes critical for synaptic function and may also regulate RNA splicing of these genes [83]. As noted previously, TET1 has been shown to positively regulate glutamatergic synaptic upscaling, though controversy has surrounded this finding as other reports have found that depletion of DNMT1/3 results in synaptic downscaling and that TET3 knockdown induces the opposite effect [84–86]. These discrepancies may be due to differences in cell type, experimental conditions, or even independent functions of different TET isoforms. No matter the explanation, as more is done to elucidate the true function of 5hmC and TET enzymes in synaptic function, the more interest will grow to explore how vitamin C plays a role in the epigenetic regulation of neurological processes.

VITAMIN C, METHYLATION DYNAMICS, AND NEUROLOGICAL DISEASE

Vitamin C has a clear physiological role in neuronal development through its function in DNA demethylation. Dysregulation of methylation/demethylation dynamics has been implicated in numerous neurological disease states. Vitamin C may therefore wield therapeutic potential to help ameliorate these conditions by inducing TET-mediated DNA demethylation to restore normal methylation profiles and thwart neurological disease. The remainder of this chapter discusses the role of vitamin C in Alzheimer disease and epilepsy, two neurological conditions whose etiologies are characterized by aberrant methylation dynamics and are influenced by vitamin C. Due to the multitude of epigenetic determinants reported to affect these diseases, only the influence of DNA methylation is discussed (Figure 12.1).

Alzheimer Disease

Alzheimer disease (AD) is the most frequent cause of dementia in the United States and confers a huge societal burden, carrying an estimated healthcare cost of $172 billion each year [87]. Although all forms of dementia exhibit cognitive decline and behavioral changes, AD is characterized by pathologic increases in extracellular deposition of amyloid-β (Aβ) peptide, accumulation of hyperphosphorylated tau protein, and widespread loss of central nervous system neurons and synapses. Although decades of research and tremendous funds have been spent to understand how these pathologic events arise and influence disease states, the question of their molecular etiology and ties to AD emergence remain a mystery.

Due to its accumulation in the brain and potent antioxidant properties, the relationship between vitamin C levels and AD has been heavily investigated. Despite the numerous difficulties with controlling patient vitamin C levels during studies and estimating daily intake via dietary journals in lieu of plasma measurements, an overwhelming body of evidence connects healthy vitamin C levels with neuroprotection from cognitive decline and AD [88]. Rodent studies first discovered hippocampal ascorbate uptake significantly decreases throughout aging, and APP/PSEN1 mutant mouse models of AD present additional cognitive impairment, increased Aβ accumulation, and higher morbidity when crossed with SVCT2-deficient mice [89,90]. Conversely, studies evaluating cognitive performance show that participants without cognitive impairment have higher mean vitamin C concentrations, and a high intake of dietary vitamin C is associated with a decreased risk for AD [91,92]. This molecular and epidemiological evidence, in addition to the important role it plays in neuronal development, suggests that vitamin C may thwart the features and onset of AD.

The epigenetic determinants surrounding AD are complex, and the role that DNA methylation/demethylation plays seems equivocal. Many studies have shown that methylation dynamics are disrupted in AD patient brains, although the exact effect this has on disease states is unclear. Increased methylation of *ABCA7*, *BIN1*, *ANK1*, and other AD susceptibility genes is associated with increased AD pathology burden [93]. This association was also observed in methylation of the *HOXA* gene cluster and can most likely be found in other neurodevelopmental genes [94]. Conversely, other studies have reported global decreases in 5mC in patient brain entorhinal cortex that negatively

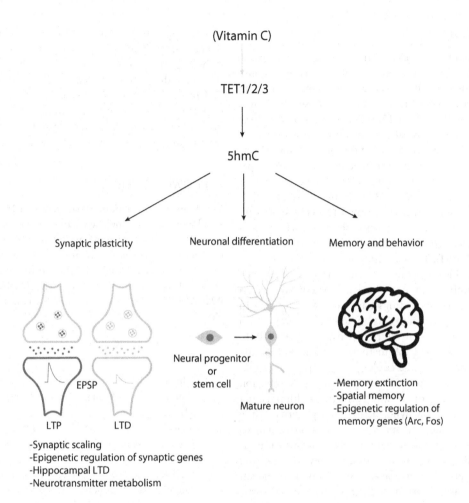

(Vitamin C)

TET1/2/3

5hmC

Synaptic plasticity   Neuronal differentiation   Memory and behavior

EPSP

Neural progenitor
or
stem cell

Mature neuron

LTP   LTD

-Memory extinction
-Spatial memory
-Epigenetic regulation of
 memory genes (Arc, Fos)

-Synaptic scaling
-Epigenetic regulation of synaptic genes
-Hippocampal LTD
-Neurotransmitter metabolism

Figure 12.1. The epigenetic role of vitamin C in neuronal function. Ten-eleven translocation (TET) enzymes have been shown to play a crucial role in many neuronal functions including synaptic plasticity, neuronal differentiation from progenitors or stem cells, and behaviors such as memory. Dysregulation of TET 1/2/3 in numerous model systems has resulted in global alterations of 5hmC content and subsequent disruption of these processes. Vitamin C has been implicated in many aspects of neuronal function and has recently been shown to propagate TET-mediated 5hmC generation and DNA demethylation. Therefore, it is plausible that vitamin C may exert its influence on neuronal function and development through TET-mediated demethylation of genes underlying these processes. (EPSP, excitatory postsynaptic potential; LTD, long-term depression; LTP, long-term potentiation.)

correlate with Aβ hippocampal load [95,96]. Work investigating 5hmC content in AD patient brains has been similarly ambiguous. Some studies have reported decreased 5hmC content in AD patient entorhinal cortex and cerebellum, while others have found increased 5hmC content in AD middle frontal gyrus and a positive correlation between 5hmC and AD pathology [95–99]. These discrepancies could be due to differences in the age of patients from which diseased brain tissue was derived. AD is a progressive multistage disease of aging, and methylation profiles could conceivably change between consecutive stages of disease severity. Furthermore, these studies examined brain

tissue from a variety of brain regions, and there is little reason to believe that methylation profiles are similar between these regions. This variability in patient age and tissue selection could explain inconsistencies in the literature and furthermore confound interpretation of promising findings. Despite the lack of a clear-cut understanding regarding methylation and AD etiology, it is plausible that dysregulation of methylation dynamics may underlie disease onset and severity. More work is needed to examine the role of 5mC and 5hmC in disease pathology and whether the neuroprotective role of vitamin C in AD is mediated by its function in regulating these marks.

Epilepsy

Vitamin C has also been implicated in mitigating epileptic seizure events. APP/PSEN1 AD mouse models exhibit altered electroencephalogram activity and are more susceptible to seizures when crossed with SVCT2-deficient mice [90]. These symptoms arise after only a 30% reduction from baseline ascorbate concentrations. Vitamin C was shown to be neuroprotective of hippocampal neurons during seizures, and in rat models of pentalenetetrazol (PTZ)-induced seizures, it has been well-demonstrated to decrease seizure intensity, prolong latency of seizure onset, suppress seizure episodes, and decrease mortality rates [100–104]. These findings have also been verified in guinea pigs who, like humans, cannot produce vitamin C and must consume it via dietary means [105]. These findings have generated interest in applying vitamin C as an adjuvant therapy with anticonvulsive drugs in epilepsy treatment [100].

A role for DNA methylation, and potentially vitamin C, in the etiology of epilepsy has begun to be elucidated in recent years. Deep methylation sequencing of chronic epileptic rat hippocampi reveals global hypermethylation patterns, while human hippocampi with hippocampal sclerosis exhibit promoter hypermethylation of protein-coding genes involved in neurodevelopment and synaptic transmission [106,107]. This phenomenon was also observed in temporal neocortical tissue from drug-refractory epileptic patients [108]. Genes previously associated with neuronal hyperactivity such as BDNF, Grin2, and Brin2b have also been observed to exhibit altered methylation profiles in epilepsy models [109–111]. Unsurprisingly, DNMTs including DNMT1 and DNMT3a have been found to be significantly upregulated in patients with temporal lobe epilepsy, and epileptic hippocampal slices exhibit reduced spontaneous excitatory transmission upon application of DNMT inhibitors [112,113]. Collectively, these studies suggest that DNA hypermethylation characterizes epilepsy and that vitamin C may act as a therapeutic by restoring normal DNA methylation profiles through TET-mediated demethylation.

CONCLUSIONS

Vitamin C has been known for decades to play a role in numerous physiologic processes and to promote normal development. The tremendously high concentration of vitamin C in the nervous system suggests an important role in neuronal development and synaptic function that is only beginning to be elucidated. The recent discovery that vitamin C is a critical regulator of TET-mediated DNA demethylation creates new opportunities to examine the role of epigenetic processes in neurological function and to discover how vitamin C may contribute to them.

ACKNOWLEDGMENTS

We thank Vladimir Camarena, Sushmita Mustafi, and David Sant for their assistance. We apologize to colleagues whose work we were not able to cite here due to space limitations. The work on the epigenomic regulation by vitamin C in the Wang lab is supported by grants (R01NS089525, R21CA191668) from the National Institutes of Health.

REFERENCES

1. Schubeler, D. 2015. Function and information content of DNA methylation. Nature. 517, 321–326.
2. Bhutani, N., Burns, D. M. and Blau, H. M. 2011. DNA demethylation dynamics. Cell. 146, 866–872.
3. Tahiliani, M., Koh, K. P., Shen, Y., Pastor, W. A., Bandukwala, H., Brudno, Y., Agarwal, S. et al. 2009. Conversion of 5-methylcytosine to 5-hydroxymethylcytosine in mammalian DNA by MLL partner TET1. Science. 324, 930–935.
4. Ito, S., Shen, L., Dai, Q., Wu, S. C., Collins, L. B., Swenberg, J. A., He, C. and Zhang, Y. 2011. Tet proteins can convert 5-methylcytosine to 5-formylcytosine and 5-carboxylcytosine. Science. 333, 1300–1303.
5. He, Y. F., Li, B. Z., Li, Z., Liu, P., Wang, Y., Tang, Q., Ding, J. et al. 2011. Tet-mediated formation of 5-carboxylcytosine and its excision by TDG in mammalian DNA. Science. 333, 1303–1307.
6. Salminen, A., Kauppinen, A. and Kaarniranta, K. 2015. 2-Oxoglutarate-dependent dioxygenases are sensors of energy metabolism, oxygen availability, and iron homeostasis: Potential role in the regulation of aging process. Cell. Mol. Life Sci. 72, 3897–3914.
7. Rose, N. R., McDonough, M. A., King, O. N., Kawamura, A. and Schofield, C. J. 2011. Inhibition of 2-oxoglutarate dependent oxygenases. Chem. Soc. Rev. 40, 4364–4397.

8. Li, D., Guo, B., Wu, H., Tan, L. and Lu, Q. 2015. TET family of dioxygenases: Crucial roles and underlying mechanisms. *Cytogenet. Genome. Res.* 146, 171–180.

9. Minor, E. A., Court, B. L., Young, J. I. and Wang, G. 2013. Ascorbate induces ten-eleven translocation (Tet) methylcytosine dioxygenase-mediated generation of 5-hydroxymethylcytosine. *J. Biol. Chem.* 288, 13669–13674.

10. Dickson, K. M., Gustafson, C. B., Young, J. I., Zuchner, S. and Wang, G. 2013. Ascorbate-induced generation of 5-hydroxymethylcytosine is unaffected by varying levels of iron and 2-oxoglutarate. *Biochem. Biophys. Res. Commun.* 439, 522–527.

11. Blaschke, K., Ebata, K. T., Karimi, M. M., Zepeda-Martinez, J. A., Goyal, P., Mahapatra, S., Tam, A. et al. 2013. Vitamin C induces Tet-dependent DNA demethylation and a blastocyst-like state in ES cells. *Nature.* 500, 222–226.

12. Yin, R., Mao, S. Q., Zhao, B., Chong, Z., Yang, Y., Zhao, C., Zhang, D. et al. 2013. Ascorbic acid enhances Tet-mediated 5-methylcytosine oxidation and promotes DNA demethylation in mammals. *J. Am. Chem. Soc.* 135, 10396–10403.

13. Chen, J., Guo, L., Zhang, L., Wu, H., Yang, J., Liu, H., Wang, X. et al. 2013. Vitamin C modulates TET1 function during somatic cell reprogramming. *Nat. Genet.* 45, 1504–1509.

14. Van Robertson, W. B. and Schwartz, B. 1953. Ascorbic acid and the formation of collagen. *J. Biol. Chem.* 201, 689–696.

15. Klose, R. J., Kallin, E. M. and Zhang, Y. 2006. JmjC-domain-containing proteins and histone demethylation. *Nat. Rev. Genet.* 7, 715–727.

16. Accari, S. L. and Fisher, P. R. 2015. Emerging roles of JmjC domain-containing proteins. *Int. Rev. Cell. Mol. Biol.* 319, 165–220.

17. Shen, X., Kim, W., Fujiwara, Y., Simon, M. D., Liu, Y., Mysliwiec, M. R., Yuan, G. C., Lee, Y. and Orkin, S. H. 2009. Jumonji modulates polycomb activity and self-renewal versus differentiation of stem cells. *Cell.* 139, 1303–1314.

18. Okada, Y., Scott, G., Ray, M. K., Mishina, Y. and Zhang, Y. 2007. Histone demethylase JHDM2A is critical for Tnp1 and Prm1 transcription and spermatogenesis. *Nature.* 450, 119–123.

19. Tateishi, K., Okada, Y., Kallin, E. M. and Zhang, Y. 2009. Role of Jhdm2a in regulating metabolic gene expression and obesity resistance. *Nature.* 458, 757–761.

20. Tsukada, Y., Fang, J., Erdjument-Bromage, H., Warren, M. E., Borchers, C. H., Tempst, P. and Zhang, Y. 2006. Histone demethylation by a family of JmjC domain-containing proteins. *Nature.* 439, 811–816.

21. Wang, T., Chen, K., Zeng, X., Yang, J., Wu, Y., Shi, X., Qin, B. et al. 2011. The histone demethylases Jhdm1a/1b enhance somatic cell reprogramming in a vitamin-C-dependent manner. *Cell Stem Cell.* 9, 575–587.

22. Ebata, K. T., Mesh, K., Liu, S., Bilenky, M., Fekete, A., Acker, M. G., Hirst, M., Garcia, B. A. and Ramalho-Santos, M. 2017. Vitamin C induces specific demethylation of H3K9me2 in mouse embryonic stem cells via Kdm3a/b. *Epigenet. Chromatin.* 10, 36.

23. Yu, X. X., Liu, Y. H., Liu, X. M., Wang, P. C., Liu, S., Miao, J. K., Du, Z. Q. and Yang, C. X. 2018. Ascorbic acid induces global epigenetic reprogramming to promote meiotic maturation and developmental competence of porcine oocytes. *Sci. Rep.* 8, 6132.

24. Harrison, F. E. and May, J. M. 2009. Vitamin C function in the brain: Vital role of the ascorbate transporter SVCT2. *Free. Radic. Biol. Med.* 46, 719–730.

25. Hughes, R. E., Hurley, R. J. and Jones, P. R. 1971. The retention of ascorbic acid by guinea-pig tissues. *Br. J. Nutr.* 26, 433–438.

26. Kratzing, C. C., Kelly, J. D. and Kratzing, J. E. 1985. Ascorbic acid in fetal rat brain. *J. Neurochem.* 44, 1623–1624.

27. Oke, A. F., May, L. and Adams, R. N. 1987. Ascorbic acid distribution patterns in human brain: A comparison with nonhuman mammalian species. *Ann. N. Y. Acad. Sci.* 498, 1–12.

28. Rice, M. E. 2000. Ascorbate regulation and its neuroprotective role in the brain. *Trends Neurosci.* 23, 209–216.

29. Padayatty, S. J., Doppman, J. L., Chang, R., Wang, Y., Gill, J., Papanicolaou, D. A. and Levine, M. 2007. Human adrenal glands secrete vitamin C in response to adrenocorticotrophic hormone. *Am. J. Clin. Nutr.* 86, 145–149.

30. Lonnrot, K., Metsa-Ketela, T., Molnar, G., Ahonen, J. P., Latvala, M., Peltola, J., Pietila, T. and Alho, H. 1996. The effect of ascorbate and ubiquinone supplementation on plasma and CSF total antioxidant capacity. *Free Radic. Biol. Med.* 21, 211–217.

31. Reiber, H., Ruff, M. and Uhr, M. 1993. Ascorbate concentration in human cerebrospinal fluid (CSF) and serum. Intrathecal accumulation and CSF flow rate. *Clin. Chim. Acta.* 217, 163–173.

32. Dhariwal, K. R., Hartzell, W. O. and Levine, M. 1991. Ascorbic acid and dehydroascorbic acid measurements in human plasma and serum. *Am. J. Clin. Nutr.* 54, 712–716.

33. Okamura, M. 1979. Uptake of L-ascorbic acid and L-dehydroascorbic acid by human erythrocytes and HeLa cells. *J. Nutr. Sci. Vitaminol. (Tokyo)*. 25, 269–279.

34. May, J. M., Qu, Z. C., Qiao, H. and Koury, M. J. 2007. Maturational loss of the vitamin C transporter in erythrocytes. *Biochem. Biophys. Res. Commun.* 360, 295–298.

35. Sotiriou, S., Gispert, S., Cheng, J., Wang, Y., Chen, A., Hoogstraten-Miller, S., Miller, G. F. et al. 2002. Ascorbic-acid transporter Slc23a1 is essential for vitamin C transport into the brain and for perinatal survival. *Nat. Med.* 8, 514–517.

36. Berger, U. V., Lu, X. C., Liu, W., Tang, Z., Slusher, B. S. and Hediger, M. A. 2003. Effect of middle cerebral artery occlusion on mRNA expression for the sodium-coupled vitamin C transporter SVCT2 in rat brain. *J. Neurochem.* 86, 896–906.

37. Castro, M., Caprile, T., Astuya, A., Millan, C., Reinicke, K., Vera, J. C., Vasquez, O., Aguayo, L. G. and Nualart, F. 2001. High-affinity sodium-vitamin C co-transporters (SVCT) expression in embryonic mouse neurons. *J. Neurochem.* 78, 815–823.

38. Rice, M. E. and Russo-Menna, I. 1998. Differential compartmentalization of brain ascorbate and glutathione between neurons and glia. *Neuroscience.* 82, 1213–1223.

39. Hediger, M. A. 2002. New view at C. *Nat. Med.* 8, 445–446.

40. Lee, S. H., Lumelsky, N., Studer, L., Auerbach, J. M. and McKay, R. D. 2000. Efficient generation of midbrain and hindbrain neurons from mouse embryonic stem cells. *Nat. Biotechnol.* 18, 675–679.

41. Lee, J. Y., Chang, M. Y., Park, C. H., Kim, H. Y., Kim, J. H., Son, H., Lee, Y. S. and Lee, S. H. 2003. Ascorbate-induced differentiation of embryonic cortical precursors into neurons and astrocytes. *J. Neurosci. Res.* 73, 156–165.

42. Shin, D. M., Ahn, J. I., Lee, K. H., Lee, Y. S. and Lee, Y. S. 2004. Ascorbic acid responsive genes during neuronal differentiation of embryonic stem cells. *Neuroreport.* 15, 1959–1963.

43. Qiu, S., Li, L., Weeber, E. J. and May, J. M. 2007. Ascorbate transport by primary cultured neurons and its role in neuronal function and protection against excitotoxicity. *J. Neurosci. Res.* 85, 1046–1056.

44. May, J. M. 2012. Vitamin C transport and its role in the central nervous system. *Subcell. Biochem.* 56, 85–103.

45. Diliberto, E. J., Jr. and Allen, P. L. 1980. Semidehydroascorbate as a product of the enzymic conversion of dopamine to norepinephrine. Coupling of semidehydroascorbate reductase to dopamine-beta-hydroxylase. *Mol. Pharmacol.* 17, 421–426.

46. Diliberto, E. J., Jr. and Allen, P. L. 1981. Mechanism of dopamine-beta-hydroxylation. Semidehydroascorbate as the enzyme oxidation product of ascorbate. *J. Biol. Chem.* 256, 3385–3393.

47. Hoehn, S. K. and Kanfer, J. N. 1980. Effects of chronic ascorbic acid deficiency on guinea pig lysosomal hydrolase activities. *J. Nutr.* 110, 2085–2094.

48. Deana, R., Bharaj, B. S., Verjee, Z. H. and Galzigna, L. 1975. Changes relevant to catecholamine metabolism in liver and brain of ascorbic acid deficient guinea-pigs. *Int. J. Vitam. Nutr. Res.* 45, 175–182.

49. He, X. B., Kim, M., Kim, S. Y., Yi, S. H., Rhee, Y. H., Kim, T., Lee, E. H. et al. 2015. Vitamin C facilitates dopamine neuron differentiation in fetal midbrain through TET1- and JMJD3-dependent epigenetic control manner. *Stem Cells.* 33, 1320–1332.

50. Patak, P., Willenberg, H. S. and Bornstein, S. R. 2004. Vitamin C is an important cofactor for both adrenal cortex and adrenal medulla. *Endocr. Res.* 30, 871–875.

51. Mefford, I. N., Oke, A. F. and Adams, R. N. 1981. Regional distribution of ascorbate in human brain. *Brain Res.* 212, 223–226.

52. Lee, L., Kang, S. A., Lee, H. O., Lee, B. H., Jung, I. K., Lee, J. E. and Hoe, Y. S. 2001. Effect of supplementation of vitamin E and vitamin C on brain acetylcholinesterase activity and neurotransmitter levels in rats treated with scopolamine, an inducer of dementia. *J. Nutr. Sci. Vitaminol. (Tokyo)*. 47, 323–328.

53. Kuo, C. H., Hata, F., Yoshida, H., Yamatodani, A. and Wada, H. 1979. Effect of ascorbic acid on release of acetylcholine from synaptic vesicles prepared from different species of animals and release of noradrenaline from synaptic vesicles of rat brain. *Life Sci.* 24, 911–915.

54. Atlante, A., Gagliardi, S., Minervini, G. M., Ciotti, M. T., Marra, E. and Calissano, P. 1997. Glutamate neurotoxicity in rat cerebellar granule cells: A major role for xanthine oxidase in oxygen radical formation. *J. Neurochem.* 68, 2038–2045.

55. Ciani, E., Groneng, L., Voltattorni, M., Rolseth, V., Contestabile, A. and Paulsen, R. E. 1996. Inhibition of free radical production or free radical scavenging protects from the excitotoxic cell death mediated by glutamate in cultures of cerebellar granule neurons. *Brain Res.* 728, 1–6.

56. Shah, S. A., Yoon, G. H., Kim, H. O. and Kim, M. O. 2015. Vitamin C neuroprotection against dose-dependent glutamate-induced neurodegeneration in the postnatal brain. *Neurochem. Res.* 40, 875–884.

57. Meadows, J. P., Guzman-Karlsson, M. C., Phillips, S., Holleman, C., Posey, J. L., Day, J. J., Hablitz, J. J. and Sweatt, J. D. 2015. DNA methylation regulates neuronal glutamatergic synaptic scaling. *Sci. Signal.* 8, ra61.

58. Zhang, R. R., Cui, Q. Y., Murai, K., Lim, Y. C., Smith, Z. D., Jin, S., Ye, P. et al. 2013. Tet1 regulates adult hippocampal neurogenesis and cognition. *Cell Stem Cell.* 13, 237–245.

59. Veenstra-VanderWeele, J., Anderson, G. M. and Cook, E. H., Jr. 2000. Pharmacogenetics and the serotonin system: Initial studies and future directions. *Eur. J. Pharmacol.* 410, 165–181.

60. Li, X., Frye, M. A. and Shelton, R. C. 2012. Review of pharmacological treatment in mood disorders and future directions for drug development. *Neuropsychopharmacology.* 37, 77–101.

61. Chang, C. W., Chen, M. J., Wang, T. E., Chang, W. H., Lin, C. C. and Liu, C. Y. 2007. Scurvy in a patient with depression. *Dig. Dis. Sci.* 52, 1259–1261.

62. Kinsman, R. A. and Hood, J. 1971. Some behavioral effects of ascorbic acid deficiency. *Am. J. Clin. Nutr.* 24, 455–464.

63. Levine, M., Conry-Cantilena, C., Wang, Y., Welch, R. W., Washko, P. W., Dhariwal, K. R., Park, J. B. et al. 1996. Vitamin C pharmacokinetics in healthy volunteers: Evidence for a recommended dietary allowance. *Proc. Natl. Acad. Sci. USA.* 93, 3704–3709.

64. Zhang, M., Robitaille, L., Eintracht, S. and Hoffer, L. J. 2011. Vitamin C provision improves mood in acutely hospitalized patients. *Nutrition.* 27, 530–533.

65. Kennedy, D. O., Veasey, R., Watson, A., Dodd, F., Jones, E., Maggini, S. and Haskell, C. F. 2010. Effects of high-dose B vitamin complex with vitamin C and minerals on subjective mood and performance in healthy males. *Psychopharmacology (Berl).* 211, 55–68.

66. Gosney, M. A., Hammond, M. F., Shenkin, A. and Allsup, S. 2008. Effect of micronutrient supplementation on mood in nursing home residents. *Gerontology.* 54, 292–299.

67. Cocchi, P., Silenzi, M., Calabri, G. and Salvi, G. 1980. Antidepressant effect of vitamin C. *Pediatrics.* 65, 862–863.

68. Brody, S. 2002. High-dose ascorbic acid increases intercourse frequency and improves mood: A randomized controlled clinical trial. *Biol. Psychiatry.* 52, 371–374.

69. Amr, M., El-Mogy, A., Shams, T., Vieira, K. and Lakhan, S. E. 2013. Efficacy of vitamin C as an adjunct to fluoxetine therapy in pediatric major depressive disorder: A randomized, double-blind, placebo-controlled pilot study. *Nutr. J.* 12, 31.

70. Fitzpatrick, P. F. 2003. Mechanism of aromatic amino acid hydroxylation. *Biochemistry.* 42, 14083–14091.

71. Ramsey, A. J., Hillas, P. J. and Fitzpatrick, P. F. 1996. Characterization of the active site iron in tyrosine hydroxylase. Redox states of the iron. *J. Biol. Chem.* 271, 24395–24400.

72. Fitzpatrick, P. F. 1989. The metal requirement of rat tyrosine hydroxylase. *Biochem. Biophys. Res. Commun.* 161, 211–215.

73. Kraus, T. F., Guibourt, V. and Kretzschmar, H. A. 2015. 5-Hydroxymethylcytosine, the "Sixth Base," during brain development and ageing. *J. Neural. Transm. (Vienna).* 122, 1035–1043.

74. Szulwach, K. E., Li, X., Li, Y., Song, C. X., Wu, H., Dai, Q., Irier, H. et al. 2011. 5-hmC-mediated epigenetic dynamics during postnatal neurodevelopment and aging. *Nat. Neurosci.* 14, 1607–1616.

75. Li, X., Yao, B., Chen, L., Kang, Y., Li, Y., Cheng, Y., Li, L. et al. 2017. Ten-eleven translocation 2 interacts with forkhead box O3 and regulates adult neurogenesis. *Nat. Commun.* 8, 15903.

76. Hahn, M. A., Qiu, R., Wu, X., Li, A. X., Zhang, H., Wang, J., Jui, J. et al. 2013. Dynamics of 5-hydroxymethylcytosine and chromatin marks in mammalian neurogenesis. *Cell. Rep.* 3, 291–300.

77. Xu, Y., Xu, C., Kato, A., Tempel, W., Abreu, J. G., Bian, C., Hu, Y. et al. 2012. Tet3 CXXC domain and dioxygenase activity cooperatively regulate key genes for Xenopus eye and neural development. *Cell.* 151, 1200–1213.

78. Tveden-Nyborg, P., Johansen, L. K., Raida, Z., Villumsen, C. K., Larsen, J. O. and Lykkesfeldt, J. 2009. Vitamin C deficiency in early postnatal life impairs spatial memory and reduces the number of hippocampal neurons in guinea pigs. *Am. J. Clin. Nutr.* 90, 540–546.

79. Rudenko, A., Dawlaty, M. M., Seo, J., Cheng, A. W., Meng, J., Le, T., Faull, K. F., Jaenisch, R. and Tsai, L. H. 2013. Tet1 is critical for neuronal activity-regulated gene expression and memory extinction. *Neuron.* 79, 1109–1122.

80. Kaas, G. A., Zhong, C., Eason, D. E., Ross, D. L., Vachhani, R. V., Ming, G. L., King, J. R., Song, H. and Sweatt, J. D. 2013. TET1 controls CNS 5-methylcytosine hydroxylation, active DNA demethylation, gene transcription, and memory formation. *Neuron.* 79, 1086–1093.

81. Li, X., Wei, W., Zhao, Q. Y., Widagdo, J., Baker-Andresen, D., Flavell, C. R., D'Alessio, A., Zhang, Y. and Bredy, T. W. 2014. Neocortical Tet3-mediated accumulation of 5-hydroxymethylcytosine promotes rapid behavioral adaptation. *Proc. Natl. Acad. Sci. USA.* 111, 7120–7125.

82. Chen, R., Zhang, Q., Duan, X., York, P., Chen, G. D., Yin, P., Zhu, H. et al. 2017. The 5-Hydroxymethylcytosine (5hmC) reader UHRF2 is required for normal levels of 5hmC in mouse adult brain and spatial learning and memory. *J. Biol. Chem.* 292, 4533–4543.

83. Khare, T., Pai, S., Koncevicius, K., Pal, M., Kriukiene, E., Liutkeviciute, Z., Irimia, M. et al. 2012. 5-hmC in the brain is abundant in synaptic genes and shows differences at the exon-intron boundary. *Nat. Struct. Mol. Biol.* 19, 1037–1043.

84. Nelson, E. D., Kavalali, E. T. and Monteggia, L. M. 2008. Activity-dependent suppression of miniature neurotransmission through the regulation of DNA methylation. *J. Neurosci.* 28, 395–406.

85. Yu, H., Su, Y., Shin, J., Zhong, C., Guo, J. U., Weng, Y. L., Gao, F. et al. 2015. Tet3 regulates synaptic transmission and homeostatic plasticity via DNA oxidation and repair. *Nat. Neurosci.* 18, 836–843.

86. Clark, E. A. and Nelson, S. B. 2015. Synapse and genome: An elusive tete-a-tete. *Sci. Signal.* 8, pe2.

87. Reitz, C., Brayne, C. and Mayeux, R. 2011. Epidemiology of Alzheimer disease. *Nat. Rev. Neurol.* 7, 137–152.

88. Harrison, F. E. 2012. A critical review of vitamin C for the prevention of age-related cognitive decline and Alzheimer's disease. *J. Alzheimers Dis.* 29, 711–726.

89. Dixit, S., Bernardo, A., Walker, J. M., Kennard, J. A., Kim, G. Y., Kessler, E. S. and Harrison, F. E. 2015. Vitamin C deficiency in the brain impairs cognition, increases amyloid accumulation and deposition, and oxidative stress in APP/PSEN1 and normally aging mice. *ACS. Chem. Neurosci.* 6, 570–581.

90. Siqueira, I. R., Elsner, V. R., Leite, M. C., Vanzella, C., Moyses Fdos, S., Spindler, C., Godinho, G. et al. 2011. Ascorbate uptake is decreased in the hippocampus of ageing rats. *Neurochem. Int.* 58, 527–532.

91. Travica, N., Ried, K., Sali, A., Scholey, A., Hudson, I. and Pipingas, A. 2017. Vitamin C status and cognitive function: A systematic review. *Nutrients.* 9.

92. Engelhart, M. J., Geerlings, M. I., Ruitenberg, A., van Swieten, J. C., Hofman, A., Witteman, J. C. and Breteler, M. M. 2002. Dietary intake of antioxidants and risk of Alzheimer disease. *JAMA.* 287, 3223–3229.

93. De Jager, P. L., Srivastava, G., Lunnon, K., Burgess, J., Schalkwyk, L. C., Yu, L., Eaton, M. L. et al. 2014. Alzheimer's disease: Early alterations in brain DNA methylation at ANK1, BIN1, RHBDF2 and other loci. *Nat. Neurosci.* 17, 1156–1163.

94. Smith, R. G., Hannon, E., De Jager, P. L., Chibnik, L., Lott, S. J., Condliffe, D., Smith, A. R. et al. 2018. Elevated DNA methylation across a 48-kb region spanning the HOXA gene cluster is associated with Alzheimer's disease neuropathology. *Alzheimers Dement.* 14, 1580–1588.

95. Chouliaras, L., Mastroeni, D., Delvaux, E., Grover, A., Kenis, G., Hof, P. R., Steinbusch, H. W., Coleman, P. D., Rutten, B. P. and van den Hove, D. L. 2013. Consistent decrease in global DNA methylation and hydroxymethylation in the hippocampus of Alzheimer's disease patients. *Neurobiol. Aging.* 34, 2091–2099.

96. Mastroeni, D., Grover, A., Delvaux, E., Whiteside, C., Coleman, P. D. and Rogers, J. 2010. Epigenetic changes in Alzheimer's disease: Decrements in DNA methylation. *Neurobiol. Aging.* 31, 2025–2037.

97. Condliffe, D., Wong, A., Troakes, C., Proitsi, P., Patel, Y., Chouliaras, L., Fernandes, C. et al. 2014. Cross-region reduction in 5-hydroxymethylcytosine in Alzheimer's disease brain. *Neurobiol. Aging.* 35, 1850–1854.

98. Coppieters, N., Dieriks, B. V., Lill, C., Faull, R. L., Curtis, M. A. and Dragunow, M. 2014. Global changes in DNA methylation and hydroxymethylation in Alzheimer's disease human brain. *Neurobiol. Aging.* 35, 1334–1344.

99. Zhao, J., Zhu, Y., Yang, J., Li, L., Wu, H., De Jager, P. L., Jin, P. and Bennett, D. A. 2017. A genome-wide profiling of brain DNA hydroxymethylation in Alzheimer's disease. *Alzheimers Dement.* 13, 674–688.

100. Sawicka-Glazer, E. and Czuczwar, S. J. 2014. Vitamin C: A new auxiliary treatment of epilepsy? *Pharmacol. Rep.* 66, 529–533.

101. Santos, L. F., Freitas, R. L., Xavier, S. M., Saldanha, G. B. and Freitas, R. M. 2008. Neuroprotective actions of vitamin C related to decreased lipid peroxidation and increased catalase activity in adult rats after pilocarpine-induced seizures. *Pharmacol. Biochem. Behav.* 89, 1–5.

102. Tome Ada, R., Ferreira, P. M. and Freitas, R. M.. 2010. Inhibitory action of antioxidants (ascorbic acid or alpha-tocopherol) on seizures and brain damage induced by pilocarpine in rats. *Arq. Neuropsiquiatr.* 68, 355–361.

103. Xavier, S. M., Barbosa, C. O., Barros, D. O., Silva, R. F., Oliveira, A. A. and Freitas, R. M. 2007. Vitamin C antioxidant effects in hippocampus of adult Wistar rats after seizures and status epilepticus induced by pilocarpine. *Neurosci. Lett.* 420, 76–79.

104. Xu, K. and Stringer, J. L. 2008. Antioxidants and free radical scavengers do not consistently delay seizure onset in animal models of acute seizures. *Epilepsy. Behav.* 13, 77–82.

105. Schneider Oliveira, M., Flavia Furian, A., Freire Royes, L. F., Rechia Fighera, M., de Carvalho Myskiw, J., Gindri Fiorenza, N. and Mello, C. F. 2004. Ascorbate modulates pentylenetetrazol-induced convulsions biphasically. *Neuroscience.* 128, 721–728.

106. Kobow, K., Kaspi, A., Harikrishnan, K. N., Kiese, K., Ziemann, M., Khurana, I., Fritzsche, I. et al. 2013. Deep sequencing reveals increased DNA methylation in chronic rat epilepsy. *Acta Neuropathol.* 126, 741–756.

107. Miller-Delaney, S. F., Bryan, K., Das, S., McKiernan, R. C., Bray, I. M., Reynolds, J. P., Gwinn, R., Stallings, R. L. and Henshall, D. C. 2015. Differential DNA methylation profiles of coding and non-coding genes define hippocampal sclerosis in human temporal lobe epilepsy. *Brain.* 138, 616–631.

108. Wang, L., Fu, X., Peng, X., Xiao, Z., Li, Z., Chen, G. and Wang, X. 2016. DNA methylation profiling reveals correlation of differential methylation patterns with gene expression in human epilepsy. *J. Mol. Neurosci.* 59, 68–77.

109. Machnes, Z. M., Huang, T. C., Chang, P. K., Gill, R., Reist, N., Dezsi, G., Ozturk, E. et al. 2013. DNA methylation mediates persistent epileptiform activity in vitro and in vivo. *PLOS ONE.* 8, e76299.

110. Martinowich, K., Hattori, D., Wu, H., Fouse, S., He, F., Hu, Y., Fan, G. and Sun, Y. E. 2003. DNA methylation-related chromatin remodeling in activity-dependent BDNF gene regulation. *Science.* 302, 890–893.

111. Ryley Parrish, R., Albertson, A. J., Buckingham, S. C., Hablitz, J. J., Mascia, K. L., Davis Haselden, W. and Lubin, F. D. 2013. Status epilepticus triggers early and late alterations in brain-derived neurotrophic factor and NMDA glutamate receptor Grin2b DNA methylation levels in the hippocampus. *Neuroscience.* 248, 602–619.

112. Levenson, J. M., Roth, T. L., Lubin, F. D., Miller, C. A., Huang, I. C., Desai, P., Malone, L. M. and Sweatt, J. D. 2006. Evidence that DNA (cytosine-5) methyltransferase regulates synaptic plasticity in the hippocampus. *J. Biol. Chem.* 281, 15763–15773.

113. Zhu, Q., Wang, L., Zhang, Y., Zhao, F. H., Luo, J., Xiao, Z., Chen, G. J. and Wang, X. F. 2012. Increased expression of DNA methyltransferase 1 and 3a in human temporal lobe epilepsy. *J. Mol. Neurosci.* 46, 420–426.

INDEX

Printed in the United States
by Baker & Taylor Publisher Services